D0071131

AUSTRALIAN MATHEMATICAL SOCIETY LEC

Editor-in-Chief: Professor S.A. Morris, Department of Mathematics, Statistics and Computing Science, The University of New England, Armidale N.S.W. 2351, Australia

Subject Editors:
Professor C.J.Thompson, Department of Mathematics, University of Melbourne, Parkville, Victoria 3052, Australia
Professor C.C. Heyde, Department of Statistics, University of Melbourne, Parkville, Victoria 3052, Australia
Professor J.H. Loxton, Department of Pure Mathematics, University of New South Wales, Kensington, New South Wales 2033, Australia

Australian Mathematical Society Lecture Series. 7

The Petersen Graph

D. A. Holton
Department of Mathematics and Statistics,
University of Otago, New Zealand

J. Sheehan
Department of Mathematical Sciences,
University of Aberdeen

CAMBRIDGE
UNIVERSITY PRESS

Published by the Press Syndicate of the University of Cambridge
The Pitt Building, Trumpington Street, Cambridge CB2 1RP
40 West 20th Street, New York, NY 10011, USA
10 Stamford Road, Oakleigh, Melbourne 3166, Australia

© Cambridge University Press 1993

First published 1993

Library of Congress cataloguing in publication data available

British Library cataloguing in publication data available

ISBN   0 521 43594 3 paperback

Transferred to digital printing 2004

Contents

Contents

Contents

Contents

Contents

Preface

The Petersen graph has fascinated many graph theorists over the years because of its appearance as a counterexample in many places. Because of its ubiquity, it seemed a natural graph to use as a central theme for a book. As a result of using this graph as our centre piece, much of this book deals with the properties of cubic 3-connected graphs and the ideas that generalize from them.

Incidentally, a biography of Julius Petersen can be found in J.Lützen, G.Sabidussi, B. Toft, Julius Petersen 1839-1910, A Biography, Preprint, Odense University, 1990.

The book has grown out of lecture courses that we have given in various places over a number of years. In all cases the audience were final honours year students, graduate students or research colleagues. Hence we would expect the current volume to be useful for senior students and research workers. Because of this we have included a reasonably large number of Exercises and an extensive set of references.

In citing references we have used the form [A-B 34] for the 1934 paper of the authors Able and Baker and [aB 34] for the 1934 (or perhaps 1834) paper of the single author A. Board. When two authors have the same initials an extra letter has been inserted to distinguish between them so [aB 89] becomes [aaB 89] if there is another author with initials A,B. References for each chapter are at the end of that chapter.

Figures, lemmas, theorems, corollaries, conjectures, etc., are numbered consecutively throughout each section of each chapter. We refer to Lemma 3.2.4, the fourth item from section two of chapter three, as such outside Chapter 3 but as Lemma 2.4 wherever it is mentioned inside Chapter 3.

In the first chapter we establish the basic definitions and concepts that are used throughout. The second chapter starts a sequence of chapters related to the Four Colour Theorem. Here we lead up to an outline of the Appel and Haken proof by way of the Five Colour Theorem. Some attention is then given to alternative approaches to the main theorem. Chapter 3 looks at the edge colouring of graphs and introduces the concept of a snark. As often occurs in this volume, the Petersen graph is an example of this concept. It also seems to be intimately related to all currently known snarks. In the next chapter we lead on from colouring to factoring and cover Tutte's characterization of perfect matchings. We also consider Tutte's notion of f-factors. The relation of flows to colourings is investigated in Chapter 5. In Chapter 6, the constraints of diameter and girth found in the

Petersen graph are extended to the concept of cages. Hypohamiltonian graphs are the subject of Chapter 7. Here we again start with the Petersen graph and end with the current state of the art. Chapter 8 takes up the symmetry side of the Petersen graph from the standpoint of automorphisms. The last chapter gives a pot-pourri to show the wide influence of our central character. Very brief mentions are made here of a number of fascinating areas of graph theory.

The pace of the chapters is a little uneven in that we do not always expect the reader to follow every step of our path through an argument, or even a section. We sometimes use a fairly broad brush and leave the details for the stimulated reader to follow up through the Exercises or the extensive references. The speed quickens particularly at the end of Chapter 4, in Sections 5.1 and 5.3, from time to time in the Chapters 7 and 8 and quite a lot in the last chapter where we are trying to give an overview of a number of interesting areas.

The development of the chapters is indicated below. While the research reader can probably dip in to the book at any stage, it is suggested that other readers should follow the sequences indicated.

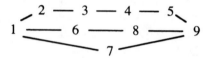

Finally we would like to thank several people who were of great assistance to us in the production of this book. First we are indebted to a number of nameless referees who did a magnificent job with drafts of this document. Their work was above and beyond the normal call on those of their profession. Second we must mention Tank Aldred for his many useful comments on the manuscript. These comments helped to reduce the number of errors we had created. Third we would like to thank those who were involved in some form or other in the production of the manuscript. Thank you Mark Borrie, Mary-Jane Campbell, Sharron Eade, David Gilbert (for most of the diagrams), Irene Goodwin (for the final draft of the manuscript), Lenette Grant, and Maree Watson. And last, but definitely not least, we would like to thank our families without whose forebearance and support we would not have been able to complete this, or any other project.

1
The Petersen Graph

0.   Prologue

The index of Capobianco and Molluzzo's book "Examples and Counterexamples in Graph Theory" [C-M 78] shows the following entry:-

"Petersen graph, 36, 54, 109, 146, 158, 164, 179, 213"

This makes it by far the most referenced graph in that volume. It also gives some idea of the importance of the Petersen graph in the study of graph theory.

1.   What is a graph?

The Petersen graph P got its name because of its appearance in 1898 in a paper by J. Petersen [jP 98] (see Fig.1.1). However, this was not the first time it had appeared in print. The graph first appeared in [aK 86] as the graph of the Desargues' configuration.

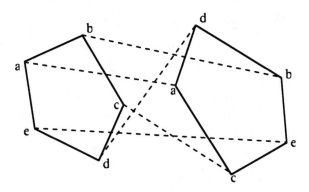

The Petersen graph P

Figure 1.1

It was appropriate that in [jP 98] the graph was used as a counterexample because, as we shall see, it continues to be a yardstick with which to measure conjectures. It also has many interesting properties. These two facts give it a unique place in the theory of graphs.

In this chapter we will introduce basic graph theoretic properties and show how they apply to the Petersen graph, P. The remainder of the book will develop a number of areas in which P plays an important role.

But what is a graph?

Precisely, it is an ordered pair $G = (VG, EG)$ where VG is the non-empty set of **vertices** of G, and EG, the set of **edges** of G, is a subset of the 2-subsets of VG. That is, the edges of G are sets of size two, whose members are elements of VG. Unless otherwise specified we will assume $|VG|$ is finite.

For instance, the following are graphs.

$G_1 = (\{1,2,3\}, \{\{1,2\}, \{2,3\}, \{3,1\}\})$

$G_2 = (\{1,2,3,4\}, \{\{1,2\}, \{1,3\}, \{1,4\}\}).$

When dealing with examples it is frequently valuable to give graphs a pictorial representation. This is done by letting the vertices of a graph G be represented by dots and the edges by lines joining the relevant dots.

The graphs G_1 and G_2 are shown in visual representations in Figure 1.2.

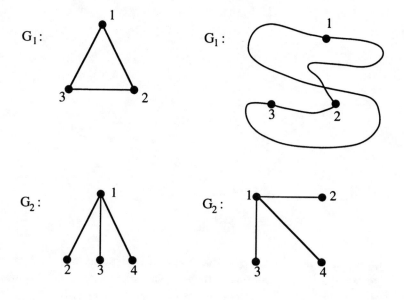

Figure 1.2

The moral to be drawn from Figure 1.2 is that there is no **unique** way of drawing a graph. Vertices may be placed arbitrarily in the plane and edges may be drawn in any convenient way. Generally we will use straight lines for edges but in the end we will be guided in the main by aesthetics and the purpose in hand.

In general we will use uv for the edge {u,v} as this will avoid a surfeit of braces. Further, in such a case we say that u and v are **adjacent**, the vertices u,v are **incident** with uv, and uv is **incident** with u and v. We sometimes write u~v to show that u is adjacent to v and u≁v when uv ∉ EG.

With this in mind the Petersen graph P, has VP = {1,2,3,4,5,1′,2′,3′,4′,5′} and EP = {12,23,34,45,51,1′3′,2′4′,3′5′,1′4′,5′2′,11′,22′,33′,44′,55′}. We show one of its common pictorial representations in Figure 1.3. Unless stated otherwise, we will always refer back to this drawing and labelling of P. It will sometimes be useful to let A = {12,23,34,45,51}, B = {11′,22′,33′,44′,55′} and C = {1′3′,2′4′,3′5′,4′1′,5′2′}.

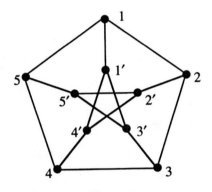

Figure 1.3

Because uv is a 2-subset of VG, then u and v are distinct vertices. Occasionally we will want to discuss 'edges' which join u to itself . These we refer to as **loops.**

Again, we defined EG to be a subset of the 2-subsets of VG. Hence there is at most one edge incident with a given pair of vertices. Sometimes we allow multiple edges between a given pair of vertices but in that case we will call the 'graphs' involved, **multigraphs.**

If we allow multigraphs to have loops, then they will be called **pseudographs.** We note that what we have defined here as graphs are elsewhere often called **simple graphs.**

A graph, multigraph and pseudograph on 5 vertices are shown in Figure 1.4.

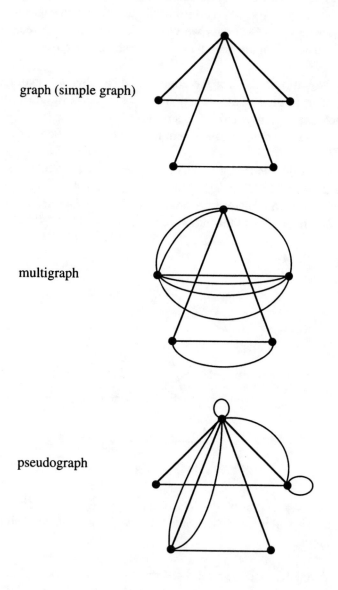

graph (simple graph)

multigraph

pseudograph

Figure 1.4

We also define a **eunegraph**. This is a pseudograph which may also contain "incomplete" edges or **splines**. That is, edges that are incident with only one vertex. We will usually convert eunegraphs to pseudographs by joining the splines to make edges, loops or multiple edges. A eunegraph will be represented as in Figure 1.5, where vertices u and v are incident to splines.

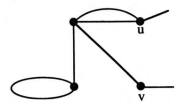

Figure 1.5

The term eunegraph is coined here for the first time. However it is a concept that we will find valuable in later chapters. Graphs with incomplete edges have proved useful in a number of places in the literature.

It is important to be able to decide when two graphs are the same.

Recall the graph G_1 of Figure 1.2. Clearly if we relabel G_1, then we should obtain essentially the same graph. Hence $G_3 = (\{u,v,w\},\{uv,\ uw,\ wv\})$ is the 'same' as G_1. We therefore say that graphs G and H are **isomorphic** if there exists a bijection $\varphi: VG \to VH$ which is also a bijection from EG to EH when we define $\varphi(uv) = \varphi(u)\varphi(v)$. The mapping φ is an **isomorphism** between G and H. We write $G \cong H$ for 'G is isomorphic to H'.

In Figure 1.6 we give four isomorphic graphs on six vertices.

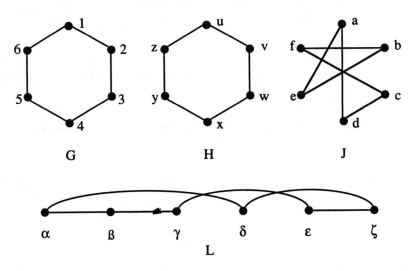

Figure 1.6

For instance, an isomorphism φ between G and L can be expressed as follows:-

$$\varphi(1) = \alpha \qquad \varphi(2) = \beta \qquad \varphi(3) = \gamma \qquad \varphi(4) = \varepsilon \qquad \varphi(5) = \zeta \qquad \varphi(6) = \delta.$$

It is straightforward to check that φ is a bijection on the edges of G and L.

Figure 1.7 shows three isomorphic copies of the same graph. These are in fact three more ways to draw P.

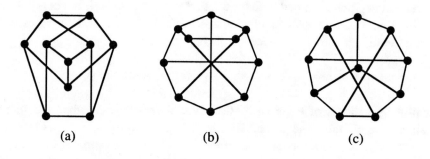

(a)   (b)   (c)

Figure 1.7

Now isomorphism is an equivalence relation on the set of all graphs. Consequently when we talk about a graph we are usually referring to it as a representative of its equivalence class. As such any labelling or names we assign to the vertices are for ease of manipulation. Unlike a road map of a country where the vertices represent towns and cities and the edges represent roads, no special significance will be attached to this labelling. If, as in the road map, there **is** good reason for assigning certain names to vertices, then we will call the graph a **labelled graph**.

In Figure 1.8 we give all the graphs on three vertices, as well as all the labelled graphs on three vertices.

Define the **order** of G to be |VG| and the **size** of G to be |EG|. Then the graphs of Figure 1.8 have order 3 and the size of graph J in Figure 1.6 is 6 and equals its order. Further, the order of the Petersen graph is 10 and its size is 15.

Now it is relatively easy to determine the number of labelled graphs of order n, but the number of graphs of order n needs special techniques involving Pólya's Theorem, which we will not go into in this book; see [H-P 73].

graphs

labelled
graphs

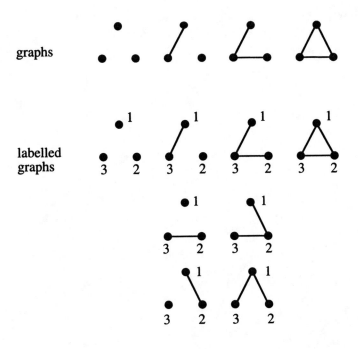

Figure 1.8

The number of edges in G incident with a vertex u (plus twice the number of loops at u, in the case of a pseudograph) is called the **degree** of u and is written $\deg_G u$. Thus $\deg_{G_2} 1 = 3$ and $\deg_{G_2} 2 = 1$ for the graph G_2 of Figure 1.2. Every vertex of P has degree 3.

Where there is no ambiguity we drop the subscript G and simply write deg u. The smallest, respectively largest, degree of any vertex in a graph G is denoted by $\delta(G)$, respectively $\Delta(G)$.

We are now able to state and prove our first result in graph theory.

Lemma 1.1
$$\sum_{u \in VG} \deg u = 2|EG|.$$

Proof: The edge uv is counted twice in the summation, once in deg u and once in deg v. Hence the result follows. □

Corollary 1.2 Every graph has an even number of vertices of odd degree. □

A **neighbour** of a vertex u ∈ VG   is any member of the set
{v ∈ VG : uv ∈ EG}, which is called the **neighbourhood of u**, $N_G(u)$. (If there
is no ambiguity we write simply N(u).) We note that |N(u)| = deg u.

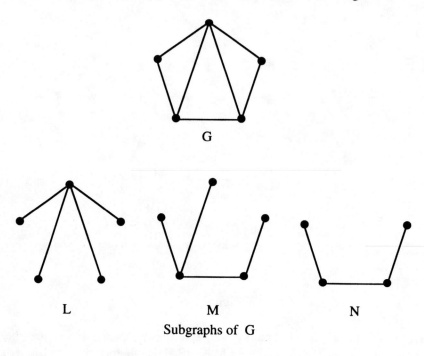

Subgraphs of G

Figure 1.9

Another common graph-theoretical notion is that of subgraph. We say that H is
a **subgraph** of G if VH ⊆ VG   and   EH ⊆ EG. A graph and some of its
subgraphs are shown in Figure 1.9.

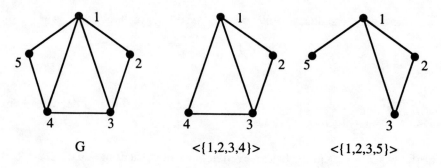

Figure 1.10

Two special types of subgraph are worth noting. We say that H is an **induced** subgraph of G, if every pair of vertices of H that are adjacent in G, are adjacent in H. If $VH = X \subseteq VG$, we write $< X >$ for the subgraph of G induced by the vertex set X. Figure 1.10 gives two examples of induced subgraphs. The labelling of G is given for ease of description.

A special case of an induced subgraph is the **vertex deleted subgraph**. This is defined to be $G_v = <VG \setminus \{v\}>$. It is the graph obtained from G by removing the vertex v and all edges of G that were incident with v. See Figure 1.11 for two examples, one of which is isomorphic to the subgraph N of Figure 1.9 and the other to the subgraph $<\{1, 2, 3, 5\}>$ of Figure 1.10.

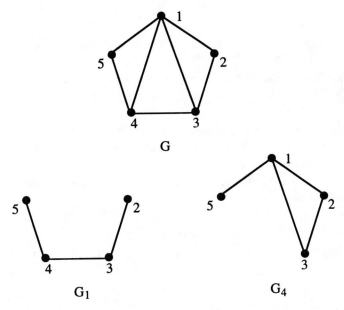

Figure 1.11

Rather than G_v we sometimes write G - v. We define $G - \{u,v\}$ to be the graph $(G_u)_v$ (or alternatively the graph $(G - u) - v$).

Figure 1.12 (a), (b) shows the vertex deleted subgraphs of P obtained by deleting vertex 1 and vertex 1'. By the symmetry of the drawing of P in Figure 1.3, $P_{1'} \cong P_{2'} \cong P_{3'} \cong P_{4'} \cong P_{5'}$. It is not difficult to show though that $P_1 \cong P_{1'}$. Hence there is essentially only one vertex deleted subgraph of P.

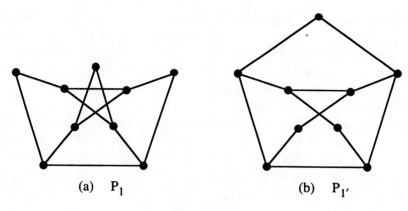

(a) P_1 (b) $P_{1'}$

Figure 1.12

There is an important open question in graph theory which deals with vertex deleted subgraphs. This is Ulam's Conjecture or the Reconstruction Conjecture, which says that any graph can be determined uniquely from its collection of vertex deleted subgraphs. Equivalently, $\{G_v : v \in VG\}$ is unique to the graph G.

Much work is currently being done on this conjecture. Unfortunately it is all beyond the scope of this book, but the interested reader should look at the excellent review articles by Bondy and Hemminger [B-H 77] and Nash-Williams [cN 78]. However we will have cause to consider vertex deleted subgraphs at several stages throughout this book.

The other useful type of subgraph which deserves mention at this stage is the spanning subgraph. Now H is a **spanning subgraph** of G if H is a subgraph of G with $|VH| = |VG|$. A graph and one of its spanning subgraphs are shown in Figure 1.13.

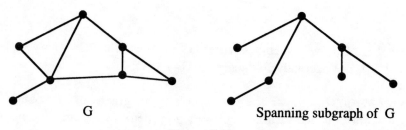

G   Spanning subgraph of G

Figure 1.13

Figure 1.14 shows a spanning subgraph of P which will be of interest later. This subgraph is taken from Figure 1.7(c). The reasons for the labelling will become clear in Section 5.

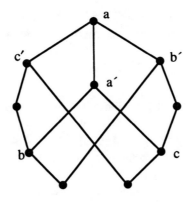

Figure 1.14

If $u,v \in VG$ and $e = uv \in EG$, then $G - uv$ is the spanning subgraph of G with edge set $EG \setminus \{uv\}$.

2. Regular graphs and connectivity

Apart from P, we encounter several graphs so often that it is worth giving them specific names.

Probably the easiest graph to deal with, is the one where $EG = \emptyset$. This is called the **empty graph** and consists solely of isolated vertices. Trivially the empty graph of order n is a spanning subgraph of every other graph of order n.

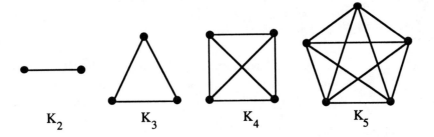

Figure 2.1

At the other end of the scale we have the **complete graph**, where EG consists of all 2-subsets of VG i.e. every vertex is adjacent to every other vertex. We denote the complete graph of order n by K_n. In Figure 2.1 we show K_2, K_3, K_4 and K_5. Clearly every graph on n vertices is a (spanning) subgraph of K_n.

If G is any graph of order n, then it is useful to define the **complement of G**,

$\overline{G}$, to be the graph with vertex set VG and such that $E\overline{G} = EK_n \setminus EG$. Hence the empty graph of order n is the complement of K_n and vice-versa.

It is worth noting that $\overline{\overline{G}} = G$.

Now K_n and $\overline{K}_n$ each have the property that every vertex has the same degree. In general if $\deg_G u = r$ for every vertex u where r is a constant, then we say that G is **regular of degree r** or **r-regular**. Hence K_n is an (n-1)-regular graph and $\overline{K}_n$ is a 0-regular graph. The Petersen graph is 3-regular.

It is clear that the only 0-regular graphs are the empty graphs. To classify all the 1-regular graphs we note from Corollary 1.2 that such graphs must have even order 2m, say. Then a 1-regular graph of order 2m consists of m edges with every vertex incident with precisely 1 edge. Two edges which are not incident on a common vertex are said to be **independent**. Hence the 1-regular graphs of order 2m consist of m mutually independent edges.

Analogously, two vertices which do not have an edge in common are said to be **independent**. A set of vertices is **independent** if the vertices are pairwise independent. The **independence number**, $\alpha(G)$ of a graph G, is the maximum cardinality of an independent set in G. For example $\alpha(P) = 4$.

What of the 2-regular graphs? The 'simplest' 2-regular graph is the **cycle on n vertices**, C_n. This is the graph with vertex set $VC_n = \{i : i = 0,1,...,n-1\}$ and $EC_n = \{i \sim i+1: i = 0,1,...,n-1\}$, where the addition is modulo n. It is now not difficult to show that any 2-regular graph of order n is made up of a collection of cycles of order less than or equal to n.

A **cycle** in a graph is a subgraph isomorphic to C_m for some m. We sometimes refer to this as an m-cycle. Clearly P has the 5-cycle (1,2,3,4,5). If m is odd, then we say that C_m is an **odd cycle**. Similarly, an **even cycle** has m even.

In view of the classification of 1-regular graphs and 2-regular graphs, it seems useful to introduce the notion of the **union** of two graphs. Now $G \cup H$ is the union of G and H if $V(G \cup H) = VG \cup VH$, where $VG \cap VH = \varnothing$, and $E(G \cup H) = EG \cup EH$. (Note that elsewhere, VG and VH are allowed to have vertices in common. This will not be the case in this book.)

Hence the 1-regular graphs of order 2m are simply $\bigcup\limits_{i=1}^{m} K_2$ and the 2-regular graphs of order n are $\bigcup\limits_{i \in I} C_i$, where I is a collection (not a set) of numbers $i \geq 3$ such that $\sum\limits_{i \in I} i = n$.

We also need the concept of the join of two graphs G and H with disjoint vertex sets. So we define G + H to be the **join** of G and H, where $V(G + H) = VG \cup VH$ and the edges of G + H are the edges of $G \cup H$ together with {uv : u ∈ VG and v ∈ VH}. So $C_4 = A + B$, where A and B are distinct graphs both isomorphic to $\overline{K}_2$.

One might now ask what about the 3-regular graphs? This turns out to be a much more difficult problem. We note first of all, that such graphs are called **cubic**. Fortunately the graph whose vertices are the vertices of a cube and whose edges are the edges of a cube, is a cubic graph (see Figure 2.2). P is also a cubic graph.

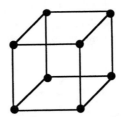

the graph of the cube

Figure 2.2

A general classification of cubic graphs is much more difficult than that of 1–regular or 2-regular graphs (so we will not even contemplate the question for r-regular graphs with r > 3). But a complete answer can be obtained for a special type of cubic graph. Before we can come to grips with this we need to delve into the concept of connectivity.

A graph G is **connected** if, for every pair of vertices u, v of VG, it is possible to find an alternating sequence of vertices and edges in G of the form $u = u_0, u_0u_1, u_1, u_1u_2, ... , u_n = v$.

So K_n, C_n and P are connected but $G \cup H$ is not, no matter what G and H are. For n > 1, $\overline{K}_n$ is also not connected.

If G is not connected, then we say it is **disconnected**. In such a case, G will be the union of connected graphs. A maximal connected subgraph of G is called a **component**. The graph G is the union of its components. So $\overline{K}_n$ is the union of n components each of which is a single vertex.

We note that the components of any 2-regular graph are cycles and of any 1–regular graph are K_2's.

But some graphs are more connected than others. Some have vertex deleted subgraphs which are disconnected; others have all their vertex deleted subgraphs connected, but not all of the next round of vertex deleted subgraphs are connected.

Consequently we say that G has **connectivity** k, if there exists a subset S of VG with |S| = k such that <VG \ S> is disconnected, and there is no subset S' of VG with |S'| < k for which <VG \ S'> is disconnected.

Looking at C_6 we see that we need to delete two vertices before the cycle is disconnected. Hence C_6 has connectivity 2. It is clear that P has connectivity at most 3 since {1',2,5} disconnects P. However, no 2-subset of VP disconnects P, so in fact it has connectivity 3.

By convention we say that K_n has connectivity n - 1 even though it cannot be disconnected by the removal of n - 1 vertices.

The set S, in the definition of connectivity, is a cutset of G. Indeed any set N of vertices whose removal disconnects G is called a **cutset**.

A less sharp version of connectivity is k-connectivity. We say that G is **k–connected** if it is of connectivity k' for some k' ≥ k.

Edge connectivity is defined in a similar way. Thus a set of edges whose removal disconnects a graph is an **edge cut** or an **edge cutset**. The **edge connectivity** of a graph is the size of the smallest edge cut whose removal disconnects the graph. And a graph is **k-edge-connected** if its edge connectivity is at least k.

We note that an edge cut of cardinality one is called a **bridge (cut-edge)**. Hence the graph of Figure 2.3 has a bridge (the edge e) and therefore has edge connectivity one.

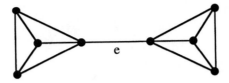

Figure 2.3

Clearly the edge connectivity of a graph, like the connectivity, cannot be greater than its minimum degree δ. This is simply because if we remove the δ edges from a vertex of minimum degree, the graph will be disconnected. Thus P has edge connectivity at most 3. But by inspection P has no bridges and by a little deeper investigation it can be seen that P has no edge cut of size 2. Hence the edge connectivity of P is 3, the same as its connectivity.

Since the graph of Figure 2.3 also has its connectivity equal to its edge connectivity it might be conjectured (on the basis of two examples) that edge connectivity and connectivity are always the same.

The graph H of Figure 2.4 shows that such a conjecture would be false. Since H_v is disconnected, the connectivity of H is one. Since $\{e_1, e_2\}$ is a cutset and H has no bridge, its edge connectivity is 2. (Note that both the connectivity and the edge connectivity are less than δ, which for H, is 3.)

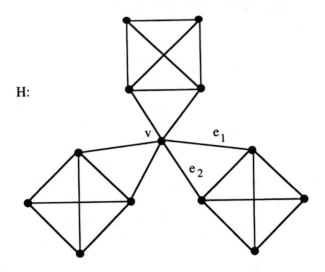

Figure 2.4

Graphs like H led Whitney [hW 32] to the following result.

Theorem 2.1   The connectivity of a graph is at most equal to its edge connectivity. The edge connectivity is at most equal to the graph's minimum degree.

Proof: Choose an edge cutset $S = \{e_1, e_2, ..., e_t\}$ and a set of vertices U such that each e_i is incident with one vertex of U. Since S is an edge cutset, U is a vertex cutset. Finally, it is clear that $|U| \leq t$.   □

We have seen that for P, edge connectivity and connectivity have the same value. This is in fact true for all cubic graphs.

Theorem 2.2 Let G be a cubic graph. Then the connectivity of G and its edge connectivity are equal.

The proof of this is quite straightforward but needs the discussion of a few cases so we omit it here; see Exercise 12.

Because of the importance to us of cubic graphs we now give characterizations of the variously connected cubic graphs. Clearly from Theorem 2.1, the connectivity of a cubic graph is at most 3.

If e and f are two distinct edges of G then a **joining** of e and f is the operation of inserting a new vertex u on e and v on f and adding the edge uv; see Figure 2.5.

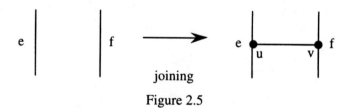

joining

Figure 2.5

Theorem 2.3 A graph G is 3-connected and cubic if and only if it is K_4 or can be obtained from K_4 by a series of joinings.

A proof of this theorem can be found on p.140 of [wT 66].

We can use Theorem 2.3 to give another proof that P is 3-connected. To see which three edges to add to K_4 it is probably easiest to start with P and remove three. However, we work constructively from K_4 as in Figure 2.6.

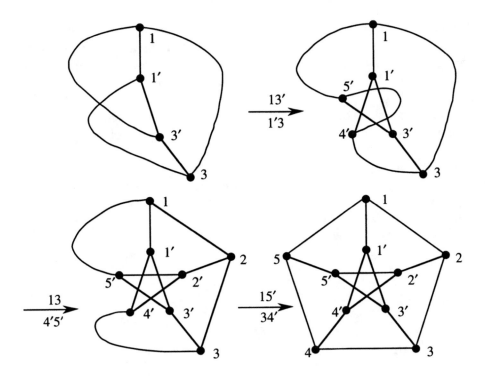

Figure 2.6

First label the vertices of K_4 as $1,1',3,3'$ as shown. Then join edges $13'$, $1'3$. This introduces the vertices $4'$, $5'$, as shown. Then join 13 and $4'5'$ which introduces vertices $2, 2'$. Finally join $15'$, $34'$ and introduce $4, 5$ to complete P.

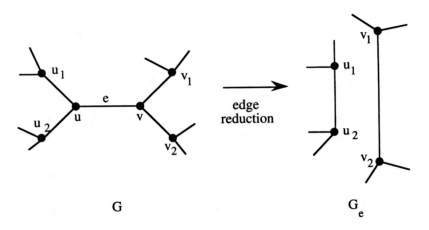

Figure 2.7

The reverse of a joining is an **edge reduction**. Let $e = uv \in EG$ with $N(u) = \{u_1, u_2, v\}$ and $N(v) = \{v_1, v_2, u\}$. The **edge reduction** G_e of G, is the multigraph obtained from G by removing the edges adjacent to u and v and adding the edges $u_1 u_2$ and $v_1 v_2$. This is shown in Figure 2.7.

To prove Theorem 2.3 by induction, it is necessary to show that every 3–connected cubic graph G has a 3-connected cubic edge reduction.

Wormald [nW 79] has produced similar results to Theorem 2.3 for 1-connected and 2-connected cubic graphs. Two constructions are needed. The **addition of** K_4 to an edge e of G is achieved by taking $G \cup K_4$ and joining e to an edge of K_4. The **insertion of two triangles** in an edge uv of G, is produced by replacing the edge uv by the graph of Figure 2.8.

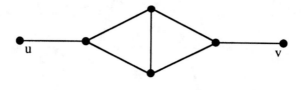

Figure 2.8

Wormald's two theorems now complete the classification of cubic graphs.

Theorem 2.4 Every connected cubic graph is either K_4 or is obtained from K_4 by a series of operations which add K_4 to an edge, insert two triangles to an edge or join two edges. □

Theorem 2.5 Every 2-connected cubic graph is either K_4 or is obtained from K_4 by a series of operations which either insert two triangles to an edge or join two edges. □

The proofs of these results may be found in [nW 79].

While these three theorems do not give a description of cubic graphs which is as precise as the classification of 1-regular and 2-regular graphs, they do in principle allow us to construct all cubic graphs.

3.   Bipartite graphs

On the theme of connected graphs one might ask what is the smallest connected graph of order n.

A moment's thought will reveal that such a graph will contain no cycles. If it did, we could remove one of the edges of the cycle and still maintain its connectivity.

Define a connected acyclic graph to be a **tree**. We show the trees on five vertices in Figure 3.1. The vertices of degree one in a tree (or any other graph) are called **endvertices**.

Figure 3.1

Theorem 3.1 Every tree contains at least one endvertex.

Proof: Suppose every vertex of the tree T has degree greater than or equal to two. Choose an arbitrary vertex $u_1 \in VT$. Let $u_2 \in VT$ such that $u_1u_2 \in ET$. Now since deg $u_2 \geq 2$, there exists $u_3 \neq u_1$ such that $u_2u_3 \in ET$. Then there exists $u_4 \in VT$ such that $u_3u_4 \in ET$. If $u_4 = u_1$, we have a cycle (u_1, u_2, u_3) in T which is a contradiction. Hence $u_4 \neq u_1$.

Continuing inductively we find $u_1, u_2, ..., u_k$ such that $u_iu_{i+1} \in ET$ for $i = 1, 2, ..., k - 1$ and these k vertices are all distinct.

But n is finite, so this inductive process must terminate for some k, so $u_{k+1} = u_{i'}$ where $i' \in \{1, 2, ..., k-1\}$. Hence T contains a cycle. This contradiction proves the result.   □

Theorem 3.2 If T is a tree then $|ET| = |VT| - 1$.

Proof: We proceed by induction on $|VT|$. Suppose that $|VT| = 1$, then $|ET| = 0$ and the result holds. Now assume that $|ET| = |VT| - 1$ for $1 \leq |VT| \leq n - 1$.

Let $u \in VT$ be an endvertex of T. Then T_u is a tree and $|ET_u| = |VT_u| - 1$. But $|VT| = |VT_u| + 1$ and $|ET| = |ET_u| + 1$ and so the result follows.   □

G:

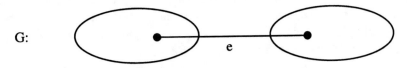

Figure 3.2

In Figure 3.2 the edge e is a bridge. It is clear that e cannot lie on a cycle since otherwise G - e would be connected.

Theorem 3.3 Every connected graph contains a spanning tree.

Proof: Let G be connected and let T be a minimal connected spanning subgraph of G. (Such a graph must exist since G itself is a connected spanning subgraph of G.) Now T is connected but for $e \in ET$, T - e is not, for otherwise T would not be minimal. Hence every edge of T is a bridge.

Suppose that T is not a tree. Then T contains a cycle and so contains an edge which is not a bridge. Hence G contains a spanning tree T.   □

As a result of Theorem 3.3, we see that trees are the smallest connected graphs of order n.

One special type of tree worthy of mention is the **path** P_n which consists of two vertices of degree one and   n - 2   vertices of degree 2 (see Figure 3.3). The **length** of P_n is n - 1.

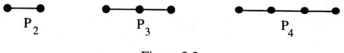

$$P_2 \qquad\qquad P_3 \qquad\qquad P_4$$

Figure 3.3

As with cycles we often refer to paths in a general graph. A **path** in G is a subgraph isomorphic to P_m for some   m. If the path is on the vertices $u_1, u_2,.., u_m$ we denote it by $u_1, u_2,..., u_m$. Further d(u,v), the **distance between u and v**, is the number of edges in the shortest path joining u and v.

Now trees have the interesting property that if we colour an arbitrary vertex black, then colour its neighbours white, then their neighbours black, and so on, every vertex is given only one colour and no edge joins two black vertices or two white vertices.

In general if the vertices of G can be divided into two sets X , Y so that no edge of G joins two vertices of X or two vertices of Y, then G is said to be a **bipartite graph.** The sets X and Y are called the **parts** of G. In particular, G is a **complete bipartite graph** if G is bipartite and every vertex of X is adjacent to every vertex of Y. If |X| = m and |Y| = n we denote this complete bipartite graph by $K_{m,n}$. Some examples of bipartite graphs are given in Figure 3.4. We note that $K_{m,n} \cong K_{n,m}$ and that $K_{m,n} = \overline{K}_m + \overline{K}_n$.

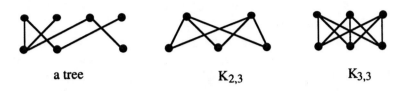

a tree $K_{2,3}$ $K_{3,3}$

Figure 3.4

An important property of bipartite graphs is given in the next theorem.

Theorem 3.4 A graph is a bipartite graph if and only if it contains no odd cycle.

Proof: Let G be bipartite with parts X and Y. Further, let $C = (u_1, u_2,...,u_m)$ be a cycle in G. Relabel, if necessary, so that $u_1 \in X$. Since G is bipartite, $u_i \in X$ for $i \equiv 1 \pmod 2$ and $u_i \in Y$ for $i \equiv 0 \pmod 2$. But $u_1 u_m \in EG$. Hence $m \equiv 0 \pmod 2$ and so C is an even cycle.

Now suppose that G contains no odd cycle. Clearly if we can show that any component of G is bipartite, then G must be bipartite. Hence, without loss of generality we may assume that G is connected.

Choose an arbitrary vertex $u \in VG$ and define $X = \{v : d(u,v)$ is even$\}$, $Y = \{w : d(u,w)$ is odd$\}$.

If $|X|$ or $|Y| \leq 2$ then the result follows trivially. Hence we may assume that $|X| > 2$. Let $v_i \in X$ and let Q_i be a shortest path from u to v_i for $i = 1, 2$. Suppose $u' \in VQ_1 \cap VQ_2$ and that u' is the furthest away from u along Q_1 of all points on Q_1 and Q_2. Since Q_1, Q_2 are shortest paths, then the distance from u to u' along Q_1 equals the distance from u to u' along Q_2. But Q_1 and Q_2 are both of even length, so the distance from u' to v_1 along Q_1 and the distance from u' to v_2 along Q_2 both have the same parity.

If $v_1 v_2 \in EG$, then the cycle consisting of Q_1 from v_1 to u', Q_2 from u' to v_2 and $v_1 v_2$ is odd. This contradiction shows that $v_1 v_2 \notin EG$.

Similarly $w_1 w_2 \notin EG$ for $w_1, w_2 \in Y$ and so X and Y are the parts of a bipartition of G.   □

Since P contains 5-cycles it is clearly not a bipartite graph.

4.   Automorphism groups

From the labelled version of the Petersen graph given in Figure 4.1(a), it is clear
that the vertices 1,2,3,4,5 are in some sense equivalent. Clearly if P is drawn
carefully we can rotate it through an angle of $2\pi/5$ so that 1 moves to 2, 2 to
3, 3 to 4, 4 to 5 and 5 to 1 and similarly 1′ to 2′, 2′ to 3′, 3′ to 4′, 4′ to
5′ and 5′ to 1′.

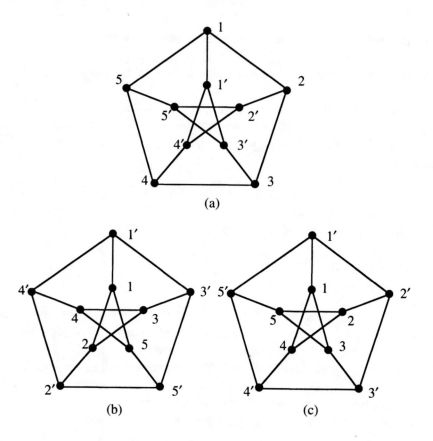

(a)

(b)   (c)

Figure 4.1

The effect of such a rotation is to send VP to VP in such a way that edges are
sent to edges. For instance, this rotation sends 11′ to 22′, 1′4′ to 2′5′ and so
on. It is therefore an isomorphism of P onto itself.

Any mapping $\varphi: VG \rightarrow VG$ which sends G isomorphically onto itself is called
an **automorphism** of G.

Thus $\varphi: VP \to VP$ such that

$\varphi(1) = 2$ $\varphi(2) = 3$ $\varphi(3) = 4$ $\varphi(4) = 5$ $\varphi(5) = 1$

$\varphi(1') = 2'$ $\varphi(2') = 3'$ $\varphi(3') = 4'$ $\varphi(4') = 5'$ $\varphi(5') = 1'$

is an automorphism of P.

Another automorphism of P is illustrated in Figure 4.1(b). Here $\varphi: VP \to VP$ is such that

$\varphi(1) = 1'$ $\varphi(2) = 4'$ $\varphi(3) = 2'$ $\varphi(4) = 5'$ $\varphi(5) = 3'$

$\varphi(1') = 1$ $\varphi(2') = 4$ $\varphi(3') = 2$ $\varphi(4') = 5$ $\varphi(5') = 3,$

where, pictorially, $\varphi(1) = 1'$ means vertex 1 moves to the position originally occupied by vertex 1', and so on. Checking all edges in the original copy of P, say in Fig 4.1(a), we see that these are sent to edges in P. For instance, $\varphi(11') = \varphi(1)\varphi(1') = 1'1$, $\varphi(15) = \varphi(1)\varphi(5) = 1'3'$ and $\varphi(14) = \varphi(1)\varphi(4) = 1'5'$.

However Figure 4.1(c) shows a map which is **not** an automorphism. Here $\varphi: VP \to VP$ is such that

$\varphi(1) = 1'$ $\varphi(2) = 2'$ $\varphi(3) = 3'$ $\varphi(4) = 4'$ $\varphi(5) = 5'$

$\varphi(1') = 1$ $\varphi(2') = 2$ $\varphi(3') = 3$ $\varphi(4') = 4$ $\varphi(5') = 5.$

Although some edges are preserved, from Figure 4.1(a), by this map, not all edges are. Clearly $\varphi(12) = \varphi(1)\varphi(2) = 1'2'$. But $1'2' \notin EP$ in the drawing of Figure 4.1(a). Hence φ is not an automorphism.

We can think of automorphisms as being permutations of VG which send edges of G into edges of G and 'non edges' of G to 'non edges'. The two automorphisms of P that have been noted from Figure 4.1 can be represented by the permutations $(12345)(1'2'3'4'5')$ and $(11')(24'53')(32'45')$.

Let's take another example. Consider the copy of K_4 shown, conveniently labelled, in Figure 4.2(a). Now $\gamma = (12)(34)$ is a permutation on VK_4 which interchanges 1 and 2 and interchanges 3 and 4. Then $\gamma(K_4)$, the image of K_4, is represented in Figure 4.2(b).

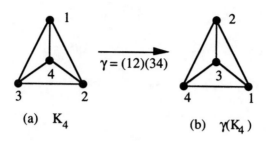

(a) K_4 (b) $\gamma(K_4)$

Figure 4.2

We note that the image of the edge 12 under γ is $\gamma(1)\gamma(2) = 21 = 12$, while the image of 24 is 13. Since it can be shown that γ preserves edges and K_4 has no 'non edges', then γ is an automorphism of K_4.

Actually the automorphisms of a graph form a group called the **automorphism group of G** and denoted by $A(G)$. For a graph with $|VG| = n$, it is clear that $A(G)$ is a subgroup of S_n, the symmetric group of degree n. (S_n is just the group of all permutations of n objects.)

Now $S_4 = \{1,(12),(13),(14),(23),(24),(34),(123),(124),(132),(134),(142),(143),$ $(234),(243),(1234),(1243),(1324),(1342),(1423),(1432),(12)(34),(13)(24),$ $(14)(23)\}$, where 1 denotes the identity permutation which fixes every number. Looking again at the drawing of K_4 in Figure 4.2(a) we see, after some effort, that every permutation of S_4 is an automorphism of K_4. Hence $A(K_4) = S_4$.

Actually, $A(K_n) = S_n$ for all n. This is readily seen by noting first that every vertex in K_n is adjacent to every other vertex and secondly that S_n just permutes the n vertices of K_n amongst themselves.

So, as we would expect, the complete graphs have a high degree of symmetry because they have a large automorphism group.

But there are graphs with no symmetries. These graphs are called **identity graphs,** since the only automorphism of their vertices which sends the graphs into themselves is the identity permutation. Hence for such graphs, $|A(G)| = 1$.

In Figure 4.3 we show the smallest non-trivial identity graph and the smallest non-trivial identity tree.

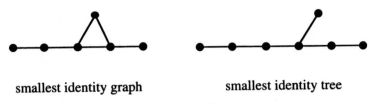

smallest identity graph   smallest identity tree

Figure 4.3

We now prove some basic results concerning automorphisms.

Lemma 4.1 If $\gamma \in A(G)$ and $u \in VG$, then deg $\gamma(u) = $ deg u.

Proof: If $uv \in EG$, then $\gamma(u)\gamma(v) \in EG$ since $\gamma \in A(G)$. Hence deg $\gamma(u) \geq$ deg u. Now every neighbour of $\gamma(u)$ comes from a neighbour of u since γ is an automorphism. We thus have the required equality.   □

Lemma 4.2 $A(G) = A(\overline{G})$.

Proof: Let $\gamma \in A(G)$. Since γ preserves adjacencies, then $\gamma(uv) \in EG$ if and only if $uv \in EG$. Hence $\gamma(uv) \notin EG$ if and only if $uv \notin EG$. Equivalently, $\gamma(uv) \in E\overline{G}$ if and only if $uv \in E\overline{G}$. So $\gamma \in A(\overline{G})$. Interchanging the role of G and $\overline{G}$ in the above gives the desired equality.   □

Now an **edge automorphism** of G is a permutation of EG which sends edges with a common endvertex into edges with a common endvertex. The set of all edge automorphisms of G forms a group called the **edge automorphism group of** G. We denote this by $A^*(G)$.

For example, $A^*(K_{1,n}) = S_n$.

Consider the graphs of Figure 4.4. Now $A^*(M) = \{1,(e_1e_4), (e_2e_3), (e_1e_4)(e_2e_3)\}$ and $A(M) = \{1,(34)\}$. Further $A^*(N) = \{1, (e_1e_3), (e_2e_4), (e_1e_2)(e_3e_4),$ $(e_1e_4)(e_2e_3), (e_1e_3)(e_2e_4), (e_1e_2e_3e_4), (e_1e_4e_3e_2)\}$ and $A(N) = \{1, (13), (24),$ $(13)(24)\}$.

So $A(G)$ and $A^*(G)$ are not always isomorphic. However the following theorem shows that there are very few exceptions.

Theorem 4.3 Let G be connected with $|VG| \geq 3$. Then $A(G) \cong A^*(G)$ if and only if G is not one of M, N or K_4.

Proof: See [B-C 71].   □

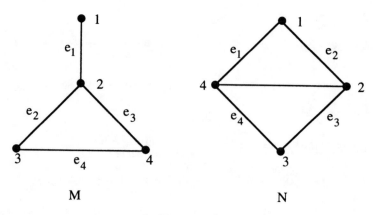

M   N

Figure 4.4

It is useful at this stage to introduce the line graph of a graph. The **line graph**, $L(G)$ of G is the graph whose vertices are the edges of G and where two such vertices are adjacent if the corresponding edges are incident to a common vertex of G. Figure 4.5 gives the line graph of K_4. We note that $\overline{L(K_4)} = K_2 \cup K_2 \cup K_2$. It turns out that $P = \overline{L(K_5)}$ (see Exercise 26).

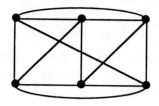

Figure 4.5

At this stage it is worth noting the following straightforward result.

Lemma 4.4 $A^*(G) \cong A(L(G))$.

We are now close to being able to find $A(P)$. Before we do though, we will need a short excursion into the theory of permutation groups.

If Γ is a permutation group acting on a set Ω and $\alpha \in \Omega$, then $\Gamma_\alpha = \{g \in \Gamma : g(\alpha) = \alpha\}$ is called the **stabilizer of** α. It consists of all the permutations of Γ which **fix** α, that is, which send α to itself.

The set $\Gamma(\alpha) = \{g(\alpha) : g \in \Gamma\}$ is the **orbit of** α in Γ. This is the subset of Ω comprising the images of α under the action of the permutations of the group Γ.

For instance, if $\Gamma = \{1,(12),(34),(12)(34)\}$ and $\Omega = \{1,2,3,4\}$, then $\Gamma_1 = \{1,(34)\}$ and $\Gamma(1) = \{1,2\}$. This leads us to the orbit-stabilizer relation.

Lemma 4.5 $|\Gamma| = |\Gamma_\alpha||\Gamma(\alpha)|$.

Proof: The result is proved by consideration of the decomposition of Γ into cosets of Γ_α; see [heW 64].   □

Now a permutation group is said to be **transitive** if $\Gamma(\alpha) = \Omega$ for some $\alpha \in \Omega$. Thus the group $\{1,(12345),(13524),(14253),(15432)\}$ is transitive while $\{1,(12),(34),(12)(34)\}$ is not. Clearly if $\Gamma(\alpha) = \Omega$ for some $\alpha \in \Omega$, then $\Gamma(\beta) = \Omega$ for all $\beta \in \Omega$. As a consequence of the definition, if Γ is transitive and $\alpha, \beta \in \Omega$, there always exists $g \in \Gamma$ such that $g(\alpha) = \beta$.

Returning to graphs, we say that G is **vertex-transitive** if $A(G)$ is transitive. Thus K_n is vertex-transitive since $A(K_n) = S_n$ and S_n is transitive. In view of Lemma 4.2, $\overline{K}_n$ is also vertex-transitive.

We can now determine $A(P)$ and show that P is vertex-transitive.

Theorem 4.6 P is a vertex-transitive graph and $A(P) \cong S_5$.

Proof: Now $P = \overline{L(K_5)}$. Hence

$$
\begin{aligned}
A(P) &= A(L(K_5)) & &\text{by Lemma 4.2,} \\
     &\cong A^*(K_5) & &\text{by Lemma 4.4,} \\
     &\cong A(K_5) & &\text{by Lemma 4.3,} \\
     &= S_5.
\end{aligned}
$$

Now from the original discussion on automorphisms relating to Figure 4.1 we know that $\gamma_1 = (12345)(1'2'3'4'5')$ and $\gamma_2 = (11')(24'53')(32'45') \in A(P)$. Using the powers of γ_1 we can see that $\{1,2,3,4,5\} \subseteq \Gamma(1)$. With γ_2 acting on $\{1,2,3,4,5\}$ we get $\{1,2,3,4,5,1',2',3',4',5'\} = \Omega = \Gamma(1)$. Hence $\Gamma(1) = \Omega$ and P is vertex-transitive.   □

This means that all the vertices of P play the same role in P and are mutually indistinguishable. This point has already been noted in Section 1.

But the symmetry of P does not stop there. In fact A(P) sends any edge of P onto every other edge. Hence the edges of P are also equivalent.

We say that G is **edge-transitive** if for any $e_1, e_2 \in EG$, there exists $\gamma \in A^*(G)$ such that $\gamma(e_1) = e_2$.

Theorem 4.7 P is edge-transitive.

Proof: Recall from p. 3 that A = {12,23,34,45,51}, B = {11′,22′,33′,44′,55′} and C = {1′3′,2′4′,3′5′,4′1′,5′2′}. If $\gamma \in A(G)$ then γ 'sends' the edge ij to $\gamma(i)\gamma(j)$. Hence γ induces a permutation $\gamma^* \in A^*(G)$. We usually identify γ and γ^*.

Clearly there is a power of γ_1 (see proof of Theorem 4.6) which sends any member of A to another member of A. The same can be said for B and C.

A power of γ_2 sends any member of A into a member of C.

Finally $\gamma_3 = (21′)(34′)(2′3′)$ sends some members of B to A some to C and fixes 55′.

Combining $\gamma_1, \gamma_2, \gamma_3$ or their powers thus sends e_1 to e_2 for any $e_1, e_2 \in EP$. Thus P is edge-transitive.   □

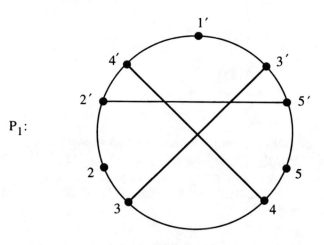

P_1:

Figure 4.6

We can use the automorphisms of P and its subgraphs to give an alternative proof of the fact that P is of connectivity 3.

First note that since deg u = 3 for any u ∈ VP, that the connectivity of P is at most 3.

Since P is vertex-transitive, all the vertex-deleted subgraphs P_v of P are isomorphic. So consider P_1 (Figure 4.6). The group of this graph has two orbits {1′,2,5} and {2′,3′,3,4,4′,5′}. Hence to see whether or not P is of connectivity two we need only see whether or not removing 1′ or 2′ from P_1 disconnects P_1. Since P - {1,1′} and P - {1,2′} are connected, then P has connectivity at least three.

5.   Hamiltonian graphs

Historically the important properties of P have been that it is neither hamiltonian nor planar nor 3-edge-colourable. In this section we discuss the hamiltonian property and in the next sections we deal with planarity and colourability.

We say that G is **hamiltonian** if there is a cycle in G which passes through every vertex. Obviously C_n and K_n are hamiltonian but no tree is. It is not difficult to see that $K_{m,n}$ is hamiltonian precisely when m = n.

Theorem 5.1 P is not hamiltonian.

Proof: Let A, B, C be the subsets of EP as defined on p. 3. Let H be a hamiltonian cycle of P.

Now B is an edge cutset of P. As such, H must use an even number of the edges of B. Hence H uses 2 or 4 edges of B.

Because of the edge-transitivity of P, we may assume that 11′ ∈ EH. Then either 12 or 15 ∈ EH. By symmetry, we may assume without loss of generality, that 12 ∈ EH.

Since P is cubic, 15 ∉ EH and hence 45, 55′ ∈ EH or else 5 is not on H.

If H uses only two edges of B, they are 11′ and 55′, so then 23, 34 ∈ EH as do 2′4′, 2′5′, 3′1′, 3′5′ and 4′1′. However, this forces two vertices (1′ and 5′) to have degree 3 in the cycle H. Hence |EH ∩ B| = 4.

By symmetry one of 22′, 44′ ∈ EH. Suppose, without loss of generality, that

$44' \in EH$. Since $34 \notin EH$ this forces 23 and $33'$ to be edges of H. Since $|EH \cap B| = 4$, $22' \notin EH$ and hence $2'4'$, $2'5' \in EH$. This, however, forces the subcycle $(2',5',5,4,4')$ in H.

Hence H does not exist.   □

If P is not hamiltonian, what is the size of the largest cycle of P ?

Lemma 5.2 Any 9 vertices of P lie on a cycle. Further, there are precisely two cycles through any 9 vertices of P.

Proof: Consider the 9 vertices of $VP \setminus \{1\}$ in the graph of P_1 in Figure 4.6. The edges $1'3'$, $1'4'$, 23, $22'$, 45, $55'$ must be in any hamiltonian cycle H of P_1.

Suppose $3'5' \in EH$. Since this implies $33' \notin EH$, we must have $34 \in EH$. But $34 \in EH$ implies $44' \notin EH$ and so $2'4' \in EH$. Hence $H = (1', 3', 5', 5, 4, 3, 2, 2', 4')$.

However, if $33' \in H$, then a similar argument forces H to be $(1', 3', 3, 2, 2', 5', 5, 4, 4', 1)$. As we have covered all possibilities these are the only two hamiltonian cycles in P_1 and hence the only two cycles containing $VP \setminus \{1\}$.

The vertex-transitivity of P shows that this is true for $VP \setminus \{v\}$ where v is any vertex of P.   □

At the other extreme we can ask for the smallest cycle in P. It is clear from the representation of P in Figure 4.1 that P contains a 5-cycle. Do 3-cycles or 4-cycles exist in P?

Because P is vertex-transitive every vertex plays the same role in P. If a 3-cycle existed in P, then one would exist at every vertex. Consider vertex 1. Since <N(1)> is the empty graph, vertex 1 is not on a 3-cycle.

Since no two neighbours of 1 have a common vertex, P contains no 4-cycles. Hence the smallest cycle in P is a 5-cycle.

In general we define the **girth**, $\gamma(G)$, of a graph G, to be the order of the smallest cycle it contains. Hence $\gamma(P) = 5$, or the girth of P is 5. On the other hand $\gamma(K_n) = 3$ and $\gamma(C_n) = n$.

Returning to P, it can be shown that it contains cycles of size 5, 6, 8 and 9.

As far as the hamiltonian concept is concerned, we still have much to learn. There is not as yet, a complete characterization of hamiltonian graphs. Indeed there is evidence to suggest that it may not be possible to find a useful characterization.

Reviews on hamiltonian graphs can be found in [jB 78] and [rG 91]. Below we list some of the main results in this topic.

Theorem 5.3 If $n = |VG| \geq 3$ and $\deg u \geq \frac{1}{2}n$ for all $u \in VG$, then G is hamiltonian.

Proof: See Dirac [gD 52]. $\square$

Theorem 5.4 Let $|VG| = n$. If $\deg u + \deg v \geq n$ for any pair of non-adjacent vertices $u, v \in VG$, then G is hamiltonian.

Proof: See Ore [oO 60]. $\square$

Another interesting approach was introduced by Fan [gF 84]; see also [hV 90]. He showed that we need not consider 'all pairs of non-adjacent vertices' but only a particular subset of pairs. Thus if for all pairs of vertices u and v which are exactly a distance two apart we have max $\{\deg u, \deg v\} \geq |VG|/2$, then G is hamiltonian. Other interesting sufficient conditions for hamiltonicity, which generalize those of Fan, have been obtained by Ainouche [aA 91].

Theorem 5.5 Let G be a graph with degrees $d_1 \leq d_2 \leq ... \leq d_n$. If $d_k \leq k < \frac{1}{2}n$ implies $d_{n-k} \geq n - k$ for each k, then G is hamiltonian.

Proof: See Chvátal [vC 72]. $\square$

If we restrict our attention to regular graphs we have the following series of results.

Theorem 5.6 Let G be a 2-connected graph which is regular of degree $n - k$, where $k \geq 3$.

If $|VG| = 2n$ and $n \geq k^2 + k + 1$, then G is hamiltonian.
If $|VG| = 2n - 1$ and $n \geq 2k^2 - 3k + 3$, then G is hamiltonian.

Proof: See Erdös and Hobbs [E-H 77]. $\square$

Theorem 5.7 Let G be 2-connected and r-regular. Then if $|VG| \leq 3r$, G is hamiltonian.

Proof: See Jackson [bJ 80]; see also [bJ 86].   ☐

This leads us to wonder what happens if $|VG| = 3r + 1$. Certainly not all such graphs are hamiltonian since for $r = 3$, $|VP| = 3r + 1$. This problem was resolved by Zhu et al and their solution puts the Petersen graph in a unique position. A simple proof of this result is contained in [B-K 88].

Theorem 5.8 Let G be 2-connected and r-regular. Then if $|VG| \leq 3r + 1$, G is hamiltonian unless $G = P$.

Proof: See [E-H 78], [Z-L-Y 85] and [B-K 88].   ☐

It would seem reasonable to try to push this line of investigation further. What can be said if $|VG| \leq 3r + a$, for various values of a? What can be said for 3–connected graphs, and so on? More complicated, but elegant, sufficient conditions for 2-connected graphs to be hamiltonian are derived in [gF 84] and [B-W 87].

We return to this discussion in Chapter 9, where the prominent role of P in the theory of hamiltonicity is pursued further.

There is a dual problem to that of finding a hamiltonian cycle. In fact it was probably **the** first graph-theoretical problem. It was posed and solved by Euler in 1736 [lE 36]; see also [B-L-W 76].

A **trail** in a graph G is an alternating sequence of vertices and edges $u_1, u_1u_2, u_2, u_2u_3, ..., u_i, u_iu_{i+1}, u_{i+1},..., u_n$. It is a **closed trail** if $u_1 = u_n$.

The problem is this. For which graphs G is it possible to find a trail (closed trail) in which every edge of G appears precisely once? Such a trail (closed trail) is called an **Euler trail (Euler tour)**.

This is equivalent to asking which graphs can be drawn without removing the pen from the paper and without tracing a given edge more than once.

The surprisingly simple answer is given in the next theorem. Euler [lE 36], [S-S-W 88] demonstrated the necessity that every vertex have even degree, but surprisingly it was not until 1873 [cH 73] that the sufficiency was established. We give a very simple proof due to Fowler [rF 88].

Theorem 5.9 Let G be a connected pseudograph. Then G has an Euler tour if and only if every vertex of G is of even degree.

Proof: The proof of sufficiency is by induction on the number of edges. Trivially the theorem is true for pseudographs satisfying the conditions of the theorem and having less than three vertices.

Now suppose that an Euler tour exists in any connected pseudograph with even degrees and less than q edges. Consider any connected even-degree pseudograph G with q edges. Since we may assume that G has at least three vertices there exists a vertex v which is adjacent to vertices x and y distinct from v; possibly x = y. Delete the edges vx and vy and (i) join x to y if x ≠ y or (ii) if x = y insert a loop at x. By induction, if the resulting pseudograph G′ is connected, then G′ has an Euler tour T′ which induces an Euler tour of G on replacing x, xy, y in T′ by x, xv, v, vy, y. If G′ is not connected then x and y belong to one component and v to the other. By induction both components have Euler tours T_1 and T_2 which together induce an Euler tour of G on replacing x, xy, y in T_1 by x, xv, T_2, vy, y.

The proof of necessity is straightforward and so is omitted.   □

Theorem 5.10 Let G be a connected pseudograph. Then G has an Euler trail if and only if it has at most two vertices of odd degree.

Proof: See [IE 36].   □

In view of the simple nature of the characterizations of the last two theorems and the apparent dual relation between closed Euler trails and hamiltonian cycles, it seems strange that a satisfactory characterization of hamiltonian graphs is so hard to find.

6.   Planar graphs

A graph is said to be **planar** if it can be drawn in the plane so that no two edges cross. It is fairly obvious that all trees are planar. The only complete planar graphs are those on four or fewer vertices. Using the Jordan curve theorem it can be shown that K_5 and $K_{3,3}$ are non-planar.

On the other hand it is possible to draw a planar graph in a non-planar fashion. Hence it is useful to call a drawing of a planar graph **plane** if it is a drawing in which no two edges cross. Figure 6.1 gives two drawings of K_4 only one of which is plane.

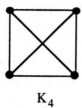
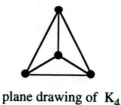

K$_4$ plane drawing of K$_4$

Figure 6.1

The plane graph G divides the plane into a number of disjoint regions called
faces. Starting from a given point in the plane, a **face** is the collection of all
points in the plane which can be reached from the given point, without crossing
an edge of G. From Figure 6.1 it can be seen that K$_4$ has four faces. (The set of
points on the 'outside' of a graph also form a face.)

There are two well-known characterizations of planar graphs which we present in
this section. The first of these is due to Kuratowski. This relies on the notion of
a subdivision.

We say that a graph is a **subdivision of G** if it can be obtained from G by the
insertion of an arbitrary number of vertices of degree two on the edges of G.
This is illustrated in Figure 6.2 where we show a subdivision of K$_4$.

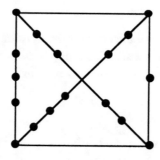

Figure 6.2

Having noted that K$_5$ and K$_{3,3}$ are non-planar we now see that they are in
some sense the only non-planar graphs. We need some special notation before
giving one of Thomassen's proofs of Kuratowski's theorem to be found in
[cT 81].

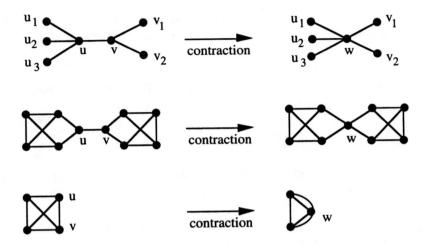

Figure 6.5

An **edge contraction** (or simply **contraction**) of a multigraph   G,   is the multigraph obtained from   G   by removing the edge   uv   and the vertices   u, v   and inserting a new vertex   w   such that w   is joined to each vertex   y   of $N_G(u) \cup N_G(v)$,   by the sum of the number of edges from   u   to   y   and from   v   to   y   in   G.   This is illustrated in Figure 6.5.

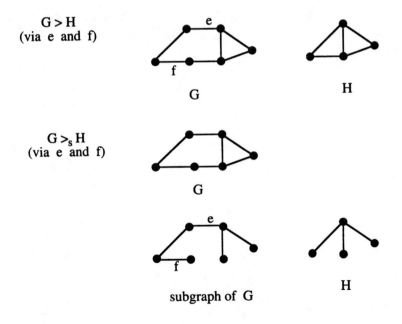

Figure 6.6

Now G is **contractible** to H if H can be obtained from G by a sequence of contractions. In such a case H will be referred to as a **contraction** of G. We write G > H.

Further, we say that H is a **subcontraction** of G if H is isomorphic to a contraction of a subgraph of G. This will be denoted by $G >_S H$.

These ideas are illustrated in Figure 6.6.

If G > H, since any graph is a subgraph of itself, then $G >_S H$. However the converse is not true in general.

Suppose that G is a graph and e ∈ EG. Define G*e to be the graph obtained from G by contracting the edge e and replacing each 2-cycle by a single edge.

The following result was proved by Kuratowski [kK 30].

Theorem 6.1 G is planar if it contains no subgraph which is a subdivision of K_5 or $K_{3,3}$.   □

The graphs K_5 and $K_{3,3}$ are easily shown to be non-planar (see Exercise 32).

Until 1981, proofs of Kuratowski's Theorem were somewhat longwinded and complicated. Then Thomassen came up with a very elegant proof based on contractions.

His first move was to note that we can restrict our attention to 3-connected graphs.

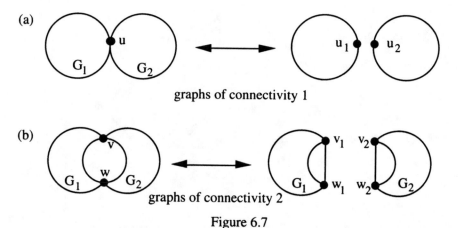

(a)

graphs of connectivity 1

(b)

graphs of connectivity 2

Figure 6.7

Consider the 1-connected graph G of Figure 6.7(a). The planarity or otherwise of G clearly depends on G_1 or G_2. By repeated reductions of the type shown in Figure 6.7(a), G will be planar if and only if certain 2-connected graphs are planar.

But Figure 6.7(b) shows us how to handle 2-connected graphs. We may break G into a collection of 3-connected graphs in this way. Hence the planarity of a 2–connected graph will depend on graphs that are 3-connected.

So we see that we now need only consider 3-connected graphs.

We say that e is a **contractible edge** in a 3-connected graph G, if G*e is 3–connected. The point is that contracting the edge e in G does not reduce the connectivity below 3. We now show that every 3-connected graph except K_4 contains a contractible edge.

Lemma 6.2 If G is 3-connected and $G \neq K_4$, then G contains a contractible edge.

Proof. Suppose e = uv is a non-contractible edge in G. Then S = {u,v,w} is a 3-cut in G for some $w \in VG \setminus \{u,v\}$. Now choose e such that $G \setminus S$ has a component U which is as large as possible. (The situation is shown in Figure 6.8).

Since S is a cutset and $G \neq K_4$, then $U' \neq \emptyset$. Furthermore, there exists $x \in N(w) \cap U'$ since G is 3-connected and not K_4.

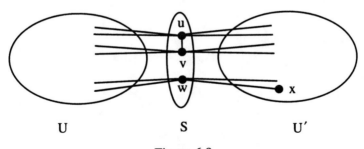

U   S   U′

Figure 6.8

Clearly U^*, the graph induced on the vertices of U and {u,v}, must be 2–connected. If not, it has a cutvertex y and then {w,y} is a cutset in G.

Now consider the edge wx. If this is non-contractible, then there exists a vertex z such that T = {w,x,z} is a 3-cut. The vertex $z \notin U^*$, since U^* is

2–connected. So U^* is contained in a component C of $G \setminus T$. Since $|VC| \geq |VU^*| > |VU|$ we have contradicted the choice of e. Hence wx is contractible. $\qquad \square$

Theorem 6.3 Let G be a 3-connected graph. Then G is planar if and only if it contains no subgraph which is a subdivision of K_5 or $K_{3,3}$.

Proof. Suppose G is planar. Then it clearly can contain no subdivision of K_5 or $K_{3,3}$.

So suppose G contains no subdivision of K_5 or $K_{3,3}$. We proceed by induction, noting that K_4 is planar.

By Lemma 6.2, G contains a contractible edge $e = uv$. Form $G*e$. A little work shows that $G*e$ contains no subdivision of K_5 or $K_{3,3}$. Hence $G*e$ is planar, by induction.

Let $e*$ be the vertex corresponding to the edge e in $G*e$. In Figure 6.9 we show a planar drawing of $G*e$. Here C is the cycle surrounding $e*$ which is a face in $(G*e) \setminus e*$. Let $u_i \in N(u) \setminus \{v\}$ in G and let Q_i be the subpaths of C joining u_i to u_{i+1} for $i = 1,2, ..., m$ with $u_{m+1} = u_1$. Let $v_j, j = 1,2, ..., n$, be the vertices of $VG \setminus \{u\}$ which are adjacent to v. There are now four cases to consider.

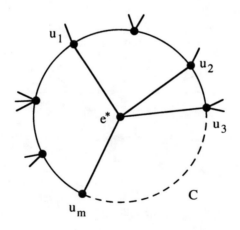

Figure 6.9

Case 1. If every v_j lies on only one Q_i, then we can insert v and its adjacencies in the plane drawing of $G*e$ to obtain a plane drawing of G; see Figure 6.10(a).

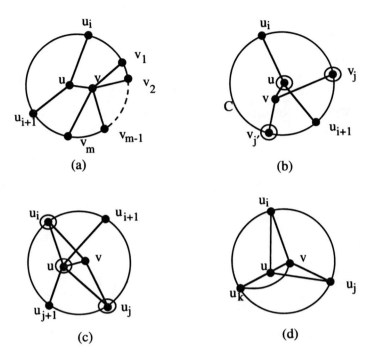

Figure 6.10

Case 2. Suppose then that v is adjacent to one interior vertex of some Q_i and to some vertex not in Q_i. This situation is shown in Figure 6.10(b). Then G contains a subdivision of $K_{3,3}$ with $\{u_i, u_{i+1}, v\}$ and $\{u, v_j, v_{j'}\}$ as the two parts of the bipartition.

So we may assume that v is only adjacent to neighbours of u and that these vertices are ends of more than one Q_i (otherwise we are in Case 1 or 2).

Case 3. Assume that v is adjacent to precisely two neighbours of u. If $m = 2$ or 3, then G is planar. Suppose, then that v is adjacent to u_i and u_j only but u_i and u_j are not endvertices of a Q_k; see Figure 6.10(c). Then G contains a subdivision of $K_{3,3}$ with parts $\{u, u_i, u_j\}$ and $\{v, u_{i+1}, u_{j+1}\}$.

Case 4. So we may assume that v has at least three such neighbours in common with u. This situation is shown in Figure 6.10(d). Then $\{u, v, u_i, u_j, u_k\}$ forms a subdivision of K_5.

This final contradiction proves the theorem.   □

Actually the proof above provides us with a much stronger result.

Corollary 6.4. If G is 3-connected and contains no subgraph which is a subdivision of K_5 or $K_{3,3}$, then G has a convex planar representation.

By a **convex planar representation** we mean that all the internal faces of the graph are convex and the edges are straight line segments. We leave this as Exercise 36.

Theorem 6.5 P is non-planar.

Proof: In Figure 1.14 we showed a spanning subgraph of P which is a subdivision of $K_{3,3}$. The vertices a, b, c and a′, b′, c′ form the two parts of $K_{3,3}$. □

The other well-known characterization of planarity is based on the concept of contraction. This is Wagner's Theorem.

Theorem 6.6 The graph G is non-planar if and only if $G >_s K_5$ or $G >_s K_{3,3}$.

Proof: See [kW 37]. □

We can now give an alternative proof to Theorem 6.5.

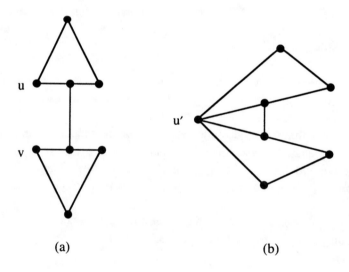

(a)   (b)

Figure 6.9

By performing contractions on the edges ii', i = 1, 2, 3, 4, 5, in P, we obtain K_5. Hence $P >_s K_5$ and is thus non-planar by Wagner's Theorem.

A similar construction to contraction is that of **identifying two vertices**. This is an edge contraction if the two vertices are adjacent. If they are not, then join them by an edge and form the edge contraction. Figure 6.9(b) shows the effect of identifying the two vertices u and v (to become u') in the graph of Figure 6.9(a).

7.   Graph colourings

An assignment of colours (elements of some set) to the vertices of a graph G, one colour to each vertex, so that adjacent vertices are assigned different colours, is called a **vertex colouring** of G. A colouring in which n colours are used is an **n-vertex-colouring**. A graph G is **n-vertex-colourable** if there exists an m–colouring of G for some $m \le n$. When it is clear we are talking about vertex colourings, the word 'vertex' will be omitted.

The minimum n for which a graph G is n-colourable is called the (vertex) **chromatic number** of G and is denoted by $\chi(G)$. If G is a graph for which $\chi(G) = n$ then G is **n-chromatic** e.g. $\chi(K_t) = t, \chi(C_{2n}) = 2$ and $\chi(C_{2n+1}) = 3$.

From the definition of bipartite graphs it should be clear that a graph is bipartite if and only if it is 2-chromatic.

Since P is not bipartite $\chi(P) \ge 3$. In Figure 7.1 we give a 3-colouring of P. Hence $\chi(P) = 3$.

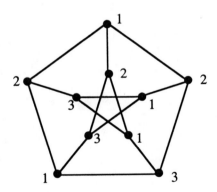

Figure 7.1

An assignment of colours to the edges of a graph G so that incident edges are assigned different colours is called an **edge-colouring** of G. Then an

n–edge–colouring and **n-edge-colourable** are defined as for vertex colourings. The (edge) chromatic number of G is often called the **chromatic index** of G and is denoted by $\chi'(G)$ e.g. $\chi'(K_{n,n}) = n$, $\chi'(C_{2n}) = 2$, $\chi'(C_{2n+1}) = 3$. In Section 3.2 we show that the chromatic index of P is 4.

Finally the **face chromatic number** of a plane graph is defined in an analogous way. In a **face-colouring**, adjacent faces must have different colours. The four colour theorem proves that 'every map (plane graph) is 4-face-colourable'. This is the theme of the next chapter.

Exercises

1. By drawing them all and checking for isomorphisms, show that there are 11 graphs of order 4, and 34 graphs of order 5.

2. Show that the 3 graphs of Figure 1.7 are all isomorphic to the drawing of P in Figure 1.3.

    In P verify that (i) for any vertex, the 6 vertices at distance 2 from it lie on a 6-cycle; (ii) for any edge there is an 8-cycle which does not pass through the endvertices of that edge; (iii) there are two 9-cycles through any specified set of 9 vertices.

3. Determine the number of labelled graphs of order n.

4. Prove Corollary 1.2.

5. Show that $P_1 \cong P_1'$.

6. Determine the unique graph whose vertex deleted subgraphs are shown below.

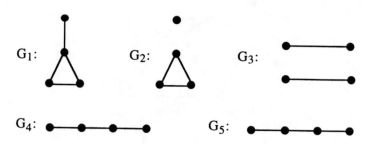

7. Prove that Ulam's Conjecture is true for regular graphs.

8.   Prove that $(\overline{G} + \overline{H}) \cong \overline{G \cup H}$.

9.   (a) Prove that $\alpha(P) = 4$.
      (b) Find $\alpha(K_n)$, $\alpha(C_n)$, $\alpha(P_n)$ for all n.
      (c) If T is a tree with $|VT| = n$, find $\alpha(T)$.

10.   How many 2-regular graphs are there of order n?

      How many connected 2-regular graphs are there of order n?

11.   Show that the connectivity of G is not greater than $\delta(G)$.

12.   Let G be cubic. Show that its connectivity and edge connectivity are equal.

13.   Show directly that P has connectivity 3.

      Show directly that P has edge connectivity 3.

      Is P the unique 3-connected cubic graph of order 10?

14.   Prove Theorem 2.2.

15.   Prove that every tree has at least two endvertices.

16.   Suppose that G is a connected graph. Prove that G is a tree if and only if $|EG| = |VG| - 1$.

17.   Let G be a bipartite graph. Is $\overline{G}$ bipartite?

18.   Find all cubic bipartite graphs on 6, 8, 10 vertices.

19.   Prove that $A(G)$ and $A^*(G)$ are groups.

20.   Show that $A(K_4) \cong A^*(K_4)$.

21.   Show that $A(K_5) \cong A^*(K_5)$.

22.   Characterize all vertex-transitive trees.

23.   Find a set of generators of $A(P)$ which consists of only two elements.

24. Determine $A(K_{m,n})$.

25. Construct a regular graph G such that G and $\overline{G}$ are connected but G is not vertex-transitive.

26. Show that $P \cong \overline{L(K_5)}$.

27. For what G is $L(G) \cong G$?

28. Is P the unique cubic non-hamiltonian graph of order 10?

     Is P the unique 3-connected cubic non-hamiltonian graph of order 10?

29. (a) Show that it is possible to form K_{2n+1} from the edge disjoint union of n hamiltonian cycles.

     (b) Show that it is possible to form K_{2n+2} from the edge disjoint union of $n + 1$ hamiltonian paths. (A **hamiltonian path** is a path which passes through every vertex of a given graph.)

     (c) Show that it is possible to form K_{2n+2} from the edge disjoint union of n hamiltonian cycles and $n + 1$ independent edges.

30. Show that P has cycles of size $5, 6, 8$ and 9 but none of size 7.

     How many cycles of each size are there?

31. Determine the girth of K_n and $K_{m,n}$.

32. Use the Jordan curve theorem to show that K_5 and $K_{3,3}$ are non-planar.

     Also prove that every subgraph of both K_5 and $K_{3,3}$ (except K_5 and $K_{3,3}$ themselves) is planar.

33. If $G >_s H$, is it true that H is isomorphic to a subgraph of a contraction of G?

34. Show that $P >_s K_{3,3}$.

35. Give an example to show that $G >_s H$ does not imply $G > H$ in general.

36. Prove Corollary 6.4.

37. A **cycle decomposition,** CD, of a graph G is a set C of cycles of G such that each edge of G belongs to exactly one member of C. A graph is **Eulerian** if every vertex is of even degree. Prove, using Theorem 5.9, that every Eulerian graph admits a CD. (We return to a discussion of CD's in Section 9.10.)

38. Define the graph $P(k,m)$ such that $VP(k,m) = \{u_i, v_i : i = 0, 1,..., k - 1\}$ and $EP(k,m) = \{u_i u_{i+1}, v_i v_{i+m}, u_i v_i : i = 0, 1,..., k - 1\}$ where addition is modulo k and $m < \frac{1}{2}k$. (See [jaB 72].)

   (a) Show that $P(5,2) \cong P$.
   (b) Find $\alpha(P(k,m))$.
   (c) Which of $P(6,2)$, $P(7,2)$, $P(11,2)$ and $P(7,3)$ are hamiltonian?
   (d) For which k, m is $P(k,m)$ non-hamiltonian? (See Section 9.13.)

   (The graph $P(k,m)$ is generally called THE **generalized Petersen graph.**)

References

[aA 91]   A. Ainouche, Four sufficient conditions for Hamiltonian graphs, *Disc. Math.*, 89, 1991, 195 - 200.

[B-C 71]   M. Behzad and G. Chartrand, *Introduction to the Theory of Graphs*, Allyn and Bacon, Boston, 1971.

[B-W 87]   A. Benhoane and A. Wojda, The Geng-Hua Fan conditions for pancyclic or hamilton-connected graphs, *J. Comb. Th.* B, 42, 1987, 167-80.

[jB 78]   J.C. Bermond, Hamiltonian graphs, in *Selected Topics in Graph Theory*, (ed. L.W. Beineke and R.J. Wilson), Academic Press, London, 1978.

[B-L-W 76] N. Biggs, E.K. Lloyd and R.J. Wilson, *Graph Theory 1736 - 1936*, Clarendon Press, Oxford, 1976.

[jaB 72]   J.A. Bondy, Variations on the hamiltonian theme, *Can. Math. Bull.*, 15, 1972, 57-62.

[B-H 77]   J.A. Bondy and B.W. Hemminger, Graph reconstruction - a survey, *J. Graph. Th*, 1, 1977, 227-268.

[B-K 88]   J.A. Bondy and M. Kouider, Hamiltonian cycles in regular 2–connected graphs, *J. Comb. Th.* B, 44, 1988, 177-186.

[B-M 76]   J.A. Bondy and U.S.R. Murty, *Graph Theory with Applications*, Macmillan, London, 1976.

[C-M 78]   M. Capobianco and J.C. Molluzzo, *Examples and Counterexamples in Graph Theory*, North Holland, New York, 1978.

[vC 72]   V. Chvátal, On Hamilton ideals, *J.Comb. Th.* B, 12, 1972, 163–168.

[gD 52]   G.A. Dirac, Some theorems on abstract graphs, *Proc. Lond. Math. Soc.*, 2, 1952, 69-81.

[E-H 77]   P. Erdös and A. Hobbs, Hamilton cycles in regular graphs of moderate degree, *J. Comb. Th.* B, 23, 1977, 139-142.

[E-H 78]   P. Erdös and A. Hobbs, A class of Hamiltonian regular graphs, *J. Graph Th.*, 2, 1978, 129-135.

[lE 36]   L. Euler, Solutio problematis ad geometriam situs pertinentis, *Comm. Acad. Sci. I. Petropolitanae*, 8, 1736, 128-140.

[rF 88]   R. Fowler, The Königsberg bridges - 250 years later. *Amer. Math. Monthly*, 95, 1988, 42-43.

[gF 84]   Geng-Hua Fan, New sufficient conditions for hamiltonian cycles in graphs, *J. Comb. Th.* B, 37, 1984, 221-227.

[rG 91]   R. Gould, Updating the hamiltonian problem - a survey, *J. Graph Th.*, 15, 1991, 121-157.

[fH 69]   F. Harary, *Graph Theory*, Addison-Wesley, Reading, 1969.

[H-P 73]   F. Harary and E. Palmer, *Graphical Enumeration*, Academic Press, New York, 1973.

[cH 73]   C. Hierholzer, Uber die Möglichkeit, einen Linienzug ohne Widerholung und ohne Unterbrechung zu umfahren, *Math. Ann.*, 6, 1873, 30-32.

[bJ 80]   B. Jackson, Hamilton cycles in regular 2-connected graphs, *J. Comb. Th.* B, 29, 1980, 27-46.

[bJ 86]   B. Jackson, Longest cycles in 3-connected cubic graphs, *J. Comb. Th.* B, 41, 1986, 17-26.

[aK 86]   A.B. Kempe, A memoir on the theory of mathematical form, *Phil. Trans. Roy. Soc. London*, 177, 1886, 1-70.

[kK 30]   K. Kuratowski, Sur la problème des courbes gauches en topologie, *Fund. Math.*, 15, 1930, 271-283.

[cN 78]   C.St.J.A. Nash-Williams, The reconstruction problem, *Selected Topics in Graph Theory*, (ed. L.W. Beineke and R.J. Wilson), Academic Press, 1978, 205-236.

[oO 60]   O. Ore, Note on Hamilton circuits, *Amer. Math. Monthly*, 67, 1960, 55.

[jP 98]   J. Petersen, Sur le théorème de Tait, *L'Intermédiare des Mathématiciens*, 5, 1898, 225-227.

[S-S-W 88] H. Sachs, M. Stiebitz and R.J. Wilson, An historical note: Euler's Königsberg letter, *J. Graph. Th.*, 12, 1988, 133-139.

[cT 81]   C. Thomassen, Kuratowski's Theorem, *J. Graph Th.*, 5, 1981, 225-241.

[wT 66]   W.T. Tutte *Connectivity in Graphs*, Toronto University Press, Toronto, 1966.

[hV 90]   H.J. Veldman, Short proofs of some Fan-type results, *Ars Combinatorica*, 29, 1990, 28-32.

[kW 37]   K. Wagner, Über eine Eigenschaft der ebenen Komplexe, *Math. Ann*, 114, 1937, 570-590.

[hW 32]   H. Whitney, Non-separable and planar graphs, *Trans. Amer. Math. Soc.*, 34, 1932, 339-362.

[heW 64]   H. Wielandt, *Finite Permutation Groups*, Academic Press, New York, 1964.

[nW 79]   N. Wormald, Classifying k-connected cubic graphs, in *Combinatorial Mathematics VI*, Lecture Notes in Mathematics No. 748, Springer Verlag, Berlin, 1979, 199-206.

[Z-L-Y 85]   Zhu Yongjin, Liu Zhenhong and Yu Zhengguang,   An improvement of Jackson's result on hamilton cycles in 2-connected regular graphs, *Annals Disc. Math.*, 27, 1985, 237-248.

2

The Four Colour Problem

0.   Prologue

"I wonder why problems about map-colourings are so fascinating? I know several people who have made more or less serious attempts to prove the Four-Colour Theorem, and I suppose many more have made collections of maps in the hope of hitting upon a counter-example. I like P.G. Tait's approach myself; he removed the problem from the plane so that it could be discussed in terms of more general figures. He showed that the Four-Colour Theorem is equivalent to the proposition that if N is a connected cubic graph, without an isthmus, in the plane, then the edges of N can be coloured in three colours so that the colours of the three meeting at any vertex are all different. It was at first conjectured that every cubical graph having no isthmus could be 'three-coloured' in this way, but this was disproved by reference to the Petersen graph, for which it may readily be verified that no three-colouring exists.

"I have often tried to find other cubic graphs which cannot be three-coloured. I do think that the right way to attack the Four-Colour Theorem is to classify the exceptions to Tait's Conjecture and see if any correspond to graphs in the plane. I did find some, but they were mere trivial modifications of the Petersen graph, obtained by detaching the three edges meeting at some vertex from one another so that the vertex becomes three vertices, and joining these three by additional edges and vertices so as to obtain another cubic graph. (Figure 0.1 is an example of such a trivial modification.)

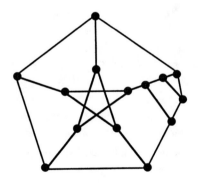

Figure 0.1

"I wondered if there could be any other exceptions to Tait's Conjecture, besides the Petersen graph... I did eventually discover one."

(Blanche Descartes [bD 48])

1. The Five Colour Theorem

On October 23rd 1852 Augustus De Morgan, Professor of Mathematics at University College, London, wrote the following to Sir William Rowan Hamilton.

'A student of mine asked me today to give him a reason for a fact which I did not know was a fact - and do not yet. He says that if a figure be anyhow divided and the compartments differently coloured so that figures with any portion of common boundary line are differently coloured, four colours may be wanted, but no more...'

The student was Frederick Guthrie who obtained the problem from his brother Francis. Francis had apparently stumbled on it while colouring in a map of the counties of England.

This then, was the origin of the famous four colour map problem which was to wait almost a century and a quarter before being solved. (Indeed there are some who believe that it is still to be solved.)

The basic problem considers the plane as being divided into regions. We are required to colour the regions so that any two regions which have a common border (containing more than a finite number of points) are coloured differently.

This map problem is often converted into a problem in graph theory. A vertex is placed in each region, including the outermost region, and two distinct vertices are then joined by an edge for each edge common to the boundaries of the two corresponding regions. In addition, a loop is added at a vertex for each bridge of the map that belongs to the boundary of the corresponding region. Suppose that the map is G then the pseudograph G^* described above is called the **dual of the map** G. G^* is drawn in the plane so that each edge crosses its associated edge of G but no other edge of G or G^*. This is always possible and so G^* is planar. An example is given in Figure 1.1. It is easy to see how to recover the map from its dual (up to a topologist's homeomorphism any way). And in fact given any plane graph we can obtain a map from it.

Francis Guthrie's conjecture (the Four Colour Conjecture) was that every map can be coloured in four or fewer colours. If we assign colours to the vertices of the dual graph and require no two adjacent vertices to have the same colour, then

Guthrie's four colour conjecture translates to 'every dual of a map is 4-vertex-colourable'.

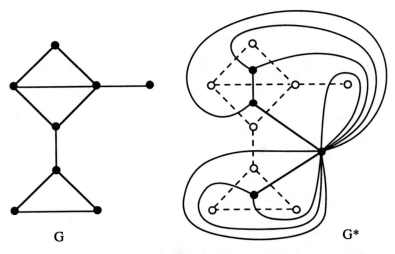

Figure 1.1

It is straightforward to see that removing all but one edge of a multiple edge between a pair of vertices or removing loops from a vertex, does not affect the chromatic number of the graph. Hence we may consider our dual graphs to be 'simple' graphs.

But we can add edges to a plane graph so that its faces are all triangles, including the outer face. An example of this is shown in Figure 1.2. The dotted lines show the edges that were added to produce the triangulated graph.

Every plane graph, all of whose faces are triangles, is called a **triangulation**. Naturally there is more than one way of producing a triangulation from a plane graph. However, if a triangulation is 4-colourable, then every spanning subgraph will be. So to prove Guthrie's conjecture it is enough to prove that the chromatic number of every triangulation is four.

We note that no triangulation contains a vertex of degree 1.

Lemma 1.1 Four colours are necessary to colour triangulations.

Proof: Consider the planar triangulation isomorphic to K_4 shown in Figure 1.3. It is clear that four colours are needed in this graph.   □

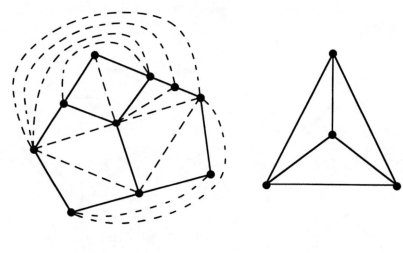

Figure 1.2   Figure 1.3

The above lemma seems to have been known to Guthrie. See [B-L-W 76] - this book gives an interesting account of the early history of the Four Colour Problem, as well as the history of other areas of graph theory. The sufficiency was not proved till 1976, when Appel, Haken and Koch, [A-H 77a], [A-H-K 77], brought the might of computer calculations to bear on the problem. In between, many false proofs were given. One of the most important of these was due to A.B. Kempe [aK 79].

There are several reasons for the importance of Kempe's work. First of all P.J. Heawood [pH 90], who pointed out the error in Kempe's paper (see [B-L-W 76]), was able to use Kempe's methods to prove the Five Colour Theorem. This shows that every triangulation is 5-colourable. Secondly, one of Kempe's basic ideas, what has become known as Kempe chains, was central to the proof obtained by Appel, Haken and Koch. We develop this notion below as we produce a proof of the Five Colour Theorem. But first we note another important idea in Kempe's work. This is a consequence of Euler's Polyhedral Formula which we prove as the next result; see [lE 50].

Theorem 1.2 If G is a connected plane graph with n vertices and e edges which divides the plane into f faces, then

$$n - e + f = 2.$$

Proof: We proceed by induction on n. If $n = 1$, then $e = 0$ and $f = 1$, so $n - e + f = 2$. So assume that the result is true for $n = k$. If we add a new vertex to a plane graph of order k to form a new plane graph, then each edge except the first, which joins this vertex to the rest of the graph, produces a new face. Hence the formula holds in general.   □

The following result was first proved by Kempe in [aK 79].

Theorem 1.3 Let n_k be the number of vertices of degree k in a triangulation T. Then

$$\sum_{k \geq 2} (6 - k)n_k = 12.$$

Proof: Now $n = \sum_{k \geq 2} n_k$. Further, if d_i is the degree of vertex i, i = 1, 2,..., n, then we know from Lemma 1.1.1 that $2e = \sum_{i=1}^{n} d_i$. Suppose that $d_i = k$ for all $i \in I \subseteq \{1,2,...,n\}$ and $d_i \neq k$ for any other i. Then $\sum_{i \in I} d_i = k|I|$. But $|I| = n_k$. Hence $\sum_{i=1}^{n} d_i = \sum_{k \geq 2} kn_k$.

But every face is bounded by 3 edges in a triangulation T. Hence if we add the number of edges surrounding all the faces we obtain a total of 3f. However, in this number, each edge is counted twice since each edge bounds two faces. Thus $3f = 2e$ which gives

$$3f = \sum_{k \geq 2} kn_k.$$

Substituting in Euler's formula we get

$$\sum_{k \geq 2} n_k - \frac{1}{2}\sum_{k \geq 2} kn_k + \frac{1}{3}\sum_{k \geq 2} kn_k = 2.$$

On simplifying we obtain the required formula.   □

Corollary 1.4 A triangulation T contains at least one vertex of degree 2, 3, 4 or 5.

Proof: From the theorem, $\sum_{k=2}^{5} (6 - k)n_k - \sum_{k \geq 6} (k - 6)n_k = 12$. Hence

$$\sum_{k=2}^{5} (6 - k)n_k = 12 + \sum_{k \geq 6} (k - 6)n_k.$$

Since the right hand side of the last equation is always positive, then so is the left hand side. Thus $n_k \geq 1$ for at least one $k = 2, 3, 4, 5$. $\qquad\square$

We are now in a position to prove the Five Colour Theorem, see [pH 90].

Theorem 1.5 Any triangulation T is 5-colourable.

Proof: First we assume that the theorem is false and that T is one of the non–5–colourable triangulations of minimum order. By Corollary 1.4 we know that T contains a vertex of degree $2, 3, 4, 5$.

If T contains a vertex of degree 2, then, since T is a simple graph, $T \cong K_3$ which is certainly 5-colourable.

If T contains a vertex v of degree 3, then part of T_v is shown in Figure 1.4(a). Since T_v is a triangulation and $|VT_v| < |VT|$, then T_v is 5-colourable. So 5–colour T_v and note that three distinct colours c_1, c_2, c_3 are used on u_1, u_2, u_3. Using this 5-colouring of T_v and placing colour c_4 on v produces a 5-colouring of T.

Suppose then that T contains a vertex v of degree 4. Then part of T_v is shown in Figure 1.4(b). This is not a triangulation so we add the edge $u_2 u_4$ to give the triangulation of Figure 1.4(b). This new triangulation can be 5-coloured and the vertices u_1, u_2, u_3, u_4 use at most the colours c_1, c_2, c_3, c_4. Using the same colouring in T, with v coloured c_5, ensures that T is 5-colourable.

Hence T must have a vertex v of degree 5 and no vertices of degree 2, 3 or 4. Now T_v contains the 5-cycle shown in Figure 1.4(c). By adding edges we turn T_v into a triangulation T_v' which is 5-colourable. If the vertices u_1, u_2, u_3, u_4, u_5 employ a maximum of four colours, then we can use this colouring to colour T, where v is given the fifth colour. Hence we may assume, without loss of generality, that u_1, u_2, u_3, u_4, u_5 are coloured c_1, c_2, c_3, c_4, c_5.

The colouring of T_v' induces a colouring of T_v. Consider the vertices of T_v which are coloured c_1 and c_3. These induce a subgraph $H_{c_1 c_3}$ of T_v. Now if u_1 and u_3 are in different components of $H_{c_1 c_3}$, we may interchange the colours c_1 and c_3 on the component of $H_{c_1 c_3}$ which contains u_3 and hence produce a 5–colouring of T_v with at most four colours on the cycle

$(u_1, u_2, u_3, u_4, u_5)$. We can then extend this 5-colouring of T_v to a 5–colouring of T by giving v the colour c_5.

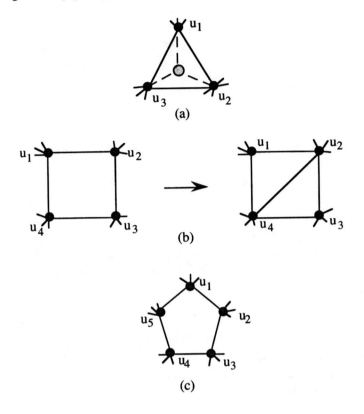

Figure 1.4

Hence there must be a 'chain' of the colours c_1, c_3 joining u_1 and u_3 in $H_{c_1 c_3}$. This is shown in Figure 1.5. Using exactly the same reasoning we may assume that in $H_{c_2 c_5}$ (the subgraph of T_v induced by the vertices coloured c_2 and c_5) there is a chain of colours c_2, c_5 joining u_2 and u_5, otherwise we may switch colours c_2 and c_5 on the component of $H_{c_1 c_3}$ containing u_5 and hence produce a 5-colouring of T.

So the $c_1 c_3$ and $c_2 c_5$ chains described above must intersect. Since T_v is a plane graph, they must intersect at a vertex, w, say. If this is the case w must be coloured in two colours since it sits on the $c_1 c_3$ chain and on the $c_2 c_5$ chain. This situation clearly cannot arise and so T can be 5-coloured.   □

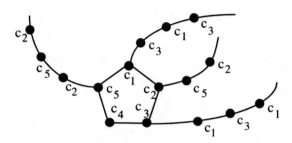

Figure 1.5

We will refer to colour chains such as the two used in the proof as **Kempe chains**.

Kempe used the idea above in his false proof of the Four Colour Theorem. He had no trouble dealing with vertices of degree 2, 3 and 4 using what has now become known as the Kempe chain argument. However he ran into difficulty with vertices of degree 5. This came about by his attempting two colour interchanges at the same time. Heawood ([pH 90] or [B-L-W 76]) pointed out the fallacy in his argument.

However there are three elements of Kempe's work which were invaluable in the Appel and Haken proof. Kempe chains have already been commented on, and the Five Colour Theorem shows their use in modifying colourings. The other two ideas which we shall enlarge on in the next sections are those of an unavoidable set and a reducible configuration.

To prove the Five Colour Theorem it was first necessary to produce a set of configurations at least one member of which occurs in **every** triangulation. Such sets are said to be **unavoidable**. Kempe showed that a vertex of degree less than six was in this sense unavoidable.

Then the unavoidable configurations had to be shown to be **reducible** in that they could not be present in any smallest counterexample.

We now pursue these two ideas in more detail but note that Kempe's strategy was to find an unavoidable set of configurations and then show these configurations to be reducible. This was exactly the same strategy which Appel and Haken adopted in their successful proof.

2. Unavoidable Configurations

A **configuration** in a triangulation is that portion of the triangulation which lies inside some cycle. This cycle is referred to as the **ring of the configuration**.

The number of vertices in the ring is called the **ring size**. Finally, in this string of definitions, we call a set of configurations **unavoidable**, if at least one member of the set is present in **every** triangulation.

By Corollary 1.4 we know that the set of vertices of degrees 2, 3, 4, 5 is an unavoidable set of configurations. The configurations here consist of eunegraphs with a single vertex of the appropriate degree and are shown dotted in Figure 2.1. Respectively, their ring sizes are 2, 3, 4, 5 and the rings are shown as solid cycles in the figure. We note here that, using the arguments of the proof of the Five Colour Theorem, vertices of degree 2, 3, 4 are easily shown not to be present in a minimum counterexample. Hence in future we will omit them from our unavoidable sets even though, strictly speaking, they should appear in all of them.

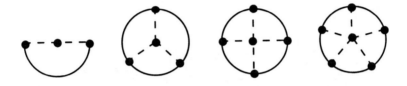

Figure 2.1

The new idea that we now need in order to produce an unavoidable set that we can handle, is that of discharging. This concept was introduced by Heesch [haH 69]. To prove that a set of configurations is unavoidable we use a proof by contradiction. So assume that there is a graph which contains no member of the set. Then assign numbers to each vertex according to some rule so that the sum of all these numbers is positive. These numbers are called the **charge** on a given vertex. Now we produce a means of moving these charges so that every vertex ends up with zero or negative charge while still, apparently, conserving charge. Thus we contradict the fact that our original set was avoidable.

The method of moving charges is called **discharging**.

We now illustrate this technique on the unavoidable set of configurations defined in Lemma 2.1 below and shown in Figure 2.2. Here vertices of degree 5 are represented by solid circles; open circles represent vertices whose degree is specified by the diagram; squares represent vertices whose degree is at least 8; and finally triangles represent vertices of unspecified degree. These configurations were originally produced by Heesch but the proof given here is due to Haken [wH 73] and is considerably shorter than the original.

Lemma 2.1 The configurations of Figure 2.2 constitute an unavoidable set.

Proof: Recall that triangulations now contain no vertex of degree less than 5. To each vertex v in the triangulation T we assign a charge of 6-deg v. Now

$$\sum_{v \in VT} (6 - \deg v) = \sum_{k \geq 2} (6 - k)n_k.$$

But by Theorem 1.3, the right hand sum is equal to 12. Hence we have assigned a total positive charge of 12 to the triangulation T.

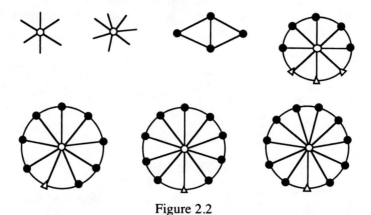

Figure 2.2

The proof now is essentially in two parts. First we show that the charge displaced from a high degree vertex u, to a neighbouring vertex x of degree 5 is always greater than a certain fraction relating to the charge. Then we show that the effect of this is to produce zero or negative charge on all vertices. This then contradicts the total positive charge of 12 on T.

We now assume that T contains none of the configurations of Figure 2.2. In particular this means that no vertex of T has degree 6 or 7. We now distribute the charges on vertices of degree 8 or more, to certain of the vertices of degree 5. To see how to do this, let $u \in VT$ with $\deg u \geq 8$. Let $N(u) = \{u_1, u_2, ..., u_r\}$ be the vertices in the neighbourhood of u, taken in cyclic order around u. If the vertex u_k has degree 5, we give it a weighting

$$w = \begin{cases} 1 & \text{if precisely one of } u_{k-1}, u_{k+1} \text{ has degree 5} \\ 2 & \text{otherwise.} \end{cases}$$

For instance, in Figure 2.3, the weighting given to vertices u_1, u_4 and u_5 is 2, while that given to u_3 and u_6 is 1.

Now let p_i be the number of vertices in $N(u)$ with weight $i, i = 1, 2$.

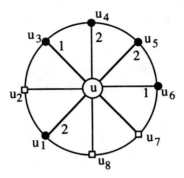

Figure 2.3

(In Figure 2.3, $p_1 = 2$, $p_2 = 3$.) We redistribute the charge on u, so that each vertex u_k of degree 5 in $N(u)$ receives a charge of

$$\frac{w(6 - \deg u)}{p_1 + 2p_2}.$$

The charge removed from u is therefore, summing over all vertices of degree 5 in $N(u)$,

$$\sum \frac{w(6 - \deg u)}{p_1 + 2p_2} = \frac{p_1 \, w(1) \, (6 - \deg u)}{p_1 + 2p_2} + \frac{p_2 \, w(2) \, (6 - \deg u)}{p_1 + 2p_2}$$
$$= 6 - \deg u.$$

where $w(i)$ indicates the vertex has weight $i, i = 1, 2$. Thus for any vertex $v \in VT$ with $\deg v \neq 5$, the new charge on v is given by

$$\begin{cases} 0 & \text{if } v \text{ has a degree 5 neighbour} \\ 6 - \deg v & \text{otherwise.} \end{cases}$$

If we can show that all vertices of degree 5 have negative charge we will have obtained our required contradiction, for now the sum of the charges taken over the whole of T cannot be positive and so cannot be 12. To do this we first show that if u is any vertex of degree greater than 5 and if x is any vertex of degree 5 with weight w_u with respect to u, then the negative charge q transferred from u to x has magnitude $|q| \geq \frac{1}{4} w_u$. We prove this in a number of steps.

(i)   Suppose that $\deg u = 8$. By the fourth configuration of our list, a vertex of degree 8 has at most four consecutive vertices of degree 5 in its ring. Hence u has at most six neighbours of degree 5. Therefore $p_1 + p_2 \leq 6$.

If $p_1 + p_2 = 6$, then the two vertices of degree greater than 5 which are adjacent to u, must separate the degree 5 vertices into two strings with weights $1,1$ and $1,2,2,1$ or weights $1,2,1$ and $1,2,1$. In both cases $p_1 = 4$, $p_2 = 2$ and $p_1 + 2p_2 = 8$.

Suppose that $p_1 + p_2 = 5$. Then, since we cannot have five consecutive degree 5 vertices, the only possibilities are illustrated in Figure 2.4.

In all cases $p_2 \leq 3$ and $p_1 + 2p_2 \leq 8$.

Now suppose that $p_1 + p_2 < 5$. Then $p_1 + 2p_2 = (p_1 + p_2) + p_2 \leq 8$.

Hence $|q| = \dfrac{w_u(\deg u - 6)}{p_1 + 2p_2} = \dfrac{2w_u}{p_1 + 2p_2} \geq \dfrac{1}{4} w_u$, since in all cases $p_1 + 2p_2 \leq 8$.

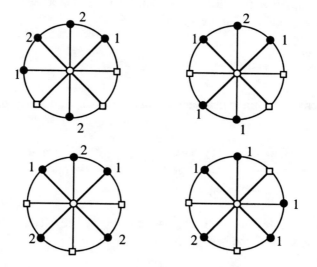

Figure 2.4

(ii)   Suppose that $\deg u = 9, 10, 11$. From the last three configurations in our list we know that $p_1 + p_2 \leq \deg u - 2$ since u must have at least two neighbours of degree greater than 5. Suppose that $p_2 \geq \deg u - 3$. There are three cases to consider: (a) $p_2 = \deg u - 3, p_1 = 0$, (b) $p_2 = \deg u - 3, p_1 = 1$ and (c) $p_2 = \deg u - 2$. Now it is easy to see that cases (b) and (c) are impossible since the two neighbours of higher degree imply in both cases that $p_1 \geq 2$. So in the remaining case, $p_1 + 2p_2 = 2 \deg u - 6$. On the other hand if $p_2 \leq \deg u - 4$ since $p_1 + p_2 \leq \deg u - 2$, $p_1 + 2p_2 = (p_1 + p_2) + p_2 \leq 2 \deg u - 6$. Then

$$|q| = \frac{w_u(\deg u - 6)}{p_1 + 2p_2} \geq \frac{w_u(\deg u - 6)}{2 \deg u - 6} \geq \frac{1}{4} w_u.$$

(iii) If $\deg u > 11$, then $p_1 + p_2 \leq \deg u$ and $p_2 \leq \deg u$. Hence $p_1 + 2p_2 \leq 2 \deg u$. Therefore

$$|q| = \frac{w_u(\deg u - 6)}{p_1 + 2p_2} \geq \frac{w_u(\deg u - 6)}{2 \deg u} \geq \frac{1}{4} w_u.$$

Hence we know that the charge displaced from a high degree vertex u, to a neighbouring vertex x of degree 5, is always greater than or equal to $\frac{1}{4} w_u$, where w_u is the weight of x with respect to the vertex u.

At this stage we note that every vertex of degree 5 possesses at least two neighbours of degree greater than 5. This is because the third configuration of Figure 2.2 is assumed not to exist in T. Further, if a vertex of degree 5 has precisely two neighbours of degree greater than 5, then these vertices are not adjacent. In fact in Figure 2.5 we show all possible configurations with a vertex of degree 5 at their centre.

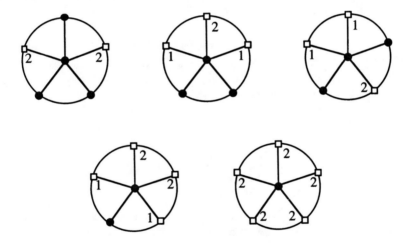

Figure 2.5

In Figure 2.5, the numbers on the edges joining vertices of degree greater than 5 to the central vertex of degree 5, are the weights given the central vertex from the vertex on the other end of that edge. Since the sum of these weights is at least four, in each case the magnitude of the total charge given to a vertex of degree five is seen to be at least one. (Recall that $|q| \geq \frac{1}{4} w_u$.) Hence the actual change in charge to the vertex x is less than or equal to -1. But each vertex x of degree 5 originally had a charge of $6 - \deg x = 1$. So after the redistribution of charge, vertex x will have zero or negative charge.

Hence all vertices now have zero or negative charge and so the total charge on the triangulation is no longer positive. This contradiction proves that the original set of configurations was unavoidable.   □

We should point out at this stage that the discharging procedure used in the previous lemma was not that used in the proof of the Four Colour Theorem. The current lemma has been used to illustrate the techniques involved and to give some idea of the form of the discharging used in the Appel and Haken proof.

3. Reducible Configurations

As we have noted earlier, a configuration in a triangulation is **reducible** if it cannot be contained in any minimum counterexample to the Four Colour conjecture. Using Kempe chains we have already seen how to 'reduce', in the Five Colour sense, vertices of degree 2, 3, 4 and 5; see Theorem 1.5. Vertices of degree 2, 3, 4 are easily shown to be reducible by exactly the same techniques. We will now show how more complicated configurations have their reducibility demonstrated.

These processes are illustrated in this section by considering a triangulation T containing the Birkhoff diamond, the third configuration consisting of four vertices of degree 5, of Figure 2.2; see also Figure 3.1. This configuration was in fact proved reducible by Birkhoff in [gB 13].

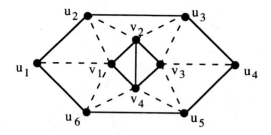

Figure 3.1

The naive method for determining reducibility is to consider the ring which contains the configuration. This is shown as the vertices $\{u_i: i=1, 2,..., 6\}$ in Figure 3.1. We have no idea what the triangulation looks like outside this ring of size 6. Hence to cover all eventualities when colouring the triangulation from the ring out, we must allow all possible colourings of the ring. If they can all be extended to a proper colouring of the vertices of the Birkhoff diamond, then T is 4-colourable and the configuration in question is reducible.

Let T' be the triangulation T with the Birkhoff diamond removed.

Suppose the vertices u_i, in order, are coloured $c_1, c_2, c_3, c_4, c_3, c_2$ in some four colouring of T'. Then we can extend this colouring to T by letting v_1, v_2, v_3, v_4 be coloured c_3, c_4, c_2, c_1, respectively.

But there are some colourings of the ring that cannot be directly extended to the whole of T. Suppose that the vertices u_i, in order, are coloured $c_2, c_1, c_3, c_1, c_3, c_1$. Now since v_2 and v_4 are both adjacent to colours c_1 and c_3, they must be coloured c_2 or c_4. However v_2 is adjacent to v_4, so they cannot both have the same colour. Without loss of generality let v_2 be coloured c_2 and v_4 be coloured c_4. But now v_3 is adjacent to a vertex of each of the four colours, so the colouring cannot be extended to the diamond configuration. At this stage we try a Kempe chain argument.

Consider $H_{c_1 c_4}$, the subgraph of T' induced by the colours c_1 and c_4. If u_4 is in a different component of $H_{c_1 c_4}$ to u_2 and u_6, then we may interchange c_1 and c_4 on the component of $H_{c_1 c_4}$ containing u_4. Then v_1, v_2, v_3, v_4 can be coloured c_3, c_2, c_1, c_4, to complete the colouring of T.

If u_4 is in the same component of $H_{c_1 c_4}$ as u_2 but in a different component to u_6, then we may interchange c_1 and c_4 on the component containing u_6 and proceed as above to produce a 4-colouring of T.

Finally then, we may assume that u_2, u_4 and u_6 are all in the same component of $H_{c_1 c_4}$. But then u_3 cannot be in the same component of $H_{c_2 c_3}$ as either u_1 or u_5. Hence we interchange c_2 and c_3 on the component of $H_{c_2 c_3}$ containing u_3 and colour v_1, v_2, v_3, v_4 in c_4, c_3, c_4, c_2, respectively, to complete the colouring of T.

When a configuration is able to be reduced by either the naive method or by the Kempe chain adjustment, we say that it is **D-reducible**.

But this is a rather inefficient way to proceed and, in addition, not all configurations are D-reducible. Since we only need to show the existence of a 4–colouring of T we do so in the following way. First remove from T the configuration C which we are trying to show is reducible. Then identify certain vertices of the ring of the configuration and add extra edges to produce a new triangulation such as T'; see Figure 3.2.

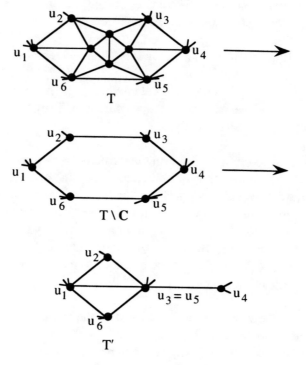

Figure 3.2

Now any colouring of T' will give a colouring of $T \setminus C$. If this in turn extends to a colouring of T we know that C is reducible. Configurations which are reducible in this manner are called **C-reducible**.

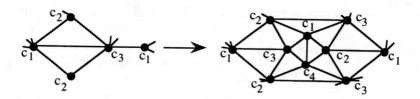

Figure 3.3

Lemma 3.1 The Birkhoff diamond is C-reducible.

Proof: We reduce the arbitrary triangulation T to the triangulation T' of Figure 3.2. There are essentially six colourings of the vertices u_1, u_2, $u_3 = u_5$, u_4, u_6 of the triangulation T'. These are c_1, c_2, c_3, c_1, c_2; c_1, c_2, c_3, c_2, c_2; c_1, c_2, c_3, c_4, c_2; c_1, c_2, c_3, c_4, c_4; c_1, c_2, c_3, c_1, c_4; c_1, c_2, c_3, c_2, c_4, where the colours are placed on the vertices u_1, u_2, u_3, u_4, u_6, respectively. All the colourings except c_1, c_2, c_3, c_2, c_2 are easily shown to extend to a colouring of T. An example is given in Figure 3.3.

However, with c_1, c_2, c_3, c_2, c_2, we can assume without loss of generality that v_2 is coloured c_1 and v_4 is coloured c_4. But then v_3 is adjacent to a vertex of each of the four colours. At this stage we have to return to our Kempe chicanery. Suppose that u_2, u_4, u_6 all belong to the same component of $H_{c_2 c_4}$. Then u_1 and u_5 belong to different components of $H_{c_1 c_3}$. Hence we can change the colour on u_5 to c_1 and complete the colouring of T by colouring v_1, v_2, v_3, v_4 in c_4, c_1, c_4, c_3, respectively. If u_2 is not in the same component of $H_{c_2 c_4}$ as either u_4 or u_6, we colour u_2 in c_4 and colour v_1, v_2, v_3, v_4 in c_3, c_2, c_1, c_4, respectively. We use a similar argument if u_6 is not in the same component of $H_{c_2 c_4}$ as either u_2 or u_4. Finally then, suppose u_4 is not in the same component of $H_{c_2 c_4}$ as either u_2 or u_6. We then colour u_2 and u_6 in c_4 and colour v_1, v_2, v_3, v_4 in c_3, c_2, c_4, c_1, respectively. $\qquad\square$

Clearly C-reducibility can be tested more quickly than D-reducibility because once vertices on the ring of a configuration have been identified, there are far fewer possible colourings available.

4. The Four Colour Theorem

In this section we will give an outline of the way in which the preceding ideas were welded together to produce a proof of the Four Colour Theorem given in [A–H 77a] and [A-H-K 77].

Theorem 4.1 Every triangulation is 4-colourable.

Proof: The general approach is exactly as in the Five Colour Theorem. First an unavoidable set of configurations is produced and then each configuration in this set is shown to be reducible.

But this cannot all be done blindly. To show that there was a strong likelihood of success for their method, Appel and Haken used a probability argument to suggest that there must exist an unavoidable set that would have no configuration of ring size greater than 14. While this argument was not watertight it was sufficient evidence to encourage them to embark on the computing required to achieve their goal. It was necessary to establish the chances of success first, because the time taken to prove the reducibility of larger rings becomes prohibitive using the techniques available to Appel, Haken and Koch.

On the other hand, they knew they would have to deal with configurations of ring size at least 12 since E.H. Moore had produced a triangulation in which the smallest C-reducible configuration had ring size 12. In fact Moore's claim about his triangulation now seems to have been false. At least, the example given in [A–H 77b] does not have the property claimed. However this error was unknown to Appel and Haken. Subsequently Moore has found another triangulation with the stated property.

So first an unavoidable set had to be produced. This was done by interaction with the computer. A set of configurations was tested for unavoidability using the discharging procedure as it was developed at that stage. What this involved is illustrated in Lemma 2.1.

The set of unavoidable configurations was assumed to be avoidable. Then given any triangulation T, with no vertex of degree less than 5, an initial charge of 60(6-deg v) was assigned to every vertex v in T. Using Theorem 1.3 we know that the total charge on T is 720. Charge can then be moved around by a computer according to the current form of the discharging algorithm. The general technique was to move charge **from** a vertex of degree 5 according to the configuration this vertex was in. In Figure 4.1 we show a situation where a charge of 10 is transferred from the vertex v of degree 5 to the vertex w of degree 7. The vertex x is a vertex of unspecified degree greater than 6.

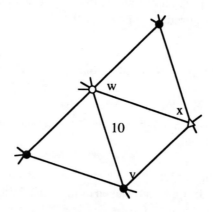

Figure 4.1

At this stage, either a total negative charge occurred, and with it the required contradiction, or there was a total positive charge. In the former case the test for reducibility could begin. In the latter case the discharging algorithm had to be modified and the computer was set to work again.

After continual mathematician-computer interaction of this sort, a discharging procedure was obtained which involved many hundreds of steps of the form shown in Figure 4.1; see [A-H 77a]. The set of unavoidable configurations finally produced contained 1879 members.

Now this set of configurations had to be tested for reducibility. First each configuration was tested for D-reducibility. If a configuration failed this test it was examined for C-reducibility. When every configuration was shown to be reducible then the theorem was proved. Should a configuration fail the reducibility tests or the tests not be completed in the time allotted, then the set of unavoidable configurations had to be modified and the whole process started again.

It perhaps is worth noting that the time limit given for rings of size 14 was 90 minutes on an IBM 370-158 or 30 minutes on a 370-168.

Thus using the basic framework of Kempe and the discharging ideas of Heesch, the proof of the Four Colour Theorem was completed some 124 years after it had been brought to the attention of de Morgan.   □

5 . Hadwiger and Hajós

Surely now mathematicians are free to turn to other problems and leave the Four Colour Theorem to amateurs who will undoubtedly still continue to find it appealing? But the proof of this famous theorem causes problems. First of all it relies greatly on the use of a computer. Because of the number of machine hours spent, it is very difficult for most mathematicians to check through the proof. They have neither that sort of access to a machine nor can they go through and check all the cases for themselves.

Traditionally all proofs of theorems have been available to all mathematicians. In the proof of the Four Colour Theorem, a certain amount has to be taken on trust.

It should be noted of course that the work has been checked. There is certainly no campaign to pull the wool over the eyes of the mathematical public. Referees checked the original proof and others, notably F. Allaire [fA 78], have proved the theorem using the same general method but using different programs.

However, there are philosophical problems with the proof. For example, Tymoczko [tT 80] argues that by allowing the computer a central role in the proof of a theorem, the essential nature of mathematics is changed.

There is also an aesthetic problem. This can be voiced by echoing the words of one H. Dumpty. 'Impenetrability! That's what I say'; see [wT 78]. The trouble is that the proof gives us no insight into **why** four colours are sufficient. What is it about the plane or about triangulations that makes four a significant number? (Halmos [paH 90] discusses the Four Colour Theorem in the context of recent advances in mathematics where he categorizes it as an 'explosion'.)

It is important at this stage to note that what we have said about the Appel and Haken proof should in no way be taken as denigrating their work. Certainly their proof is a fine mathematical achievement and is something which eluded many many mathematicians over a period of more than 100 years.

But it is a different kind of proof from almost all previous ones. Possibly the mathematical community will have to come to grips with this 'new-fangled' type of proof. In the meantime there is still the desire to find a different, more 'elegant', way to solve the problem.

There have been a number of other methods of attack suggested for the Four Colour Theorem. If one of these is successful it may provide the light we need.

The first of these that we want to discuss is Hadwiger's Conjecture [hH 43]. This is that $\chi(G) \geq k$ implies that G is subcontractible to K_k $(G >_s K_k)$ for any graph G.

Lemma 5.1. Hadwiger's Conjecture is true for $k \leq 4$.

Proof: For $k = 1$ we note that if $\chi(G) \geq 1$, then G contains a vertex and so $G >_s K_1$. If $k = 2$, then G contains an edge and so $G >_s K_2$.

For $k = 3$ we note that if $G \not>_s K_3$, then G is a forest and so $\chi(G) = 2$. This contradiction gives the result we are seeking.

The proof for $k = 4$ is not complicated and can be found in Dirac [gD 52].   □

Lemma 5.2 Hadwiger's Conjecture for $k = 5$ is equivalent to the Four Colour Theorem.

Proof: Suppose that G is planar with $\chi(G) \geq 5$. By the Hadwiger Conjecture $G >_s K_5$ and so by Wagner's Theorem (Theorem 1.6.6), G is non-planar.

For the converse, suppose that $\chi(G) \geq 5$. By the Four Colour Theorem we know that G is non-planar. Hence by Wagner's Theorem, $G >_s K_5$ or $G >_s K_{3,3}$. If we can dispose of the latter possibility the result follows. The complete proof of this is to be found in Wagner [kW 64].   □

Corollary 5.3 Hadwiger's conjecture is true for $k = 5$.   □

Theorem 5.4 If $\chi(G) = 6$, then G contains a subgraph which is contractible to K_6 minus two edges.

Proof: See [gD 63].   □

Now the fact that every 5-colourable graph is subcontractible to K_5 seems to give more insight into colourings of planar graphs than does the Four Colour Theorem. The hope then, is that a direct proof of Hadwiger's Conjecture for $k = 5$ may be found that does not rely heavily on the computer.

As an example of this approach Woodall [dW 87] proves that a graph that does not have K_{m+1} as a subcontraction must contain an independent set consisting of at least $\frac{1}{2} (m - 1)$ of its vertices.

We have seen in Chapter 1 that if 'subcontractible' is changed to 'subdivision', Wagner's Theorem, Theorem 1.6.6, is converted to Kuratowski's Theorem, Theorem 1.6.1. Using the same change of words, we produce Hajós' Conjecture from Hadwiger's Conjecture. Hence Hajós' conjectured [gH 61] that if $\chi(G) \geq k$, then G contains a subdivision of K_k.

Lemma 5.5 Hajós' Conjecture is true for $k \leq 4$.

Proof: The result for $k = 1, 2$ is immediate. For $k = 3$, we note that if $\chi(G) \geq 3$, then G must contain an odd cycle. Hence G contains a subdivision of K_3. Again the proof for $k = 4$ can be found in [gD 52].   □

The conjecture is still open for $k = 5, 6$. A partial result in this direction is given by Pelikàn in [jP 69].

Theorem 5.6 If $\chi(G) = 5$, then G contains a subdivision of K_5 minus an edge.   □

However for $k \geq 7$, the Hajós' Conjecture is definitely false. The original counterexamples were given by Catlin [pC 79]. We present one such below.

Let $\Sigma(G)$ denote the **subdivision number** of G, that is, the largest m such that G contains a subdivision of K_m. Using this terminology Hajós' Conjecture asserts that $\Sigma(G) \geq \chi(G)$.

Now define $_rG$ to be the multigraph formed from G by replacing every edge wx by r edges between w and x. Then Catlin showed that $\Sigma(L(_3C_5) \setminus \{u,v\}) = 6$ and $\chi(L(_3C_5) \setminus \{u,v\}) = 7$ where u and v are non-adjacent vertices; see Exercise 11. Here $L(G)$ denotes the line graph of G; see Section 1.4.

In addition, since $\Sigma(G + v) = \Sigma(G) + 1$ and $\chi(G + v) = \chi(G) + 1$, if Hajós' Conjecture is false for G, then it is false for $G + v$. Hence Hajós' Conjecture is false for all $k \geq 7$.

Interestingly, Bollobàs, Catlin and Erdös [B-C-E 80] proved that Hadwiger's conjecture holds for almost all graphs, while a result due to Erdös and Fajtlowicz [E-F 81], shows that Hajós' conjecture is false for almost all graphs.

6. Tait's Method

In the prologue to this chapter we quoted Blanche Descartes as saying 'I do think the right way to attack the Four-Colour Theorem is to classify the exceptions to Tait's Conjecture and see if any correspond to graphs in the plane'.

Tait approached the problem of colouring the faces of a map by colouring the boundaries between them. We saw in Section 1 how to convert the problem of colouring the regions of a map into colouring the vertices of a triangulation. Let us say that the boundaries of a map are the edges of a graph and that these edges meet in vertices of the graph. Then if we go back via duality from triangulations to these map graphs, we see that the map graphs we need to consider are connected graphs in which every vertex is of degree 3. Further, every edge lies on the boundary of two faces, so the graph is bridgeless. Figure 6.1 shows the dual graph of a triangulation.

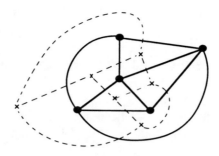

Figure 6.1

Hence Tait set about edge-colouring bridgeless cubic planar graphs. He conjectured that every such graph was 3-edge-colourable.

Now in 1880, Tait [pT 80] showed that his conjecture was equivalent to the Four Colour Theorem. Indeed he had gone so far as to produce a proof of the Four Colour Theorem. Unfortunately this proof was based on the assumption that the graphs with which he was dealing (3-connected cubic planar graphs) were hamiltonian. The error in this assumption was not pointed out until 1946 when Tutte [wT 46], produced the famous 3-connected cubic planar, but non-hamiltonian, graph of Figure 6.2.

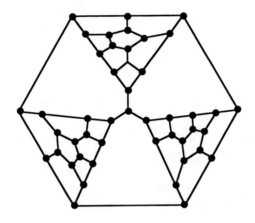

Figure 6.2

We now prove the equivalence of the Tait conjecture and the Four Colour Theorem.

Theorem 6.1. Every plane graph is 4-face-colourable if and only if every bridgeless cubic planar graph is 3-edge-colourable.

Proof: Let G be a bridgeless cubic planar graph and let G' be a plane representation of G. Assume that the faces of G' can be 4-coloured in the colours $(0,0),(0,1),(1,0),(1,1)$. Consider these colours as being elements in $Z_2 \times Z_2$ which is a group under coordinate-wise addition. Now if the edge e lies between colours c_1 and c_2, we assign the colour $c_1 + c_2$ to e; see Figure 6.3. Since G' is bridgeless, the colours c_1, c_2, c_3 around any vertex v, must all be distinct. Hence because the colours are chosen from the group $Z_2 \times Z_2$, each edge incident with v has a distinct colour. Finally, since every element in $Z_2 \times Z_2$ is self-inverse, the bridgelessness of G' guarantees that the colour $(0,0)$ is never used.

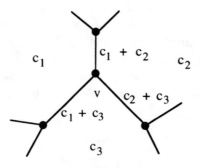

Figure 6.3

We now assume that every bridgeless cubic planar graph is 3-edge-colourable. Suppose that there exists a plane graph G which is not 4-face-colourable. Clearly G*, the dual of G, is not 4-vertex-colourable, but it is plane. Furthermore there is no loss of generality in supposing that G* is a simple graph; this does not affect the (vertex) chromatic number of G*. By adding edges to G* we may obtain a graph T* all of whose faces are triangles. In a vertex colouring, T* must use at least as many colours as G* and so T* is not 4-vertex-colourable.

Now form the dual of T*, to give T which is a bridgeless cubic planar graph. It is planar because T* is, and cubic because each face of T* is a triangle. If it contained a bridge, then some face of T* would not be a triangle. Since T is the dual of T*, it is not 4-face-colourable.

However, by hypothesis, T is 3-edge-colourable. Suppose that it can be edge coloured using the colours e_1, e_2, e_3. Let T_{ij} be the subgraph of T induced by those edges coloured e_i and e_j, for $i \neq j$. Since e_i and e_j meet every vertex of T, T_{ij} must be the union of cycles and so T_{ij} is 2-face-colourable. But each face of T is the intersection of a face of T_{12} and a face of T_{23}, say. Hence if f_1, f_2 are the face colours used for T_{12} and if g_1, g_2 are the face colours used for T_{23}, we can colour the faces of T in colours (f_i, g_j), $i, j = 1, 2$. This is done by assigning the colour (f_i, g_j) to a face which is the intersection of a T_{12} face coloured f_i and a T_{23} face coloured g_j. This gives a 4-face-colouring of T and so contradicts the assumption that G was not 4-face-colourable. □

Corollary 6.2 Every bridgeless cubic planar graph G is 3-edge-colourable.

Proof: This follows immediately from Theorems 4.1 and 6.1. □

Now naturally the Petersen graph is not 3-edge-colourable as we shall see in Section 3.2. And now we take up the theme of the book, for Tutte makes the following conjecture [wT 66].

Conjecture 6.3 (Tutte) If G is a bridgeless cubic graph with $\chi'(G) \geq 4$, then $G >_s P$.

If this conjecture is true, see Section 9.10, then it is the absence of the Petersen graph which makes for the 3-edge-colourability of cubic planar graphs.

Actually we can strengthen Corollary 6.2. We define the **crossing number** $c(G)$, of the graph G, to be the smallest number of edge intersections of any embedding of G in the plane. Hence the crossing number of a plane graph is zero and it is not difficult to show that $c(K_5) = 1$ and $c(P) = 2$. Incidentally, Bloom, Kennedy and Quintas [B-K-Q 83] conjecture that if G is a graph such that $c(G) \geq 2$, then G has a subgraph H with $c(H) = 2$. Richter [bR 88] proved this conjecture if G is cubic.

Notice that the standard drawing of P is far from the most efficient relative to crossings. However, this drawing is not just aesthetically picturesque - it leads to generalizations. See [E-H-K 81] where crossing numbers for some of these generalizations are determined.

Suppose that Tutte's Conjecture, Conjecture 6.3, is true. Then if G is a 3–connected cubic graph with $c(G) \leq 1$, $\chi'(G) = 3$. In fact this result has been proved independently of Tutte's Conjecture by Jaeger [fJ 80].

Theorem 6.4 Every bridgeless 3-connected cubic graph with crossing number at most 1 has chromatic index 3.

Tutte's Conjecture is at least consistent with Jaeger's result. In the next chapter we consider more evidence in favour of the conjecture.

Exercises

1. Prove that the only simple graph triangulation containing a vertex of degree 2 is the graph K_3.

2. Prove that $(G^*)^* \cong G$ if and only if G is connected.

3. Prove that if a planar graph contains no vertex of degree 5, then it is 4–colourable.

4.   We attempt a proof of the Four Colour Theorem using the same approach as
      the Five Colour Theorem. Clearly there is no problem with vertices of
      degree 2, 3 or 4. So suppose v has degree 5 and the neighbours of v use
      all four colours in the colouring of T_v. Hence two of the neighbours of v
      have the same colour.

      The two vertices with the same colour are not adjacent. Without loss of
      generality suppose that $N(v) = \{v_1, v_2, v_3, v_4, v_5\}$, v_1, v_3 are coloured c_1,
      v_2 is coloured c_2, v_4 is coloured c_3 and v_5 is coloured c_4. It is easy to
      see that difficulties arise when v_2 and v_5 are in the same component of
      $H_{c_2 c_4}$ and again when v_2 and v_4 are in the same component of $H_{c_2 c_3}$.

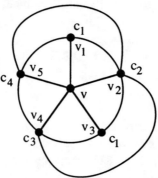

      As v_1 and v_4 are not in the same component of $H_{c_2 c_3}$, we may change the
      colour of v_1 to c_3, Similarly we can change the colour of v_3 to c_4. The
      neighbours of v now have colours c_2, c_3, c_4 and so we can use c_1 on v
      to give a 4-colouring of the graph.

      This proves the Four Colour Theorem.

      Any comments?

5.   (a)   A 4-graph is a connected plane graph in which all faces are 4-cycles.
            Construct 4-graphs on 4, 5 and 6 vertices.

      (b)   Prove that if n_k is the number of vertices of degree k in a 4-graph, then

            $$\sum_{k \geq 2} (4 - k)n_k = 8.$$

      (c)   Give an unavoidable set of configurations for a 4-graph.

(d)   Is it true that any 4-graph is 3-(vertex) colourable?

(e)   Are all 4-graphs 2-colourable? Justify your answer.

6.   Determine the ring size of the configurations of Figure 2.2.

7.   Assume that the configuration below is reducible.

     Prove that any configuration consisting of a vertex of degree 5 with 3 neighbours of degree 5 and one of degree 6 is reducible.

8.   Check all cases of Lemma 3.1.

9.   Can the Birkhoff diamond be shown to be C-reducible by identifying u_3 and u_6 in $T \setminus C$?

10. Prove that the configuration consisting of a single vertex of degree 5 is not D-reducible.

11. Prove that $L(3C_5) \setminus \{u,v\}$ is a counterexample to Hajós' Conjecture. Is it a counterexample to Hadwiger's Conjecture?

12. Prove that $\chi'(P) = 4$ and $c(P) = 2$.

13. Show that there are non-planar graphs which have chromatic number less than 5.

14. Is it true that every cubic planar graph is 3-edge-colourable?

15. Show that $c(K_5) = 1$. Determine $c(K_6)$, $c(K_7)$, $c(K_{3,3})$, $c(K_{3,4})$. In general $c(K_n)$ and $c(K_{m,n})$ are unknown. Show that

$$c(K_n) \leq \tfrac{1}{4} \lfloor \tfrac{1}{2}n \rfloor \lfloor \tfrac{1}{2}(n-1) \rfloor \lfloor \tfrac{1}{2}(n-2) \rfloor \lfloor \tfrac{1}{2}(n-3) \rfloor$$

and $c(K_{m,n}) \leq \lfloor \tfrac{1}{2}m \rfloor \lfloor \tfrac{1}{2}(m-1) \rfloor \lfloor \tfrac{1}{2}n \rfloor \lfloor \tfrac{1}{2}(n-1) \rfloor$; see [W-B 78] p.391.

References

[fA 78] F. Allaire, Another proof of the Four Colour Theorem: Part I, Proc. 7th. Manitoba Conf. Num. Math. and Computing, *Congr. Num.* XX, 1978, 3-72.

[A-H 77a] K. Appel and W. Haken, Every planar map is four colorable: Part I, Discharging, *Illinois J. Math.*, 21, 1977, 429-490.

[A-H 77b] K. Appel and W. Haken, The solution of the four-color-map problem, *Scientific American*, 237(4), 1977, 108-121.

[A-H-K 77] K. Appel, W. Haken and J. Koch, Every planar map is four colorable: Part II, Reducibility, *Illinois J. Math.*, 21, 1977, 491–567.

[B-L-W 76] N.L. Biggs, E.K. Lloyd and R.J. Wilson, *Graph Theory 1736–1936*, Clarendon Press, Oxford, 1976.

[gB 13] G.D. Birkhoff, The reducibility of maps, *Amer. J. Math.*, 35, 1913, 114-128.

[B-K-Q 83] G. Bloom, J. Kennedy and L. Quintas, On crossing numbers and linguistic structures, *Graph Theory* (Langów 1981) Lecture Notes in Maths., No.1018, Springer-Verlag, Berlin, 1983.

[B-C-E 80] B. Bollobás, P. Catlin and P. Erdös, Hadwiger's conjecture is true for almost every graph, *European J. Comb.*, 1, 1980, 195-199.

[pC 79] P.A. Catlin, Hajós graph-coloring conjecture: variations and counterexamples, *J. Comb. Th.*, 26B, 1979, 268-274.

[bD 48] B. Descartes, Network colourings, *Math. Gazette*, 32, 1948, 67–69.

[gD 52] G.A. Dirac, A property of 4-chromatic graphs and some remarks on critical graphs, *J. London Math. Soc.*, 27, 1952, 85-92.

[gD 63] G.A. Dirac, Generalizations of the five color theorem, *Theory of Graphs and Its Applications*, Academic Press, New York, 1963, 21-27.

[E-F 81]   P. Erdös and S. Fajtlowicz, On the conjecture of Hajós, *Combinatorica*, 1, 1981, 141-143.

[lE 50]   L. Euler, Private communication to Christian Goldbach, 1750 (see [B-L-W 76] p.76).

[E-H-K 81]   G. Exoo, F. Harary and J.A. Kabell, The crossing numbers of some generalized Petersen graphs, *Math. Scand.*, 48, 1981, 184-188.

[hH 43]   H. Hadwiger, Über eine Klassifikation der Streckenkomplexe, *Vierteljschr. Naturforsch. Gesellsch. Zurich*, 88, 1943, 133-142.

[gH 61]   G. Hajós, Uber eine Konstruktion nicht n-farbbaren Graphen, *Wiss. Z. Martin-Luther. Univ. Halle-Wittenberg. Math.-Nat. Reihe*, 10, 1961, 116-117.

[wH 73]   W. Haken, An existence theorem for planar maps, *J. Comb. Th. B*, 13, 1973, 180-184.

[paH 90]   P. Halmos, Has progress in mathematics slowed down?, *Amer. Math Monthly*, 97, 1990, 561-588.

[pH 90]   P.J. Heawood, Map-colour theorem. *Quart. J. Pure Appl. Math.*, 24, 1890, 332-338.

[haH 69]   H. Heesch, Untersuchungen zum Vierfarben problem, *B.I.HochSchulscripten*, 810/810a/810b, Bibliographisches Institut, Mannheim-Vienna-Zurich, 1969.

[fJ 80]   F. Jaeger, Tait's theorem for graphs with crossing number at most one, *Ars. Combinatoria*, 9, 1980, 283-287.

[aK 79]   A. Kempe, On the geographical problem of the four colours, *Amer. J. Math.*, 2, 1879, 193-200.

[jP 69]   J. Pelikán, Valency conditions for the existence of certain subgraphs, *Theory of Graphs*, Academic Press, New York, 1969, 251-258.

[bR 88]   B. Richter, Cubic graphs with crossing number two, *J. Graph Th.*, 12, 1988, 363-374.

[pT 80]   P.G. Tait, Remarks on the colouring of maps, *Proc. Royal Soc. Edin.*, 10, 1878-80, 729.

[tT 80]   T. Tymoczko, Computers, proofs and mathematicians: a philosophical investigation of the Four-Color proof, *Math. Mag.* 53, 1980, 131-138.

[wT 46]   W.T. Tutte, On hamiltonian circuits, *J. London Math. Soc.*, 21, 1946, 98-101.

[wT 66]   W.T. Tutte, On the algebraic theory of graph colourings, *J. Comb. Th.*, 1, 1966, 15-50.

[wT 78]   W.T. Tutte, Colouring problems, *The Math. Intelligencer*, 1, 1978, 72-75.

[kW 64]   K. Wagner, Beweis einer Abschwächung der Hadwiger-Vermutung, *Math. Ann.*, 153, 1964, 139-141.

[W-B 78]   A.T. White and L.W. Beineke, Topological graph theory, *Selected Topics in Graph Theory*, (ed. L.W. Beineke and R.J. Wilson), Academic Press, 1978, 15-50.

[dW 87]   D.R. Woodall, Subcontraction-equivalence and Hadwiger's conjecture, *J. Graph Th.*, 11, 1987, 197-204.

3
Snarks

0. Prologue

Erect and sublime, for one moment of time.
In the next, that wild figure they saw
(As if stung by a spasm) plunge into a chasm,
While they waited and listened in awe.

'It's a Snark!' was the sound that first came to their ears,
And seemed almost too good to be true.
Then followed a torrent of laughter and cheers:
Then the ominous words, 'it's a Boo -'

(From 'The Hunting of the Snark' by Lewis Carroll [cD 57])

The term "snark" that is defined in this chapter owes its origins to the Snark of Lewis Carroll fame. When the graph theoretical term was first used, it seemed that snarks were very rare and unusual creatures. It goes without saying that P is a snark.

1. Vizing's Theorem

We recall from p. 42 that the number of colours required to colour the edges of a graph G so that no two edges incident with a vertex have the same colour, is the **chromatic index** of G. This is denoted by $\chi'(G)$.

We note that from Exercise 3, $\chi'(P) = 4$, which is one more than the degree of the graph. We shall prove in this section the somewhat surprising result that the chromatic index of a graph G is either equal to its maximum degree $\Delta(G)$, or to one more than its maximal degree. It is immediately obvious that $\chi'(G) \geq \Delta(G)$ since every edge incident with a vertex of maximum degree must be given a different colour. But the upper bound of $1 + \Delta(G)$ found by Vizing [vV 64] is far from obvious. For an up-to-date account on recent work in this area see [hY86] and for generalizations see [F-W 77] and [B-F 91].

Theorem 1.1 For any graph G, $\Delta(G) \leq \chi'(G) \leq 1 + \Delta(G)$.

Proof: Let $\Delta = \Delta(G)$.

Our method of proof is constructive in that we give an algorithm which will enable
the edges to be coloured in $1 + \Delta$ colours.

First colour as many edges as possible arbitrarily in $1 + \Delta$ colours so that no two
incident edges share a colour. Our strategy is to show that either all the edges are
coloured or we can "extend" this maximal colouring. Suppose that as a result
there is an edge adjacent to a vertex u which is not yet coloured. Let this edge be
uv_1.

Since $\deg u \leq \Delta$ and $1 + \Delta$ colours are being used, there is a colour c which is
not used on an edge incident with u. We say c is **missing** at u and denote it by
a circled c as in the diagram of Figure 1.1(a).

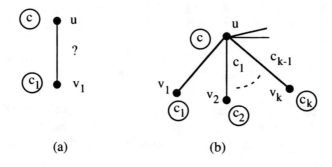

(a)   (b)

Figure 1.1

Using the same argument we see that there is a colour missing at v_1. Call it c_1.

Suppose there is an edge uv_2 coloured c_1. Again there is a colour missing at v_2,
call it c_2. Suppose there is an edge uv_3 coloured c_2. Then there is a colour
missing at v_3, call it c_3.

In this way we may construct the sequence of edges $uv_1, uv_2 ,..., uv_k$ such that
colour c_i is missing at vertex v_i for $i = 1, 2,..., k$ and colour c_j is on edge
uv_{j+1} for $j = 1, 2,..., k - 1$. See Figure 1.1(b).

If at any stage c is missing at vertex $v_m, m \in \{1,2,...,k\}$, then we cascade the
colours $c_1, c_2,..., c_{m-1}$ from $uv_2, uv_3,..., uv_m$ to $uv_1, uv_2,..., uv_{m-1}$ to give
another colouring of G with $1 + \Delta$ colours. Since c is missing at both u and
v_m we colour uv_m in colour c.

Hence, we may assume that c is not missing from any vertex $v_i, i = 1, 2,..., k$.

Now we need to know why the sequence of edges, uv_i, $i = 1, 2, ..., k$, stopped. There are two possible reasons. Either there is no edge coloured c_k adjacent to u or the colour $c_k = c_j$ for some $j < k - 1$ and has already been used. (Clearly the sequence must stop because of the finiteness of deg u but it may well stop before all the edges adjacent to u are exhausted.)

If c_k is missing at u, then cascade all the colours as before so that uv_i has colour c_i for $i = 1, 2,..., k - 1$. Now colour uv_k in colour c_k.

We must therefore suppose that $c_k = c_j$ for some $j < k - 1$. In this case cascade the colours so that uv_i has colour c_i for $i = 1, 2,..., j$ and leave uv_{j+1} uncoloured as in Figure 1.2.

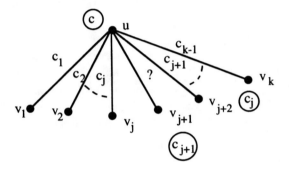

Figure 1.2

Now consider H_{cc_j}, the subgraph of G induced by the edges of colour c and c_j. Clearly every vertex in H_{cc_j} has degree at most 2, so it is the union of cycles, paths and isolated vertices.

By a previous argument, c is not missing at v_{j+1} or v_k. On the other hand c_j is missing at these vertices, since we have just moved c_j to uv_j and $c_k = c_j$. Hence in H_{cc_j}, the degree of v_{j+1} and v_k is one.

Further, c is missing at u and c_j is not missing. So again, the degree of u in H_{cc_j} is one.

But the structure of H_{cc_j} ensures that no three vertices of degree one are in the same component.

If u and v_{j+1} are in different components of H_{cc_j}, then interchange the colours of the component containing v_{j+1}. Then c is missing at both u and v_{j+1}. If we colour uv_{j+1} in c we are done.

Suppose then, that u and v_{j+1} are in the same component of H_{cc_j}. Necessarily v_k is not in this component, so interchange c and c_j in the component containing v_k. In this case further cascade the colours so that uv_i has colour c_i for $i = 1, 2, \ldots , k\text{-}1$. Now colour uv_k with colour c.

Thus in every case we have extended our colouring with $1 + \Delta$ colours to one more edge of G. But our original colouring was maximal. Hence G can be coloured in $1 + \Delta$ colours to give $\chi'(G) \leq 1 + \Delta$.   □

Hence we know that any cubic graph has chromatic index 3 or 4. It is an interesting problem then, to try to determine all 4-edge-colourable cubic graphs because this might lead to an independent proof of the Four Colour Theorem. And, after all, this is what sent Tait off in the direction of edge colouring cubic graphs.

But 4-edge-colourable cubic graphs do not appear to be thick on the ground. Indeed, as we noted in the Prologue to Chapter 2, Blanche Descartes has "often tried to find" such graphs and, apart from the Petersen graph, only managed to discover one on 210 vertices. Isaacs [rI 75], who found two infinite classes said that anyone seeking a 4-edge-colourable cubic graph would be "vividly impressed by the maddening difficulty of" such a task. In fact, until Isaacs' 1975 paper, there were only four kinds of graphs of this type known. These were Petersen's graph, Blanuša's graphs on 18 vertices [dB 46], Szekeres' graph on 50 vertices [gS 73], and Descartes' graph on 210 vertices [bD 48].

As a result, Martin Gardiner [mG 76] called them **snarks** after the elusive figment of the vivid imagination of the Rev C. Dodgson [cD 76] (see Prologue to this chapter).

But we need to take a little care in defining precisely what we mean by a snark. As Ms. Descartes has pointed out in the beginning of Chapter 2, there are some trivial ways that 4-edge-colourable cubic graphs can be produced. We now present a series of results which will enable us to define snarks precisely. But first we need a result of Blanuša [dB 46].

Lemma 1.2 Let G be a cubic multigraph which is 3-edge-colourable with colours $c_i, i = 1, 2, 3$. Let Z be any edge cut in G and n_i the number of edges of Z coloured c_i. Then the n_i are congruent modulo 2.

Proof: The removal of the edges of Z from G will result in two disjoint sets of vertices A, B with $A \cup B = VG$ and such that the endvertices of the edges of Z are in A or B. Further there is no edge joining a vertex of A to a vertex of B.

Now when we colour G, each vertex in A is one end of an edge coloured c_i. The vertices of A are split by this colour into two sets, one being the endvertices of the edges of Z which were coloured c_i and the other being pairs of vertices joined by an edge coloured c_i. Hence the number of vertices in A is equal to the number of endvertices of the edges of Z which were coloured c_i plus a multiple of two. Hence n_i is congruent to $|A|$ modulo 2.   □

This will turn out to be a very powerful lemma. As a first simple consequence of this we note the following.

Corollary 1.3 Let G be a cubic graph containing a bridge. Then $\chi'(G) = 4$.
                                                                               □

Hence we can construct 4-edge-colourable graphs from two 3-edge-colourable graphs, simply by adding a new vertex on an arbitrary edge of each graph and joining the new vertices by an edge.

Thus it can be seen that an infinite number of 4-edge-colourable graphs can be constructed if we allow bridges in our graphs. But the motivation for considering edge colourings was the Four Colour Problem. The cubic graphs derived from triangulations do not have bridges; see p. 71. Hence we will not include graphs with bridges in the class of snarks.

We now continue our catalogue of 'trivial modifications' without comment.

Lemma 1.4 Let G be a cubic graph containing the triangle (a_1, a_2, a_3). Let G' be the multigraph obtained from G by contracting this triangle to a single vertex a; see Figure 1.3. Then $\chi'(G) = 3$ if and if only $\chi'(G') = 3$.

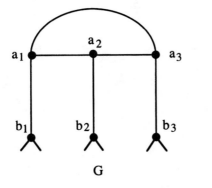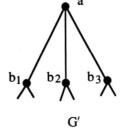

Figure 1.3

Proof: Suppose that $\chi'(G) = 3$. Then by Lemma 1.2, each edge $a_ib_i, i = 1, 2, 3$ has a different colour c_i. If c_i is now assigned to ab_i in G' we obtain a 3–edge-colouring of G'. Given a 3-edge-colouring of G' we can easily produce a 3-edge-colouring of G.   □

Lemma 1.5 Let $G^{\square}$ be obtained from the cubic graph G by inserting vertices a_1, a_2 and a_3, a_4 in the distinct edges e_1 and e_2 of G, and adding the edges a_1a_4, a_2a_3 (see Figure 1.4). If $\chi'(G^{\square}) = 4$ then $\chi'(G) = 4$.

Figure 1.4

Proof: Suppose that $\chi'(G) = 3$ and e_1, e_2 have the same colour c_1. Then we colour $u_1a_1, a_2v_1, u_2a_3, a_4v_2$ in c_1, a_1a_2, a_3a_4 in c_2 and a_1a_4, a_2a_3 in c_3. This gives a 3-edge-colouring of $G^{\square}$.

If e_1 and e_2 have colours c_1, c_2 we can similarly produce a 3-edge- colouring of $G^{\square}$.   □

Note that the converse of Lemma 1.5 is false. Figure 1.5 shows a $P^{\square}$ which is 3-edge-colourable.

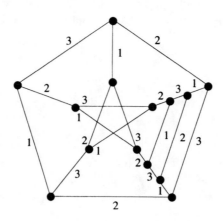

Figure 1.5

Lemma 1.6 Let G be a cubic bridgeless graph which contains an edge cut $\{u_1v_1, u_2v_2\}$ of size two, where u_1 and v_2 are not in the same component of $G - \{u_1v_1, u_2v_2\}$. To this latter graph add the edges u_1u_2, v_1v_2. Let the component of the new multigraph which contains u_1u_2 be G_1 and the remaining component be G_2.

Then $\chi'(G) = 3$ if and only if G_1 and G_2 both have chromatic index 3.

Proof: Since G is bridgeless $u_1 \neq u_2$ and $v_1 \neq v_2$. If G is 3-edge-colourable, then by Lemma 1.2 we know that the edges u_1v_1, u_2v_2 must both have the same colour c_1, say. Then colour u_1u_2 in G_1 and v_1v_2 in G_2, in colour c_1, to give G_1 and G_2 a 3-edge-colouring.

Conversely suppose that $\chi'(G_1) = \chi'(G_2) = 3$. In any 3-edge-colouring of G_1 and G_2 we may permute the colours so that v_1v_2 has the same colour c_1, say, as u_1u_2. Now in G we use the same colouring for all the edges of G_1 and G_2 which are in common with G and use c_1 for the edges u_1v_1, u_2v_2, to give a 3–edge-colouring of G.   □

We note here that if G contains a square, then G_1 and G_2 may both be multigraphs. However, if G does not contain a square, at most one of G_1, G_2 is a multigraph.

Let G be a graph and A be an edge cut. A is **trivial** if there exists a vertex of degree $|A|$ which is incident to all the edges of A. Otherwise A is a **proper** edge cut.

Let G be a 3-edge-connected cubic graph which contains a proper edge cut $A = \{u_iv_i : i = 1, 2, 3\}$. Label the edges so that no v_i is in a component of $G - A$ which contains u_1, u_2 or u_3. To $G - A$ add vertices u, v and edges uu_i, vv_i, $i = 1, 2, 3$. Label the component of $G - A$ which contains u, G_1, and the other component G_2. G_1 and G_2 are called **3-cut reductions** of G.

Lemma 1.7 G is 3-edge-colourable if and only if both 3-cut reductions G_1 and G_2 are.

Proof: By Lemma 1.2, if $\chi'(G) = 3$, then each of the edges u_iv_i has a different colour in any 3-edge-colouring. Using the colours of G together with the colours c_i for the edges uu_i and vv_i we obtain 3-edge-colourings of both G_1 and G_2. Notice here that because G is 3-edge-connected and A is proper, G_1 and G_2 are not multigraphs.

The converse follows after permuting the colours of G_1 and G_2 so that uu_i and vv_i are coloured c_i. Then colour u_iv_i with c_i in G.   □

Clearly every cubic graph contains an edge cut A of size 3, we simply choose three edges incident to a given vertex. However we are mainly interested in the situation where G - A has two components, which each contain a cycle. Let A be a **cyclic cutset** if G - A has two components each of which contains a cycle. The **cyclic edge-connectivity** of G, $\lambda_c(G)$, is the size of the smallest cyclic cutset of G. We say that G is **cyclically k-edge-connected,** if $\lambda_c(G) \geq k$. In other words, a cutset of size at least k has to be deleted before the graph is broken into two components each of which contains a cycle.

The graph shown in Figure 1.6(a) is cyclically 3-edge-connected, while that of Figure 1.6(b) is not. Clearly any graph containing a bridge is not cyclically 3–edge-connected.

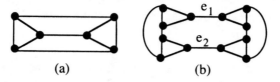

(a)   (b)

Figure 1.6

Of particular interest to us here are cyclically 4-edge-connected graphs. We note that P is such a graph. (In fact $\lambda_c(P) = 5$). Clearly P must be at least cyclically 3-edge-connected since it is 3-edge-connected. If P is not cyclically 4–edge–connected and A is an edge cut of P containing 3 edges, then P - A must consist of two components with a cycle. Since $\gamma(P) = 5$, these components have 5 vertices. However, only three of the vertices in either component have degree 2 and so these components contain a cycle on fewer than 5 vertices. Hence P is cyclically 4–edge-connected.

We are now in a position to give a precise definition of the term "snark".

We wish to discover the essence of 4-edge-colourability, so we ignore chromatic index 4 graphs with a bridge, because Corollary 1.3 shows us that such graphs have no well defined structure. They are also graphs which can be avoided in the context of the Four Colour Theorem.

Lemmas 1.4 and 1.5 show that 4-edge-colourable graphs with girth 3 or 4 can be obtained from smaller 4-edge-colourable graphs. Hence graphs of girth less than 5 are not minimal in the sense of chromatic index 4.

Lemma 1.6 shows that graphs with a 2-edge-cut which are 4-edge-colourable come from smaller 4-edge-colourable multigraphs. Since we may now assume that the graphs with a 2-edge-cut do not have girth 4, at least one of these smaller multigraphs is simple.

Lemma 1.7 shows us that graphs with proper 3-cuts which are 4-edge-colourable, come from smaller 4-edge-colourable graphs. Hence we may discard graphs with proper 3-cuts from our basic 4-edge-colourable collection.

Hence we define a **snark** to be a cyclically 4-edge-connected cubic graph of girth at least five, which has chromatic index four.

Lemma 1.8 The Petersen graph is the smallest snark and is the unique snark on 10 vertices.

Proof: Suppose that G is a snark. Let A be any cyclic edge-cut of G. Then G - A has two components which contain cycles of size at least 5, since the girth of G is 5. Hence $|VG| \geq 10$. Since P is a snark, then the smallest snark(s) must have order 10.

Let G be a snark of order 10 and let A be a cyclic edge-cut. Suppose $|A| = 4$. Then G - A contains two components each with 5 vertices or else $\gamma(G) < 5$. However in either of these components there must be four vertices of degree 2 and one of degree 3. This is not possible.

If $|A| > 5$, then $|VG| > 10$. So let $A = \{ii',: i = 1, 2, 3, 4, 5\}$. Then G - A consists of two 5-cycles. Let one of these cycles be $(1,2,3,4,5)$. If $1' \sim 2'$ or $1' \sim 5'$ then G contains a 4-cycle. Hence $1' \sim 3'$ and $1' \sim 4'$. Similarly, $2' \sim 4', 5'$ and $3' \sim 5', 1'$. So G = P.   □

In the light of the above discussion we rephrase Conjecture 2.6.3.

Conjecture 1.9 All snarks are subcontractible to P.

Can it really be that the essential nature of 4-edge-colourability is to be found in P?

The evidence of small graphs supports this conjecture; see [C-W 81], [jF 82] and [A–B–H–R87]). There are no snarks on 12, 14 or 16 vertices. On 18, there are precisely two; see Figure 2.2. Six snarks exist of order 20 and 20 of order 22. These 26 snarks are shown in Appendices 1 and 2. Obviously, since a snark is cubic, there are no odd snarks.

Interestingly (see [mW 69] and [C-P 72]) P = P(5,2) is the only **generalized**
Petersen graph to have chromatic index equal to 4. See Exercise 38 Chapter 1 for
the definition of the generalized Petersen graph.

2 . Isaacs' Snarks

Up until 1975 only four types of snarks were known. These were P on 10
vertices, Blanuša's graphs on 18 vertices, Szekeres' graph on 50 vertices and
Descartes' graph on 210 vertices. In 1975, Isaacs [rI 75] produced two infinite
families of snarks, as well as a completely separate new snark. One of the families
(using the 'dot product operation') included all of the previously known snarks
with the exception of the Petersen graph. The other family, consisting of the
flower snarks, had been independently discovered by Grinberg in 1972, but never
published. Also in some unpublished work in 1976, F. Loupekhine described a
construction which produces another infinite family of snarks. Very readable
surveys on snarks are given in [C-W 81] and [bS 87]. For the moment we
concentrate on the work of Isaacs.

The dot product

One of the two infinite families of snarks constructed by Isaacs is obtained by
using an operation called the dot product. A **dot product** of two connected cubic
graphs L and R denoted by L·R, is defined as follows:

(1)   remove any pair of adjacent vertices x and y from L;
(2)   remove any two independent edges ab and cd from R;
(3)   join {r,s} to {a,b} and {t,u} to {c,d} or join {r,s} to {c,d} and
       {t,u} to {a,b}.

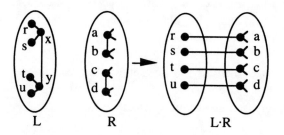

Figure 2.1

We assume that L and R are large enough so that L·R exists. We notice that
there are several different dot products L·R, depending on the particular choice of
edges and vertices to be removed and how in (3) we join corresponding vertices.
One possible dot product is shown in Figure 2.1. We will not comment on the
non-uniqueness of L·R again.

Clearly if L and R have orders m and r respectively, then any dot product L·R has order m + r - 2.

Theorem 2.1 Suppose $\chi'(L) = \chi'(R) = 4$. Then $\chi'(L \cdot R) = 4$.

Proof: Suppose that L·R is 3-edge-colourable. By Lemma 1.2, up to relabelling, the edges ar, bs, ct, du can be coloured respectively in one of the following ways: (i) (1,1,1,1) ; (ii) (1,1,2,2) and (iii) (1,2,1,2). For (i) colour the edges ab, cd with colour 1 and preserve the colours on the other edges of L·R in R. This gives a 3-edge-colouring of R. Similarly the other two cases give 3-edge-colourings of L or R.   □

By this theorem, if L and R are snarks then L·R must be a snark. (It is easy to check that $\gamma(L \cdot R) \geq 5$ and L·R has cyclic edge-connectivity at least 4.) As an example take L ≅ P ≅ R. This gives the Blanuša snarks shown in Figure 2.2.

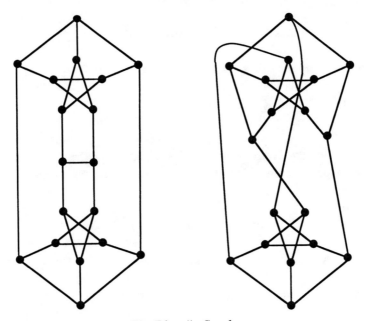

The Blanuša Snarks

Figure 2.2

Another example is given by the Szekeres snark, see Figure 2.3, which appears as a P·(P·(P·(P·(P·P)))); see Exercise 9(b).

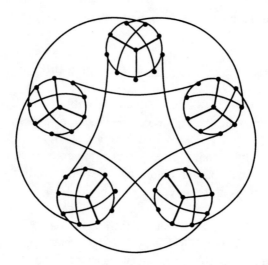

Szekeres' Snark

Figure 2.3

We have been very careful to use $\chi'(L) = 4$ and $\chi'(R) = 4$, rather than call L and R snarks. This is because it is possible for an L·R to be a snark even though not both L and R are snarks.

There are in fact, other constructions which produce snarks using graphs which do not all have chromatic index 4. To see this we introduce the star product. Let L and R be connected cubic graphs then this product is defined below.

The star product
A **star product**, L*R, of two cubic graphs L and R is obtained by

(1) deleting vertices u from L and v from R together with their incident edges;
(2) joining {r,s,t} to {a,b,c}.

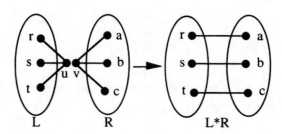

Figure 2.4

We note that L*R is not uniquely defined. One L*R is illustrated in Figure 2.4. Observe that a star product is the reverse of a 3-cut reduction.

We now show how to construct snarks from non-snarks. Let S_1, S_2 be snarks and A_1 and A_2 be connected cubic graphs. Let $G = S_2 \cdot (A_1 * (S_1 * A_2))$, where in the dot product we delete two adjacent vertices from S_2 and edges from A_1 and A_2 and in the star products we delete two distinct vertices from S_1 and one each from A_1 and A_2. The graph G is illustrated in Figure 2.5.

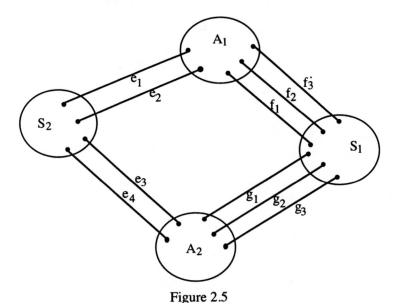

Figure 2.5

Now, assuming that $\chi'(G) = 3$, take any 3-edge-colouring of G. By Lemma 1.2 the edges $\{e_1, e_2, e_3, e_4\}$, up to colour relabelling, are coloured in the ordering (e_1, e_2, e_3, e_4) in the following possible ways:

(i) (1,2,1,2); (ii) (1,1,1,1); (iii) (1,1,2,2).

Clearly (i) would induce a 3-edge-colouring of S_2 which is impossible. Hence we can be sure that e_1 and e_2 are coloured the same and e_3 and e_4 are coloured the same. It follows from Lemma 1.2, considering the edge cut $\{e_1, e_2, f_1, f_2, f_3\}$, that f_1, f_2 and f_3 have distinct colours. Similarly g_1, g_2 and g_3 have distinct colours. This induces a 3-edge-colouring of S_1, which is impossible. Hence the assumption that $\chi'(G) = 3$ is false. Hence $\chi'(G) = 4$. We can thus produce new snarks from possibly non-snarks A_1 and A_2 provided only that A_1 and A_2 are chosen to produce the necessary girth and connectivity requirements for G.

Now we can repeat the process used in constructing $G = S_2 \cdot (A_1 * (S_1 * A_2))$. Let $G' = G \cdot (A_3 * (S_3 * A_4))$. In this dot product construction we use the edge e_4, see Figure 2.5, in G and delete one edge from A_3 and one from A_4. G' is illustrated in Figure 2.6.

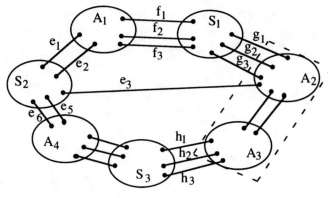

Figure 2.6

In Figure 2.7 the eunegraph illustrates in more detail the construction of G' localized at S_2. The dotted lines indicate the edges which have been deleted from S_2.

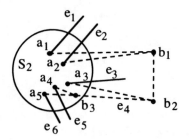

Figure 2.7

Now, assuming that $\chi'(G') = 3$, take any 3-edge-colouring of G'. Using Lemma 1.2, if e_1 and e_2 are coloured the same then f_1, f_2 and f_3 receive distinct colours and, again using Lemma 1.2, g_1, g_2 and g_3 receive distinct colours. Then clearly this colouring of G' induces a 3-edge-colouring of the snark S_1, which is impossible. A similar argument shows that e_5 and e_6 must receive distinct colours. Finally by Lemma 1.2 , the edges $\{e_1, e_2, e_3, e_5, e_6\}$, up to colour relabelling can only be coloured $(1,2,1,1,3)$ in the order $(e_1, e_2, e_3, e_5, e_6)$. Clearly, see Figure 2.7, such a colouring induces a 3-edge-colouring of S_2 which is impossible. Hence $\chi'(G') = 4$.

So, provided we are careful about girth and connectivity G' is a snark. Furthermore, subject to these constraints also, the structure of the section of the graph G', indicated by the dotted boundary in Figure 2.6, can be chosen quite arbitrarily i.e. the component of $G' - \{g_1, g_2, g_3, e_3, h_1, h_2, h_3\}$, not containing e_1 say, may be chosen arbitrarily. We shall use this idea when we construct the Descartes snark.

The Descartes snark. Let A_i $(i = 1, 2,..., 30)$ be connected cubic graphs. Let $G_1 = P \cdot (A_1 * (P * A_2))$ where an edge from each of A_1 and A_2 is used in the dot product. So assuming G_i is defined $(1 \le i \le 15)$, let $G_{i+1} = G_i \cdot (A_{2i-1} * (P * A_{2i}))$. Always in the construction of G_{i+1} we use an edge from A_{2i-1} and an edge from A_{2i}. The adjacent vertices which we delete from G_i in the dot product construction of G_{i+1} are best shown by example. In the construction of G_1, see Figure 2.8(a), we delete an arbitrary pair of adjacent vertices in P and an arbitrary edge in each of A_1 and A_2. In the construction of G_2, see Figure 2.8(b), we delete the endvertices of e_1 and in the construction of G_3, see Figure 2.8(c), we delete the endvertices of e_2, and so on.

Now by our previous discussion, we can replace the subgraphs indicated by the dotted lines in Figure 2.8(b) and Figure 2.8(c), by any graph at all without affecting the chromatic index of G_2 or G_3 respectively, providing that the graphs G_2 and G_3 remain cubic. In fact we replace the indicated subgraph in G_3 by the cycle C_9. By continuing this construction for 'each edge' of the original P used in the construction, we obtain a graph in which each vertex of P is replaced by C_9 and every edge of the original P is replaced by the eunegraph of Figure 2.9.

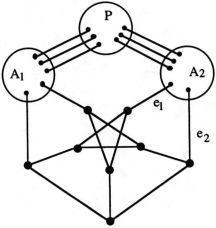

(a) $G_1 = P \cdot (A_1 * (P * A_2))$

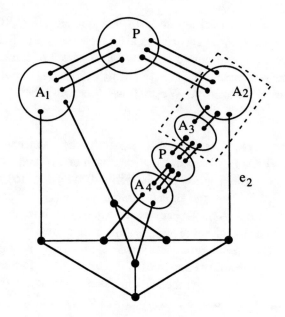

(b) $G_2 = G_1 \cdot (A_3 * (P * A_4))$

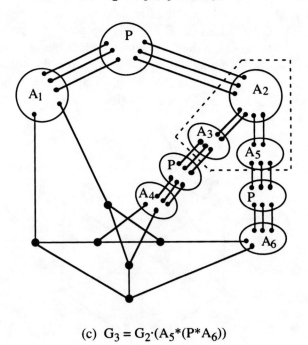

(c) $G_3 = G_2 \cdot (A_5 * (P * A_6))$

Figure 2.8

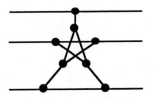

Figure 2.9

The resulting graph has order 210 and is called a Descartes snark (notice that our construction does not lead to a unique graph). Incidentally, the Descartes after whom this snark is named, is Blanche whom we referenced in the prologue of Chapter 2. (We leave the reader the interesting task of tracking down Blanche's parents.)

We have now shown that Isaacs' products include all the snarks which had previously been known except for P. Now any element of a dot product L·R is only cyclically 4-edge-connected. Essentially they all seem to be of chromatic index 4 because P is of chromatic index 4. One can certainly ([gR 87]) produce huge numbers of not very informative examples using the dot product and in some ways this is because they all have cyclic 4-edge cuts. This idea is made more concrete by the next theorem which is statement 5 of [maG 81] and Theorem 1 of [C-C-W 87].

Theorem 2.2 Let G be a snark with an edge cutset S of size 4. Let G_1, G_2 be the two eunegraphs produced by cutting the edges of S. Then either
(i)   one of G_1 or G_2 is not 3-edge-colourable; or
(ii) both G_1 and G_2 can be "extended" to snarks by adding two vertices in such a way that G is their dot product.

Proof: We may assume that G_1 and G_2 are both 3-edge-colourable. Let the cutset be $\{u_iv_i: i = 1,2,3,4\}$. Since Lemma 1.2 applies to the splines of G_1 and G_2 as if they were a cutset, we may assume that the splines of G_1 and G_2 can be coloured in order (1,1,1,1), (1,1,2,2), (1,2,1,2) or (1,2,2,1). Suppose the splines of G_1 can be coloured (1,1,1,1). Using a Kempe chain argument on a 1-2 coloured chain we see that the splines of G_1 can also be coloured in one of the other three ways. Since $\chi'(G) \neq 3$, the colours of the splines on G_1 and G_2 cannot match up. Hence the splines of G_1 can be coloured in two of the four ways above and the splines of G_2 in the other two ways.

Suppose the splines of G_1 are coloured (1,1,1,1) or (1,1,2,2). Then introduce the two vertices v,w and form $\widetilde{G}_1$ by adding u_1v, u_2v, u_3w, u_4w, vw to the edges of G. The graph $\widetilde{G}_2$ is formed by joining the splines at v_1 and v_2 together and by joining the splines at v_3 and v_4 together; see Figure 2.10.

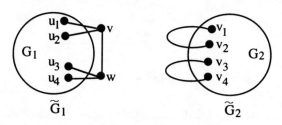

Figure 2.10

Neither of $\widetilde{G}_1$ and $\widetilde{G}_2$ is 3-edge-colourable and $G = \widetilde{G}_1 \cdot \widetilde{G}_2$. We leave the checking of the remaining snark properties of $\widetilde{G}_1$ and $\widetilde{G}_2$ to Exercise 10.

Similar constructions for $\widetilde{G}_1$ and $\widetilde{G}_2$ can be obtained if G_1 is coloured (i)(1,1,1,1) or (1,2,1,2) or (ii) (1,1,1,1) or (1,2,2,1).   □

The following result was also proved in [C-C-W 87].

Theorem 2.3 Let $G \neq P$ be a snark with an edge cutset S of size 5. Let G_1, G_2 be the two eunegraphs produced by cutting the edges of S. Then either
(i) one of G_1 or G_2 is not 3-edge-colourable; or
(ii) both G_1 and G_2 can be 'extended' to cubic graphs that are not 3-edge-colourable by adding at most five vertices; at least one of these extended graphs is smaller than G.

The importance of these results is that they suggest that to get to the heart of snarkness one needs to consider cyclically 6-edge-connected cubic graphs. Snarks with smaller connectivity are supposedly captured by Theorems 2.2 and 2.3. The point is that if either G_1 or G_2 in both theorems is not 3-edge-colourable, then it is easy to construct a snark smaller than G via $\widetilde{G}_1$ or $\widetilde{G}_2$. If both G_1 and G_2 are not 3-edge-colourable, then G is a dot product or there are smaller snarks than G from which G can be produced.

It does seem to be possible though, that snarks may exist whose only cyclic 5-edge cutset is such that every edge of the cutset is incident to exactly one vertex of a 5-cycle. The flower snark J_5 (see below) is such a graph. Here the eunegraphs obtained by cutting the cyclic 5-edge cutset can only be extended to non-3-edge-colourable graphs by the addition of a further 5 vertices in each case. Although Theorem 2.3 is true for J_5 (and potentially other such snarks) the extended graph which is smaller than J_5 is P. It would seem that this situation needs to be incorporated into the essence of snarkness before attacking cyclically 6-edge-connected snarks.

The flower snark. We define the graph J_k, where k is an odd number greater than 3, as follows. Write $I(m) = \{0, 1, \ldots, m-1\}$. Let

$VJ_k = \{u_i, v_i : i \in I(k)\} \cup \{w_j : j \in I(2k)\}$ and let
$EJ_k = \{u_i u_{i+1}, u_i v_i, w_j w_{j+1} : i \in I(k), j \in I(2k)\} \cup \{v_i w_j : i \equiv j \pmod{k}, i \in I(k),$
$\quad j \in I(2k)\}$,

where the addition on subscript i is modulo k and on subscript j is modulo $2k$.

We show J_5 and J_7 in Figure 2.11. The representations shown there make it clear why the name 'flower snark' is applied to this class.

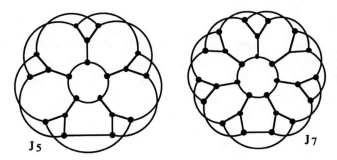

J_5 J_7

Figure 2.11

Notice that if we allow $k = 3$, then J_3 contains a triangle. Hence it is not a snark but if we contract the triangle to a single vertex then we obtain P.

Theorem 2.4 J_k is a snark.

Proof: Assume that $\chi'(J_k) = 3$. Take any 3-edge-colouring of J_k. Now consider the subgraph of J_k shown in Figure 2.12.

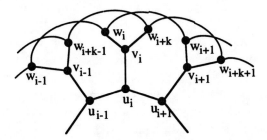

Figure 2.12

Let $E(i) = \{u_{i-1}u_i, w_{i-1}w_i, w_{i+k-1}w_{i+k}\}$, where integers are modulo k on the 'u' subscripts and modulo 2k on the 'w' subscripts. Because the edges incident to v_i receive different colours, not all the edges of $E(i)$ can be coloured the same. Suppose two of the edges are coloured the same. Suppose that $u_{i-1}u_i, w_{i-1}w_i, w_{i+k-1}w_{i+k}$ are coloured, in that order, (1,1,2). If u_iv_i is given colour 2, then v_iw_i has to be coloured 3 and v_iw_{i+k} coloured 1. This gives the colouring (3,2,3) in order, for $E(i+1) = \{u_iu_{i+1}, w_iw_{i+1}, w_{i+k}w_{i+k+1}\}$. On the other hand if u_iv_i is coloured 3, then we get the colouring (2,3,3) for $E(i+1)$.

The same argument shows that any colouring of the form (1,1,2) (in any order) for the edges of $E(i)$ gives colourings of the form (2,3,3) (in some order) for $E(i+1)$. Consequently, $E(i+2)$ is forced to be coloured (1,1,2) and so on. However $E(0) = E(k)$ and, since k is odd, this means the edges of $E(0)$ are coloured with (1,1,2) **and** (3,3,2). This contradiction implies that the edges of $E(i)$ are all coloured differently.

So now suppose $u_{i-1}u_i, w_{i-1}w_i, w_{i+k-1}w_{i+k}$ are coloured (1,2,3). Then the edges of $E(i + 1)$ in the order $u_iu_{i+1}, w_iw_{i+1}, w_{i+k}w_{i+k+1}$ are coloured either (2,3,1) or (3,1,2) i.e. the colouring is a cyclic permutation of the colouring of the edges of $E(i)$. Hence $E(k)$ (= $E(0)$) has its edges coloured in a cyclic permutation of the colouring of $E(0)$. However, and this is where we have to be careful as to which modulus we take our subscripts, $E(k)$ is a transposition of the edge–colouring of the edges in $E(0)$.

This final contradiction implies that $\chi'(J_k) = 4$. It is then straightforward to show that J_k is cyclically 5-edge-connected and of girth 6 unless k = 5 when its girth is 5.   ☐

A graph is said to be **hamiltonian-connected** if, for each pair of distinct vertices u and v, there is a hamiltonian path with ends u and v. In [K-S 90], the graphs J_k, for k $\geq$ 7 and k odd, are shown to be not hamiltonian-connected but the addition of any edge produces a graph which is. The Coxeter graph is a graph on 28 vertices, which shares this property with these J_k. The Petersen graph, see Exercise 18, is the smallest cubic graph with this property.

The double star snark ([rI 75]). As well as his two infinite families of snarks, Isaacs' discovered the double star snark; see Figure 2.13.

Isaacs' constructed this snark as part of a general scheme of constructing further infinite families of snarks. Roughly sketched the idea is this. Given a eunegraph with five splines which is potentially a part of a 3-edge-colourable graph, in any

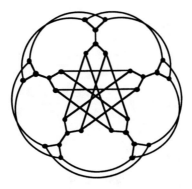

Figure 2.13

such colouring the five splines must be coloured so that one colour is repeated three times and the other two colours once each. Now assume that the structure of the eunegraph determines that the colouring of the splines are equivalent up to a 'cyclic permutation' of the splines. (This rather vague statement will become clear from the example below.) Now either the three appearances of the one colour are consecutive, with respect to some fixed ordering of the splines, or they are not. Finally if there exist two eunegraphs such that one of them is necessarily of the 'consecutive type' in any 3-edge-colouring and the other eunegraph is necessarily of the 'nonconsecutive type' in any 3-edge-colouring then, provided we match the splines in an obvious way, we can construct a graph which is not 3–edge–colourable.

We illustrate this situation with the eunegraph of Figure 2.14(a). Using the obvious cyclic permutation of the colours c_i, $i = 1, 2, 3$ we try to colour the splines so that the three splines which would be required to use c_3 are not placed consecutively around the pentagon. This colouring forces uv and vw to be coloured c_2 which is not possible in a 3-edge-colouring. Hence the eunegraph E of Figure 2.14(a) cannot have its splines coloured as shown. They must be coloured consecutively c_3, c_3, c_3, c_1, c_2 to within a permutation of the colours or a cyclic permutation of the splines. Applying this argument to the eunegraph of Figure 2.14(b) gives the colouring shown in that diagram, again to within a permutation of the colours or a cyclic permutation of the splines.

If we now try to join the splines from Figure 2.14(a) and (b) to give edges uu',vv', ww', xx', yy', we cannot obtain a 3-edge-colouring. This is because if there were a 3-edge-colouring, the colouring of Figure 2.14(b) would force a colouring of Figure 2.14(a) where the three occurrences of c_3 were not consecutive.

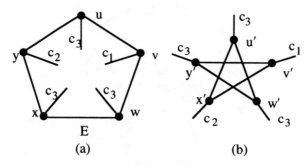

Figure 2.14

Thus we have proved the following.

Lemma 2.5 $\chi'(P) = 4$.   ☐

A similar pentagonal situation arises if we sever the edges $u_i v_i$, $i = 1, 2, 3, 4, 5$ of J_5. Here we obtain the eunegraph E mentioned above and a eunegraph which Isaacs calls V*. We have seen how joining two copies of E gives P. If we produce our mismatching of coloured spines using E and V* we obtain J_5. But in addition we may perform the same mismatch on two copies of V*. When we do, we obtain the double star snark of Figure 2.13.

Loupekhine's Snarks. This infinite family gives a further method of constructing new snarks from old. Suppose we take a path u_1, u_2, u_3, u_4, u_5 in a snark S and cut the edges so as to remove the eunegraph induced by u_2, u_3, u_4, see Figure 2.15.

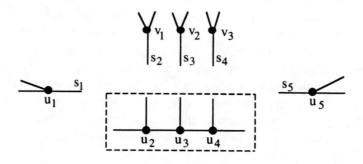

Figure 2.15

Now by Lemma 1.2 we know that if the splines s_i, $i = 1, 2, 3, 4, 5$ were joined to other splines to produce a 3-edge-colouring of some graph G, then the s_i would require three of one colour and one each of the other two. But conceivably

s_2, s_3, s_4 may be coloured c_1, while s_1 and s_5 are coloured c_2 and c_3, respectively. In this case we can take the colouring of G and produce a 3–edge–colouring of S. To see this, we colour every edge of ES $\cap$ EG with the same colour it had in G, colour $u_1u_2, u_2v_1, u_3v_2, u_4v_3, u_4v_5$ in the same colour that the splines s_1, s_2, s_3, s_4, s_5 were given in G, and colour u_2u_3 in c_3 and u_3u_4 in c_2. This contradiction shows that s_2, s_3, s_4 cannot all have the same colour. Similarly we can obtain a contradiction if both s_1, s_2 and s_4, s_5 are coloured so that each pair of splines has two distinct colours associated with it. Hence one of the pairs s_1, s_2 or s_4, s_5 must have the same matching colour, while the other pair must consist of two distinct colours.

To use this result, we first take an odd number of snarks and operate on them as above to give an odd number of eunesnarks of the type shown in Figure 2.15. Now suture together, in an obvious way, the s_1, s_2 pairs and s_4, s_5 pairs from the different eunesnarks. Join the splines s_3 to each other, or to a new vertex of degree three. By what has already been said, the new graph is not 3-edge-colourable. If we have been careful to avoid cycles of length smaller than five, we will have produced a new snark. A possible 3-edge-colouring would need to go from matched to unmatched pairs of splines as we move from eunesnark to eunesnark. Because of the odd number of eunesnarks, this forces three edges of type s_2, s_3, s_4 to have the same colour. This contradiction shows that our new graph is not 3-edge-colourable.

The Loupekhine construction is not unique because, among other things, we may cross or uncross edges between pairs of splines. It is also worth noting that this method cannot produce snarks from 3-edge-colourable graphs.

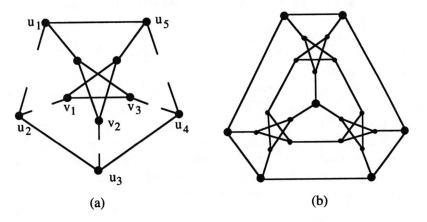

(a)   (b)

Figure 2.16

The smallest snarks in the Loupekhine family are of order 22 and are produced from the Petersen snark. We illustrate one of these beasts in Figure 2.16(b). In Figure 2.16(a) we show how the Petersen graph has been doctored to obtain the eunegraph required for a Loupekhine snark.

Loupekhine also gives a variation of this construction using an even number of snarks. This work is reported by Isaacs in an unpublished technical report [rI 76]. Yet another slight variation, which as usual is derived from copies of P, is given in Figure 2.17. This is one of the smallest snarks which can be derived in this way and has only 26 vertices.

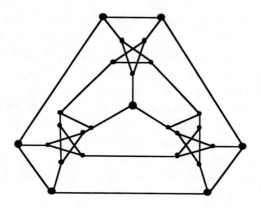

Figure 2.17

Many authors [jJ 74], [C-S-F 80], [maG 81], [mP 82], [jW 83], [gR 87], [gR88], [bS 87] to name but a few, have continued to pursue the elusive snarks. Other constructions and more information on snarks may be found in the survey article of Chetwynd and Hilton [C-H 88].

3 . k-snarks (k > 3)

We have seen that there are a large number of snarks. For example, the dot product operation itself produces infinitely many new snarks. However these all have cyclic edge-connectivity at most 4. Nevertheless we still obtain a large number of snarks with cyclic edge-connectivity at least five using other constructions. Possibly snark hunters should be looking for generalizations of Goldberg's theorem, Theorem 2.2. There may be a similar result if we assume the cyclic edge connectivity is even rather than simply 4.

However as far as we know, snarks are a heavily inbred species. All the known constructions produce new snarks from old snarks and they all find their family tree leading back to that grand old patriarch the Petersen graph. All of which lends credence to Tutte's Conjecture advanced at the end of Chapter 2, that every bridgeless cubic graph of chromatic index 4 is subcontractible to P. Indeed, assuming the truth of this conjecture, snarks are simply 'generalizations' of P.

Moreover, a variant of this conjecture may well be true if one delves into the domain of the supersnark. From Vizing's theorem we know that if G is a k–regular graph (k ≥ 3), then $\chi'(G) = k$ or k + 1. The analogue to Lemma 1.2 for k-regular graphs is:

Lemma 3.1 Suppose G is a k-regular graph (k ≥ 3) such that $\chi'(G) = k$. In any k-edge-colouring of G with colours c_i, i = 1, 2,..., k and for any edge cut Z, the numbers n_i of edges in Z coloured c_i are all congruent modulo 2.

Proof: The argument (Exercise 14) is a simple generalization of the argument of Lemma 1.2.   □

There are similar k-regular graph analogues for Corollary 1.3 and Lemmas 1.4, 1.5, 1.6 and 1.7. Some of these analogues are slightly more complicated. In particular one must be careful as to the parity of k. So in the same spirit of the definition of a snark we define a **k-snark** to be a k-regular cyclically (k+1)–edge-connected graph of even order which does not contain a clique of order k + 1 minus a 1-factor or a clique of order k.

Notice that in this definition the reference to a clique of order k + 1 minus a 1–factor is a generalization of the property, see Lemmas 1.4, 1.5, that if G is a snark then $\gamma(G) \geq 5$ i.e. G does not contain a triangle and G does not contain an induced 4-cycle - a K_4 minus a 1-factor. See Lemma 1.4 and Exercise 12 for the reason for excluding k-cliques in this definition. It then follows that a 3–snark is a snark.

It is of interest to note that the Meredith graph [gM 73] of degree k, see Figure 3.1 for k = 4, has chromatic index k + 1. However its cyclic edge-connectivity is at most k and so it is not a k-snark.

One way of deriving new k-snarks from old ones is by using the next result.

Lemma 3.2 If S is a snark of order n ≡ 0 (mod 4) then L(S) is a 4-snark.

Proof: See [fJ 74].   □

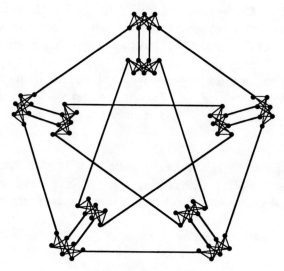

The Meredith graph

Figure 3.1

From Lemma 3.2 we know that $L(J_5)$ is a 4-snark. Figure 3.2 shows a subgraph of $L(J_5)$ which is contractible to P. Tutte's conjectures in both the subcontractible and homeomorphic cases, generalize to k-snarks.

Hence we have the next conjecture.

Conjecture 3.3: Every k-snark contains a subgraph contractible to P.

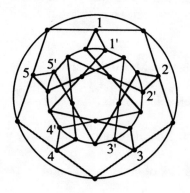

Figure 3.2

Appendix 1 The snarks of order 20

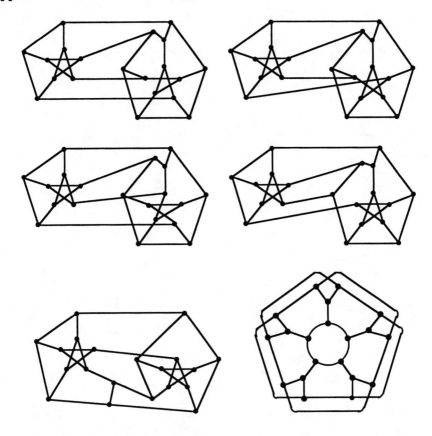

Appendix 2 The snarks of order 22

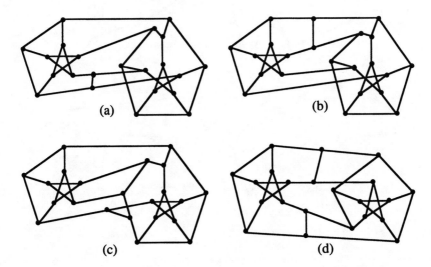

(a)

(b)

(c)

(d)

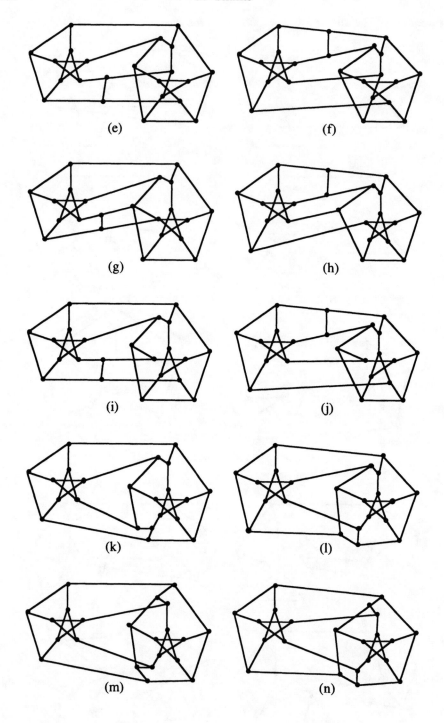

(e)

(f)

(g)

(h)

(i)

(j)

(k)

(l)

(m)

(n)

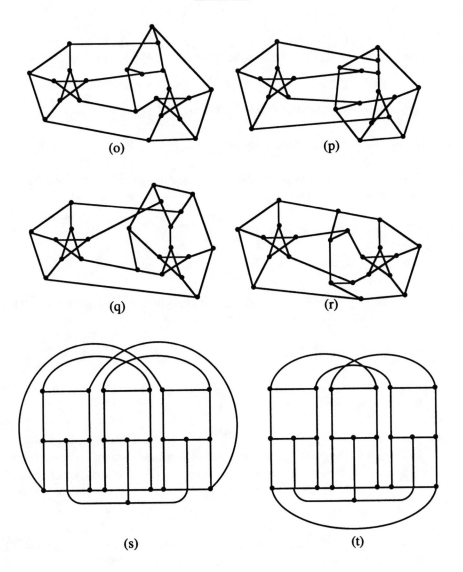

(o)   (p)

(q)   (r)

(s)   (t)

Exercises

1. Find the chromatic index of $K_n, K_{m,n}$ and an arbitrary tree T.

2. Assuming that all 3-connected planar cubic graphs are hamiltonian, prove the Four Colour Theorem.

3. Use Lemma 1.2 to prove that $\chi'(P) = 4$.

4. Show that P is cyclically 5-edge-connected.

5.   Let H be formed from cubic graphs G_1 and G_2 by adding a vertex to an arbitrary edge of G_1 and another to an arbitrary edge of G_2 and joining these vertices by an edge.

Without using Lemma 1.2 prove that $\chi'(H) = 4$.

6.   Show that there are no snarks of order 12, 14 or 16.

7.   Construct snarks of all possible orders from 18 to 30. Show that Tutte's conjecture is true for these graphs.

8.   Show that P(10,3) is not a snark.

9.   (a)   Find non-snarks L, R such that L·R is a snark.

     (b)   Show that one of the graphs P·(P·(P·(P·(P·P)))) is the Szekeres snark.

10.   Complete the proof of Theorem 2.2.

11.   Show that $L(J_k)$ contains a subgraph which is a subdivision of P. Repeat by replacing 'subdivision' by 'contractible'.

12.   Let G be any k-regular graph with k = 4 or 6. Choose (if possible) $\frac{1}{2}k$ independent edges of G and introduce two new vertices on these edges. Add edges between these new vertices so that they add a clique of order k to G. Denote the new graph by G(k). Prove that if $\chi'(G(k)) = k + 1$ then $\chi'(G) = k + 1$.

13.   Let G be 4-regular and contain a clique of order 4. Let the edges joining this clique to the rest of the graph be e_1, e_2, e_3, e_4. If the edges e_i are all given a distinct colour in some edge colouring of G, show that this forces a 5-edge-colouring of G.

14.   Prove Lemma 3.1.

15.   Prove that if G is k-regular $(k \geq 3)$ and $|VG|$ is odd then $\chi'(G) = k + 1$.

16.   Is L(P) a 4-snark?

17. The **total chromatic number**, $T(G)$, is the minimum number of colours needed to colour $VG \cup EG$ so that no two elements of $x, y \in VG \cup EG$ have the same colour if they are either incident or adjacent.

Then Behzad [mB 71] conjectured that, $\Delta(G) + 1 \leq T(G) \leq \Delta(G) + 2$.

If $T(G) = \Delta(G) + 1$, then G is of type 1. Prove that P is of type 1; see [hY 86, p. 78].

18. Prove that P is the smallest cubic edge-minimal non-hamiltonian-connected graph.

References

[A-B-H-R 87] R.E.L. Aldred, Bau Sheng, D.A. Holton and Gordon F. Royle All the small snarks, submitted.

[mB 71] M. Behzad, The total chromatic number of a graph: A survey, Combinatorial Mathematics and its Applications, (Proc Conference, Oxford, 1969), Academic Press, London 1971, 1-8.

[dB 46] D. Blanuša, Problem Cetiriju Boja (Croatian), *Hrvatsko Priordoslorno Društvo Glasnik, Mat-Fiz., Astr, Ser. II* , 1, 1946, 31-42.

[bS 87] Bau Sheng, Cycles in graphs, Masters Thesis, University of Otago, 1987.

[B-F 91] C. Berge and J. Fournier, A short proof of a generalization of Vizing's theorem, *J. Graph Th.*, 15, 1991, 333-336.

[C-C-W 87] P.J. Cameron, A.G. Chetwynd and J.J. Watkins, Decomposition of snarks, *J. Graph Th.*, 11, 1987, 13-19.

[cD 67] L. Carroll, *The Hunting of The Snark*, (Annotated by Martin Gardner), Penguin, 1967.

[C-P 72] F. Castagna and G. Prins, Every generalised Petersen graph has a Tait coloring, *Pac. J. Math.*, 40, 1972, 53-58.

[C-S-F 80] U.A. Celmins, E.R. Swart and J.L. Fouquet, The construction and characterization of snarks, *Research Report*, Univ.Waterloo, 1980.

[C-W 81]   A.G. Chetwynd and R.J. Wilson, Snarks and supersnarks, The
            Theory and Applications of Graphs, *Proc. Fourth International
            Graph Theory Conf.* (ed. by G. Chartrand et al), John Wiley and
            Sons, 1981, 215-241.

[C-H 88]   A.G. Chetwynd and A.J. Hilton, Snarks and k-snarks, *Ars
            Combinatoria*, 25C, 1988, 39-54.

[bD 48]   B. Descartes, Network-colourings, *Math. Gazette*, 32, 1948,
            67–69.

[F-W 77]   S. Fiorini and R.J. Wilson, *Edge-Colourings of Graphs*,
            Research Notes in Maths., 16, Pitman, London, 1977.

[jF 82]   J-L. Fouquet, Note sur la non existence d'un snark d'ordre 16,
            Disc. Math, 38, 1982, 163-171.

[mG 76]   M. Gardner, Mathematical games, *Scientific American*, 234,
            No.4, 1976, 126-30, and No 9, 1976, 210-211.

[maG 81]   M.K. Goldberg, Construction of class 2 graphs with maximum
            vertex degree 3, *J. Comb. Th.* B, 31, 1981, 282-291.

[rI 75]   R. Isaacs, Infinite families of nontrivial trivalent graphs which are
            not Tait colourable, *Amer. Math. Monthly*, 82, 1975, 221-239.

[rI 76]   R. Isaacs, Loupekhine's snarks: a bifamily of non-Tait-colorable
            graphs, *Technical Report* No. 263, Dept. of Math. Sciences, John
            Hopkins University, 1976.

[fJ 74]   F. Jaeger, Sur l'indice chromatique du graphe représentatif des
            arêtes d'un graphe regulier, *Disc. Math.*, 9, 1974, 161-172.

[jJ 74]   J.T. Jakobsen, On critical graphs with chromatic index 4, *Disc.
            Math.*, 9, 1974, 265-276.

[K-S 90]   R. Kalinowski and Z. Skupien, Large Isaacs' graphs are maximally
            non–hamiltonian-connected, *Disc. Math.*, 82, 1990, 101-104.

[gM 73]   G.H.J. Meredith, Regular n-valent n-connected non-hamiltonian
            non n-edge-colorable graphs, *J. Comb. Th.* B, 14, 1973, 55-60.

[mP 82]   M. Preissmann, Snarks of order 18, *Disc.Math.*, 42, 1982, 125–126.

[gR 87]   G. Royle, Constructive enumeration of graphs, Ph.D. Thesis, University of Western Australia, 1987.

[gR 88]   G. Royle, Constructive enumeration of graphs, Thesis, *Bull. Austral. Math. Soc.*, 38, 1988, 159.

[gS 73]   G. Szekeres, Polyhedral decompositions of cubic graphs, *Bull. Austral. Math. Soc.*, 8, 1973, 367-387.

[vV 64]   V.G. Vizing, On an estimate of the chromatic class of a p-graph (Russian), *Diskret Analiz.*, 3, 1964, 25-30.

[jW 83]   J. Watkins, On the construction of snarks. *Ars Combinatoria*, 16 B, 1983, 111-124.

[mW 69]   M.E. Watkins, A theorem on Tait colorings with an application to the generalised Petersen graphs, *J. Comb. Th.*, 6, 1969, 152-164.

[hY 86]   H.P. Yap, *Some Topics in Graph Theory*, L.M.S. Lecture Note Series 108, C.U.P., 1986.

4

Factors

0. Prologue

In the court of King Arthur there dwelt 150 knights and 150 ladies-in-waiting. The king decided to marry them off, but the trouble was that some pairs hated each other so much that they would not even get married, let alone speak! King Arthur tried several times to pair them off but each time he ran into conflicts. So he summoned Merlin the Wizard and ordered him to find a pairing in which every pair was willing to marry. Now Merlin had supernatural powers and he saw immediately that none of the 150! possible pairings was feasible, and this he told the king. But Merlin was not only a great wizard, but a suspicious character as well, and King Arthur did not quite trust him. "Find a pairing or I shall sentence you to be imprisoned in a cave forever!" said Arthur. Fortunately for Merlin, he could use his supernatural powers to find the reason why such a pairing could not exist. He asked a certain 56 ladies to stand on one side of the king and 95 knights on the other side, and asked: "Is any one of you ladies, willing to marry any of these knights?", and when all said "No!", Merlin said: "O King, how can you command me to find a husband for each of these 56 ladies among the remaining 55 knights?" So the king, whose courtly education did include the pigeonhole principle, saw that in this case Merlin had spoken the truth and he graciously dismissed him.

(Part of a fairy tale by the Brothers Lovász and Plummer in [L-P 86])

In trying to prove the Four Colour Theorem Tait [pT 80] stated a theorem which he believed to be 'universally true'. The theorem said that 'the edges of the polyhedron, which has trihedral summits only, can be divided into three groups, one from each group, ending in each summit'. Petersen showed his confusion by stating Tait's 'theorem' as essentially 'every bridgeless cubic graph is 1–factorable'. It is as a counterexample to this that the famous Petersen graph appears. One of the very reasons that the Petersen graph first attracted attention was because of the peculiarity of its 1-factors. The Petersen graph, it could be argued, led to the interest in and development of, the theory of factors.

The Four Colour Problem of course initiated the interest in factors as well as in hamiltonian cycles. These themes have remained the most central topics in the theory of graphs. There is now an enormous body of results on factorization; see [A-K 85] and [L-P 86]. We will just hint at some of them by concentrating on Tutte's 1-factor and f-factor theorems.

1.   Tait's conjecture

We recall that Tait conjectured that every planar cubic bridgeless graph was 3–edge-colourable. But this idea of colouring edges can be thought of in a different light. Let   G   be a planar cubic bridgeless graph which is 3–edge–colourable and let H_i be the subgraph of G induced by colour c_i. Then H_i consists of independent edges. In fact there are half as many independent edges as there are vertices of G. See Figure 1.1 for example.

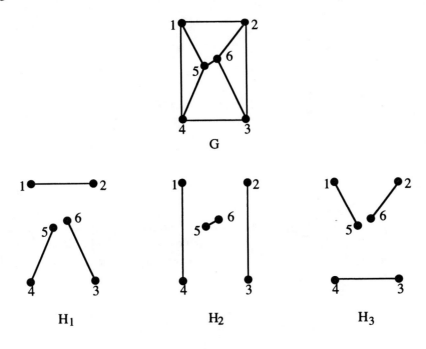

Figure 1.1

Such a subgraph, made up of independent edges so that every vertex of the original graph is a vertex of the subgraph is called a **1-factor** or a **perfect matching**.

Some graphs are the union of disjoint 1-factors as is the graph of Figure 1.1. Such graphs are said to be **1-factorable** or to contain a **1-factorization**. The graphs H_1, H_2, H_3 provide a 1-factorization of G, the graph of Figure 1.1.

Tait's conjecture can therefore be rephrased in the language of factors. It now says that every planar cubic bridgeless graph is 1-factorable. As a result of Corollary 2.6.2 we know that this is true.

In this chapter we pursue the idea of factors. The concept can be pushed in a number of directions. First a subgraph of G consisting solely of independent edges and isolated vertices is known simply as a **matching**. Any vertex incident to one of these independent edges is said to be **matched** or **covered** in the matching. If the number of isolated vertices is specified as d, say, then the subgraph is called a **defect-d matching**. The graph G of Figure 1.2 has a matching H which is a defect-1 matching.

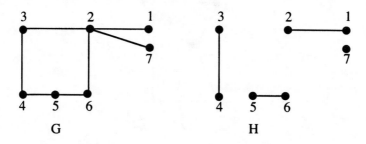

Figure 1.2

If every edge of a graph is in some 1-factor, then the graph is said to be **factor covered.** Note that this is not the same as being 1-factorable. A 1-factorable graph is factor covered but the converse does not hold. Perhaps not surprisingly, the Petersen graph provides a counterexample.

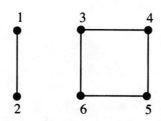

Figure 1.3

But Figure 1.3 gives a simpler example. This graph is certainly factor covered, since 12, 34 and 56 are in the 1-factor {12, 34, 56}, while 36 and 45 are in the 1-factor {12, 36, 45}. However the graph is not 1-factorable, since 12 must be in **every** 1-factor.

One can also observe that in a 1-factorable graph the number of edges is divisible by half the size of the vertex set. This is not the case with the graph of Figure 1.3.

To generalize the concept of factor covered we note that a graph is **defect-d covered** if every edge is in some defect-d matching. We note that the graph G of Figure 1.2 is not defect-1 covered, since the edge 23 is in no defect-1 matching. But the graph obtained from G by adding the edge 17 is defect-1 covered.

The concept of 1-factor can be generalized to any regular subgraph of fixed degree which spans a given graph. Thus G is said to have an **r-factor** if G contains a regular spanning subgraph of degree r. The graph of Figure 1.4 has a 2-factor. (In fact it has more than one 2-factor.)

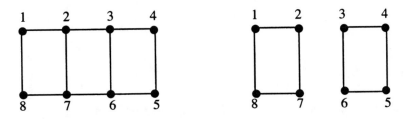

A graph and one of its 2-factors

Figure 1.4

Later in the chapter we introduce the notion of an f-factor, where f is a function. By the context there will be no confusion between r-factors and f-factors as r will always be an integer and f a function.

In the balance of this chapter then, we develop some of the theory of factors and note the place of the Petersen graph in this theory.

2.   Tutte's characterization.

In this section we give Tutte's characterization of those graphs that have a 1–factor. This theorem is clearly one of the basic theorems in graph theory and it is certainly the most important tool in the whole study of the factors of graphs. Indeed, without this theorem there would be no theory. Much of what follows in the rest of this chapter is a consequence of this key result.

What could possibly stop a graph from having a 1-factor? Well, clearly if |VG| is odd, then there is no chance of G possessing a 1-factor. So suppose |VG| is even and consider the situation of Figure 2.1. The graph G has a 1-factor, the edges of this 1-factor being indicated on the diagram. The set S is a cutset of G. Edges not in the 1-factor, including edges joining the vertices of S to the components of G \ S, are not shown.

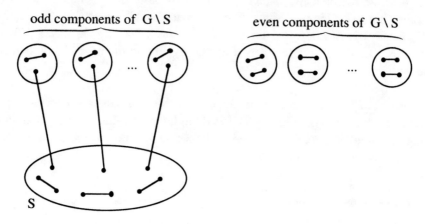

Figure 2.1

First define an **odd component** to be a component which has an odd number of vertices. Now if, in Figure 2.1, we added more odd components to G \ S so that the number of odd components exceeded |S|, the resulting graph could not possibly contain a 1-factor. This is readily seen, since at best the even components of G \ S will each have a 1-factor while the odd components will have at least one vertex not in a 1-factor. These vertices would have to be paired with vertices of S for the graph to have a hope of containing a 1-factor. But |S| is now not large enough.

Let $h_G(S)$ be the number of odd components of G \ S. If $h_G(S) > |S|$ for some $S \subseteq VG$, then we say that S is a **1-barrier**.

We have just shown that if G has a 1-barrier then it cannot have a 1-factor. Actually this situation also covers the case when |VG| is odd, since then we simply take $S = \emptyset$.

So if G has a 1-factor it is necessary that G has no 1-barrier. It is perhaps a little surprising that the non-existence of a 1-barrier is also sufficient for the existence of a 1-factor.

This is Tutte's characterization of graphs which contain 1-factors. The original proof given in [wT 47] used Pfaffians. Here we give a more direct proof due to Lovász [IL 75].

Theorem 2.1 A graph either has a 1-factor or a 1-barrier but not both.

Proof: From the discussion above we know that if a graph has a 1-barrier then it cannot have a 1-factor. Hence if it has a 1-factor it cannot have a 1-barrier. We must now show that a graph has to have one or the other.

Now suppose that G' is a graph which has neither a 1-factor nor a 1-barrier. So for all $S \subseteq VG'$, $h_{G'}(S) \leq |S|$. Choose G so that G' is a spanning subgraph of G and such that G does not contain a 1-factor. Subject to these constraints choose G so that it is edge-maximal i.e. if $uv \notin E(G)$ then $G + uv$ contains a 1-factor for any such choice of $u,v \in V(G)$. Since G' is a spanning subgraph of G, if $S \subseteq VG = VG'$ then $h_G(S) \leq h_{G'}(S) \leq |S|$. So G has no 1-barrier. Also if $S = \emptyset$ then $h_G(S) \leq 0$ i.e. $|VG|$ is even.

Let $H = \{v \in VG : \deg v = n - 1\}$ where $|VG| = n$. If $H = VG$ then $G \cong K_n$ which has a 1-factor. So $H \neq VG$.

Suppose now that $G \setminus H$ is the disjoint union of complete graphs. If there are exactly k of these complete graphs which are odd, then if $|H| \geq k$ we can construct a 1-factor of G as in Figure 2.2. We simply select a vertex u_i from each of the odd components. Since $|VG|$ is even, $|H \setminus \{v_1, v_2, ..., v_k\}| = |H| - k$ is even as $|VG| \equiv |H| - k \pmod 2$. Hence since $\langle H \setminus \{v_1, ..., v_k\} \rangle$ is a complete even graph, our matching can be extended to the whole of G. Therefore $|H| < k$ and so G has a 1-barrier. This is a contradiction. Hence at least one component of $G \setminus H$ is not complete.

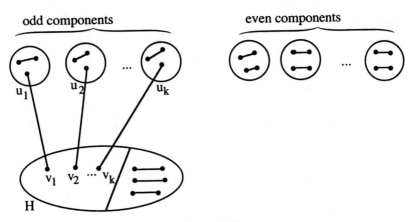

Figure 2.2

Let K be a component of $G \setminus H$ which is not complete. Hence we may choose $u,v \in V(K)$ with $d(u,v) \geq 2$. Let Q be a shortest path between u and v and let x,y,z be consecutive vertices on Q. Since Q is a shortest path, $xz \notin E(K)$. Since $y \notin H$ there exists a vertex w such that $yw \notin E(G)$. By the maximality of G we may choose 1-factors F_1 and F_2 in $G + \{xz\}$ and $G + \{yw\}$ respectively. Further, let F be the subgraph of $G + \{xz,yw\}$ induced by the symmetric difference $F_1 \Delta F_2$. A little thought, see Figure 2.3, then leads to the conclusion that F is the disjoint union of even cycles. Suppose that xz and yw are in different components C_1, C_2 of F. Then construct a 1-factor L of G consisting of the edges of F_1 in C_2 and the edges of F_2 elsewhere. Figure 2.3 indicates why L is a 1-factor in F. This is because L is a 1-factor in $G+\{xz,yw\}$ which does not include the edges xz and yw.

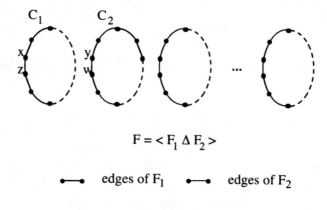

$$F = <F_1 \Delta F_2>$$

•——• edges of F_1 •——• edges of F_2

Figure 2.3

Hence xz and yw are edges in the same component C of F. Using the same notation as Figure 2.3, Figure 2.4 illustrates how the edges of F_1 in that part of C running from y through w to z, together with yz and the edges of F_2 in the rest of G including the rest of C, give rise to a 1-factor in G. This final contradiction proves the theorem.   ◻

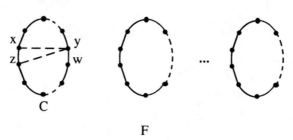

Figure 2.4

The above theorem is usually stated in the following way.

Corollary 2.2 The graph G has a 1-factor if and only if $h_G(S) \leq |S|$ for all $S \subseteq VG$. ▢

We now give some applications of Tutte's theorem. The first is a result of Petersen [jP 91]. This is not the paper from which the graph got its name but it is the first paper concerned with the general consideration of factors of graphs.

Theorem 2.3 Let G be a cubic graph with no 1-factor. Then G has at least three bridges, not all of them contained in the same path.

Proof: Suppose G has no 1-factor. Choose a 1-barrier S of G. Let K be an odd component of $G \setminus S$ and let $q(K,S)$ denote the number of edges $ks, k \in K$ and $s \in S$. Then, since $|VK|$ is odd,

$$q(K,S) = \sum_{k \in K} \deg k - \sum_{k \in K} q(k,K) = 3|VK| - 2|EK| \equiv 1 \pmod 2.$$

Let a be the number of odd components K such that $q(K,S) = 1$ and let b be the number of odd components K such that $q(K,S) \geq 3$. Let E be the set of edges sk where $s \in S$ and $k \in K$ for any odd component K. Then, since G is a cubic graph,

$$3|S| \geq |E| \geq a + 3b. \tag{1}$$

Since S is a 1-barrier, $h_G(S) > |S|$. Furthermore, since $h_G(S) \equiv |S| \pmod 2$; see Exercise 2,

$$h_G(S) \geq |S| + 2. \tag{2}$$

Therefore, from (1) and (2),

$$3(a + b) = 3h_G(S) \geq a + 3b + 6. \tag{3}$$

Hence $a \geq 3$ i.e. G has at least 3 bridges. Clearly these bridges are not contained in the same path. ▢

It is worth noting that Fleischner [hF 92] has proved that if G is a bridgeless graph with minimum degree 3, then G has a 1-factor.

Now Theorem 2.3 was proved by Petersen in 1891 by a different method. Nash-Williams [cN 82] writes 'in that paper Petersen expressed a feeling that his theorem could be generalized to regular graphs of any odd valency but said he found the probable labour involved too daunting to attempt this task. In [dK 31] König quoted this remark with the somewhat sceptical comment that no such generalization had been achieved despite the passage of 45 years during which Petersen's original proof had undergone considerable simplifications. Some years later, however, Tutte [wT 47] proved his 1-factor theorem which tells us, without restriction to regular graphs of odd valency, which finite graphs have perfect matchings'.

Petersen's result can be partially rephrased to claim that any 2-edge-connected cubic graph contains a 1-factor. Instead of stating this as a corollary to Theorem 2.3 we state and prove a generalization.

Theorem 2.4 Let G be an $(r-1)$-edge-connected, r-regular graph with $r \geq 2$ and odd. Then G has a 1-factor.

Proof: Let $S \subseteq VG$. If K is any odd component of $G \setminus S$, then as in the proof of Theorem 2.3,

$$q(K,S) = r|VK| - 2|EK| \equiv r|VK| \equiv r \ (\text{mod } 2). \tag{1}$$

Since G is $(r-1)$-edge-connected,

$$q(K,S) \geq r - 1. \tag{2}$$

From (1) and (2),

$$q(K,S) \geq r. \tag{3}$$

Therefore, using the same notation as in Theorem 2.3, and remembering each vertex of S has degree r,

$$r|S| \geq |E| \geq r \ h_G(S). \tag{4}$$

So S is not a 1-barrier. Hence G has a 1-factor.   □

In this theorem, r is necessarily odd since otherwise G is Eulerian and cannot be $(r-1)$-edge-connected.

Since an $(r-1)$-connected graph is $(r-1)$-edge-connected, the corollary below follows immediately.

Corollary 2.5 Let G be an $(r-1)$-connected r-regular graph with $r \geq 2$ and r odd. Then G has a 1-factor.   □

The next application is the famous Marriage Theorem proved by Hall in [pH 35]. Suppose a group of boys each makes a list of the girls he would be prepared to marry. Under what conditions can all the boys be married so that each boy is married to a girl on his list? Hall's theorem answers this question completely.

Because of the sexism explicit in the statement of the problem we rephrase it in terms of bipartite graphs. We note that here we are interested in a matching, not a perfect matching (1-factor).

Once again we have a theorem in which the obvious necessary condition is also sufficient. We use $N(S)$ to denote $\bigcup_{s \in S} N(s)$. We say that a matching M in a graph G **covers (matches)** a set $X \subseteq VG$ if for each $x \in X$ there exists an edge $e \in EM$ such that x is incident with e.

Theorem 2.6 Let G be a bipartite graph with parts X and Y. Then G contains a matching which covers X if and only if

$$|N(S)| \geq |S| \text{ for all } S \subseteq X.$$

Proof: Let M cover every vertex of X. Suppose $S \subseteq X$. Define

$$S' = \{y : y \in Y, sy \in EM \text{ for some } s \in S\}.$$

Then $|S| = |S'|$ and $|N(S)| \geq |S|$ since $S' \subseteq N(S)$. Hence for all $S \subseteq X$, $|N(S)| \geq |S|$.

Now assume that $|N(X')| \geq |X'|$ for all $X' \subseteq X$. Form the graph G' from G by adding all possible edges between vertices of Y. If $|VG|$ is odd then also add a new vertex v to G which is joined to every vertex in Y. Hence $|VG'|$ is always even. If G' has a 1-factor, then G has a matching which covers every vertex of X. This follows since in G' no two vertices in X are adjacent, so any edge of a 1-factor of G' which has one end-vertex in X must have the other end-vertex in Y. Such an edge must also belong to G.

Suppose then that G' does not have a 1-factor. Then by Tutte's theorem, it contains a 1-barrier S. Write $S_X = S \cap X$ and $S_Y = S \cap Y$. Denote by X_1 the subset (possibly empty) of vertices of X such that their neighbourhoods are entirely in S_Y i.e. $N(X_1) \subseteq S_Y$; see Figure 2.5.

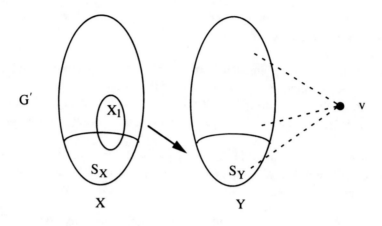

Figure 2.5

Since S is a 1-barrier - remember $h_{G'}(S) \equiv |S| \pmod 2$,

$$h_{G'}(S) \geq |S| + 2.$$
(1)

By the definition of X_1, and since $<Y>$ is complete, $G' \setminus S$ consists of the $|X_1 \setminus S|$ isolated vertices of $X_1 \setminus S$ and just one other component - which includes v if $|VG|$ is odd. So from (1),

$$|X_1| + 1 \geq h_{G'}(S) \geq |S| + 2.$$
(2)

However, by hypothesis, $|X_1| \leq |N(X_1)|$. Since $N(X_1) \subseteq S_Y \subseteq S$ it follows that

$$|X_1| \leq |N(X_1)| \leq |S_Y| \leq |S|.$$
(3)

The desired contradiction then follows from (2) and (3).   □

This theorem first appeared in [pH 35].

Corollary 2.7 Let G be an r-regular bipartite graph. Then G contains a 1–factor.

Proof: Suppose G has bipartition (X,Y). Let $S \subseteq X$. The number of edges going from S to N(S) is r|S|. However, the number of edges going from N(S) to S cannot exceed r|N(S)|. Therefore $r|S| \leq r|N(S)|$. So Hall's condition holds and G has a 1-factor.   □

There is now a vast literature on results related to the Marriage Theorem. We discuss just two results in this area. The reader should refer to Lovász and Plummer [L-P 86] to see the extent of the literature. Many of these theorems go back as far as König [dK 31] and Egerváry [eE 31]; see [iA 89] page 37 for an English translation. Undoubtedly one of the areas where progress has been made over the last few years is in generalizations of the Marriage Theorem for infinite graphs.

In 1935, P. Hall gave his criterion for deciding whether a bipartite graph with parts X and Y contained a matching which covered every vertex of X. The criterion of Theorem 2.6, namely

$$|N(S)| \geq |S| \text{ for all } S \subseteq X,$$

is known as **Hall's condition**.

Put differently, Hall's Theorem states that the graph does not contain a matching covering every vertex of X if and only if there exists a set $S \subseteq X$ such that $|N(S)| < |S|$. Such a set is of course, a 1-barrier.

Hall's criterion has been generalized by Enomoto, Ota and Kano, [E-O-K 88], to establish a sufficient condition for a bipartite graph to have a k-factor.

Until 1983 [A-N-S 83] it remained an open problem to generalize Hall's Theorem for all infinite bipartite graphs. The first progress was made by M. Hall [mH 48] who proved that Hall's criterion holds also for infinite bipartite graphs each of whose vertices has finite degree. The question was next settled for the case of countable bipartite graphs ([D-M 74], [P–S76], [cN 78]). The form of the Aharoni, Nash-Williams, Shelah solution, follows that of P. Hall's theorem: A bipartite graph with parts X and Y has a matching which covers each vertex of X if and only if it does not contain one of a set of 'barriers'; see Exercises 16, 17.

In a similar vein, Aharoni [rA 84], [rA 88], has established criteria for characterizing graphs of any cardinality which possess a 1-factor.

If a graph G contains a 1-factor then it frequently contains more than one 1–factor. We now present some results which give some bounds for $f(G)$, the number of 1-factors of G. It is perhaps a little surprising that $f(G)$ is as large as it is.

Theorem 2.8 Let G be a bipartite graph which contains a 1-factor and let δ be the minimum degree of G. Then $f(G) \geq \delta!$.

Proof: We present a proof in the spirit of Halmos and Vaughan [H-V 50]. Since G contains a 1-factor, if X and Y are parts of the bipartition, then $|X| = |Y|$. Let $|X| = m$. We proceed by induction on m. The inductive hypothesis is that if G is a bipartite graph which contains a 1-factor and δ is the minimum degree of a vertex in the part X, then $f(G) \geq \delta!$.

If $m = 1$, then the result clearly follows. So assume that the result is true for all values of $|X|$ less than or equal to $m - 1$ $(m \geq 2)$.

First suppose that every subset $T \subseteq X$ is such that $|N(T)| > |T|$. Then remove xy from G, where $x \in X$ and $y \in Y$. In $G - \{x,y\} = G'$, the conditions of the Marriage Theorem hold and G' has two parts of order $m - 1$. Now $\delta(G') \geq \delta - 1$, so by the induction hypothesis $f(G') \geq (\delta-1)!$. Since there are at least δ choices of an edge xy adjacent to x we see that $f(G) \geq \delta!$.

We may therefore suppose that there exists $T \subset X$ such that $N(T) = |T|$. Let $H = <T \cup N(T)>$ and let $H' = <(X \setminus T) \cup (Y \setminus NT)>$. These are bipartite graphs in which the conditions of the Marriage Theorem hold and so each contains a 1–factor. By the induction hypothesis H contains at least $\delta!$ 1-factors. Taking the union of these 1-factors with any 1-factor of H' gives $\delta!$ 1-factors in G. $\square$

There are many less elementary results concerning $f(G)$. For example, see [IL72].

Theorem 2.9 Suppose that G is a k-connected graph $(k \geq 1)$. Suppose that $f(G) \geq 1$. Then either $f(G) \geq k!$ or for all $u, v \in V(G), u \neq v, f(G \setminus \{u,v\}) > 0$. $\square$

3. Petersen's graph

This section is the hub of the book, for here we see the context in which Petersen's graph was born. Not that the graph which bears his name, first appeared in print in the way in which we shall describe. In fact it would seem that if precedence were all, then Petersen's graph should be called Kempe's graph. This is because the representation of P shown in Figure 3.1 was given in [aK86], where it was derived from the Desargues configuration of projective geometry.

We recall that in trying to prove the Four Colour Theorem, Tait hit on the idea of edge-colouring cubic graphs. However in his paper [pT 80], he clouds the issue by stating as a theorem that every cubic graph is 3-edge-colourable. He then added

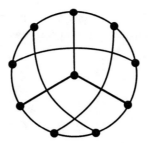

Figure 3.1

'The difficulty of obtaining a simple proof of this theorem originates in the fact that it is not true without limitation'. So he provided the counterexample of Figure 3.2.

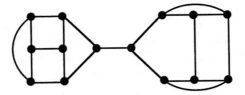

Figure 3.2

Tait went on to state his theorem in a form which he believed to be 'universally true'. The new theorem said that 'the edges of the polyhedron, which has trihedral summits only, can be divided into three groups, one from each group ending in each summit'.

Petersen then showed his confusion by stating Tait's 'theorem' as essentially 'every bridgeless cubic graph is 1-factorable'. It is as a counterexample to this that the famous Petersen graph appears.

Although Petersen was dealing with a factor problem, it is interesting to note that his proof is given in the language of colourings; see [jP 98]. There is a marked similarity between his proof that the Petersen graph is not 1-factorable and some of the constructions given in Chapter 3 to produce snarks. In fact it is identical to the proof of Lemma 3.2.5.

Theorem 3.1 The Petersen graph is not 1-factorable.

Proof: Suppose that P is 1-factorable or equivalently, that $\chi'(P) = 3$. Consider any 3-edge-colouring with colours c_1, c_2 and c_3. Then, up to a relabelling of the colours, Petersen demonstrated that the edges emanating from a 'pentagon' of P would have three edges coloured c_1, one edge coloured c_2 and the other edge coloured c_3. (Here Petersen just proves a special case of Lemma 3.1.2.)

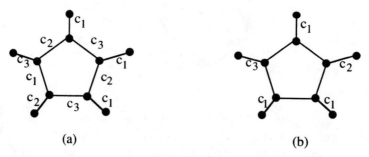

(a)   (b)

Figure 3.3

This choice of colours can essentially be made in only two ways as shown in Figure 3.3. We can extend the colouring of Figure 3.3(a) to a colouring of the pentagon, but this cannot be done for the colouring of Figure 3.3(b). Petersen now observed that P consists of two pentagons joined by five edges in such a way that if these five edges were coloured with respect to one pentagon as in Figure 3.3(a), they were coloured à la Figure 3.3(b) with respect to the other pentagon. Hence $\chi'(P) \neq 3$. □

Petersen's drawing of P in [jP 98] is shown in Figure 3.4.

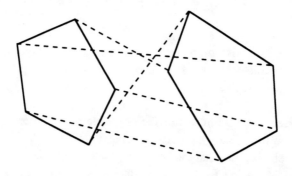

Figure 3.4

It is interesting to note however, that $\overline{P}$ **is** 1-factorable.

Armed with the concept of 1-factorability we can extend the result of Corollary 2.7.

Theorem 3.2 Let G be an r-regular bipartite graph $(r \geq 1)$. Then G is 1–factorable.

Proof: We proceed by induction on r. The result is obviously true for $r = 1$. Hence assume that the result holds for $r \leq k$. If $r = k + 1$, then by Corollary 2.7 we know that G contains a 1-factor, F, say. Now $G \setminus F$ satisfies the induction hypothesis, so adding F to the disjoint 1-factors of $G \setminus F$ gives the required disjoint union of 1-factors for G.   □

Using this theorem, we can give a nice proof of the following theorem of Petersen [jP 91].

Theorem 3.3 Every 2r-regular graph is 2-factorable $(r \geq 1)$.

Proof: A graph of even degree has an Euler tour; see Theorem 1.5.9. That is, starting at any vertex we may travel continuously over all the edges of G, without repetition, until we return to the original vertex. So if G is 2r-regular it has such an Euler tour.

Take any Euler tour of G and, as we traverse an edge of G, orient it in the direction of travel to obtain a directed graph $\vec{G}$. This graph has r edges oriented toward each vertex and   r   edges oriented away from each vertex.   If $VG = \{v_1, v_2, ..., v_n\}$, form the bipartite graph H with parts $\{u_1, u_2, ..., u_n\}$, $\{u'_1, u'_2, ..., u'_n\}$ and join u_i to u'_j in H if and only if $\overrightarrow{v_i v_j} \in E\vec{G}$.

Now deg $u_i = r$ and so H is an r-regular bipartite graph. Hence by Corollary 2.7, H has a 1-factor F. Suppose that $F = \{..., u_i u'_j, u_j u'_k, ..\}$. Then in G, the edges corresponding to F are $K = \{..., v_i v_j, v_j v_k, ...\}$. The subgraph of G induced by K is the union of disjoint cycles. Now any other factor F' of H which is disjoint from F gives rise to a union of cycles K' which are edge disjoint from the cycles of K. Hence H is 1-factorable if and only if G is 2–factorable. Theorem 3.2 guarantees that H is 1-factorable.   □

4 . Factor-covered
Recall that G is factor-covered if every edge of G is contained in a 1-factor.

Now P is an example of a factor-covered graph. This result can be proved directly or by noting that P contains a 1-factor and that every edge of P is equivalent.

We can now generalize Tutte's characterization of graphs which contain a 1-factor, Theorem 2.1. The result is due to Little, Grant and Holton [L-G-H 75].

Theorem 4.1 Suppose that $|VG|$ is even. Then G is factor-covered if and only if for all $S \subseteq V$

(i) $h_G(S) \leq |S|$; and

(ii)   if $h_G(S) = |S|$ then $<S> \cong |S|K_1$.

Proof: Suppose that G is factor-covered. Then G contains a 1-factor and (i) is satisfied from Corollary 2.2. Now choose $S \subseteq VG$. Assume that there exist $u,v \in S$ with $e = uv \in EG$. Since G is factor-covered, $G - e$ has a 1-factor. Write $S_0 = S \setminus \{u,v\}$. Then, using Corollary 2.2,

$$h_G(S) \leq h_{G-e}(S_0) \leq |S_0| = |S| - 2 < |S|. \tag{1}$$

So if $h_G(S) = |S|$, from (1) it follows that $E(<S>) = \varnothing$. Therefore (ii) holds.

Conversely suppose (i) and (ii) are true. Assume that G is not factor-covered. Then we may choose $u,v \in VG$, $uv \in EG$, so that $G - \{u,v\}$ has no 1-factor. Write $K = G - \{u,v\}$. Since K has no 1-factor we may choose $S_0 \subseteq VK$ satisfying

$$h_K(S_0) \geq |S_0| + 2. \tag{2}$$

Write $S = S_0 \cup \{u,v\}$. Then from (i), (2) and the definition of K and S_0,

$$|S| \geq h_G(S) = h_K(S_0) \geq |S_0| + 2 = |S|. \tag{3}$$

So from (3), $h_G(S) = |S|$. Since $uv \in E(<S>)$ it follows that condition (ii) is violated. This contradiction proves the theorem.   □

As a consequence of the above we may now generalize Theorem 2.4.

Theorem 4.2 Let G be an $(r-1)$-edge-connected r-regular graph with $r \geq 2$ and r odd. Then G is factor-covered.

Proof: From Theorem 2.4, G has a 1-factor. Thus $h_G(S) \leq |S|$ for all $S \subseteq VG$. Suppose now that for some $S \subseteq VG$, $h_G(S) = |S|$. Then equation (4) in the proof of Theorem 2.4 implies that $|E| = r|S|$. Hence $E<S> = \varnothing$. Therefore conditions (i) and (ii) of Theorem 4.1 both hold and consequently G is factor–covered.   □

Corollary 4.3 Every $(r-1)$-connected r-regular graph with $r \geq 2$ and r odd is factor-covered.   □

Corollary 4.4 Every cubic bridgeless graph is factor-covered. ▫

Corollary 4.5 Every cubic bridgeless graph G has $f(G) \geq 3$.

Proof: There are three edges incident with a given vertex and every edge is in a 1–factor. ▫

Corollary 4.6 If G is (r-1)-connected r-regular with $r \geq 2$ and r odd, then $f(G) \geq r$.

The connectivity restriction is best possible as far as the theorem above goes, in that there exist (r-2)-connected r-regular graphs G with $r \geq 3$ and r odd, which are not factor-covered. Indeed they do not even contain a 1-factor; see Figure 4.1 for an example with $r = 3$.

If we drop the connectivity restriction of Theorem 4.2 but demand a certain amount of symmetry in our graphs, then we obtain the next result.

Theorem 4.7 Every connected even vertex-transitive graph is factor-covered.

Proof: See [L-G-H 75]. The proof is by induction and relies heavily on the notion of atomic parts introduced in [mW 70] . ▫

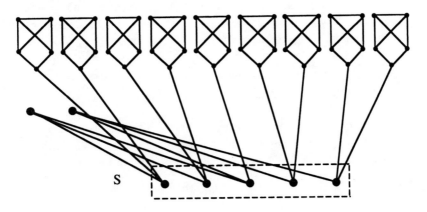

Figure 4.1

If every edge in a graph is in some defect-d matching, then we say that the graph is **d-covered**. The results of this section along with Corollary 2.2 can all be generalized to d-covers; see the Exercises.

We note finally that there exist (r - 2)-connected graphs which are r-regular for
r≥ 3 and |VG| ≡ d (mod 2), such that G is not d-covered, nor does it have a
defect-d matching. An example for r = 3, d = 4 is shown in Figure 4.1. In the
figure, S is a 1-barrier.

5. f-factors

In this section we generalize the idea of r-factors. It is a generalization which only
in the last few years has been fully appreciated. Tutte's theorem for 1-factors was
the first real advance in the theory since the days of Petersen over a half century
before. The 1-factor theorem was easy to state and easy to appreciate. Tutte's
theorem for f-factors [wT 52] took some 30 years before it became fully
understood and appreciated, even though Tutte himself went to some trouble to
show how it could be used and how indeed it could be proved using the 1-factor
theorem. The theorem for f-factors is harder to state than the 1-factor theorem but
this can only in part explain the time lag between its proof and its application by
mathematicians other than Tutte himself.

To begin with, we must introduce some notation. Let G be a graph. Suppose
X, Y ⊆ VG. Let $q(X,Y) = |\{(x,y) : xy \in EG, x \in X, y \in Y\}|$. For brevity
$q(x,Y) = q(\{x\},Y)$ and we write $q(X) = q(X,X)$. So $q(X) = 2|EX|$ and
$q(x,VG) = \deg x$.

Let $f : VG \to N$ be a function from VG into the set N of non-negative integers.
An **f-factor** of G is a spanning subgraph F of G such that $\deg_F u = f(u)$ for
all $u \in VG$. For convenience we let $f(S) = \sum_{s \in S} f(s)$, where $S \subseteq VG$.

The concept of f-factor extends the notion of r-factors by allowing irregular
spanning subgraphs of G.

There is an obvious conflict with the earlier notation for an r-factor i.e. an f-factor
where f is the constant function $f(u) = r$ for all $u \in VG$. We shall resolutely
only use the symbols f (and g) when we are dealing with the more general
concept.

The problem then is, given the function f, when does G have an f-factor? The
short answer, in line with Theorem 2.1, given by Tutte is "when it doesn't have an
f-barrier". So what is an f-barrier? We need a few more ideas before we can
answer that question.

A **graph triple** $B = (S,T,U)$ **for** G is an ordered triple (S,T,U) such that
{S,T,U} is a partition of VG. Let C be a component of <U>. Write

$J_G(B;C) \equiv J(B;C) = f(VC) + q(VC,T).$

Then C is said to be an **odd** or **even component** of $<U>$ according as $J(B;C)$ is odd or even. We illustrate the situation so far in Figure 5.1.

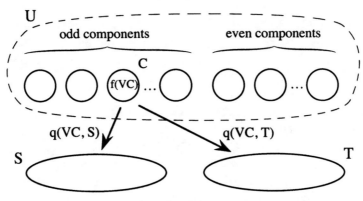

Figure 5.1

Let $h_G(B) \equiv h(B)$ denote the number of odd components of $<U>$. As an example suppose f is the function defined by $f(u) = 1$ for all $u \in VG$. So an f–factor is a 1-factor. Then $J(B;C) = |VC| + q(VC,T)$. In the particular case when $T = \varnothing$ then $J(B;C) = |VC|$ and $J(B;C)$ is an odd or even component of $<U>$ according as C has an odd or even number of vertices. Hence when $B=(S,\varnothing,U)$, $h_G((S,\varnothing,U)) = h_G(S)$ i.e. the number of odd components (in our earlier sense) of $G \setminus S$. Incidentally, as in the 1-factor theorem, the even components of $<U>$ can almost be ignored. In all the counting arguments the presence of even components merely 'helps'.

One other notational device which is very convenient is to write $\deg(T)$ rather than $\sum_{t \in T} \deg t$. Then, for example

$\deg(T) = q(T) + q(T,S \cup U).$

The **deficiency** $\delta(B)$ **of the graph triple** $B = (S,T,U)$ is defined by

$$\delta(B) = h(B) - f(S) + f(T) - \deg(T) + q(S,T). \tag{*}$$

This definition is constantly referred to below. We say that B is an f-**barrier** if $\delta(B) > 0$ and B is a **maximum f-barrier** if $\delta(B)$ is as large as possible among all f-barriers.

Examples

(i)   When B = (S,∅,U), as in the discussion above, and the f-factor is a
1–factor, (*) becomes

$$\delta(B) = h(B) - f(S) = h_G(S) - |S|.$$

Hence B is an f-barrier if and only if $h_G(S) > |S|$ i.e. if and only if S is a
1–barrier. So an f-barrier is a generalization of the concept of a 1-barrier.
Furthermore, see Exercise 14, in the same way that $h_G(S) \equiv |S|$ (mod 2), so
$\delta(B) \equiv 0$ (mod 2).

(ii)   Suppose f(VG) is odd. Then B = (∅,∅,VG) is an f-barrier. This is
because from (*), $\delta(B) = h(B)$. Now for any component C of <U> (i.e. for
any component C of G) J(B;C) = f(VC). However since f(VG) is odd, f(VC)
must be odd for at least one component C of G. Hence $h(B) \geq 1$ and from (*) ,
$\delta(B) > 0$. Therefore B is an f-barrier. In particular therefore, in the case when
the f-factor is a 1-factor, f(VG) = |VG|. Thus when |VG| is odd, VG itself is a
1–barrier.

(iii)   Suppose there exists $v \in VG$ such that deg v < f(v). Then of course G
cannot have an f-factor - remember that an f-factor is a spanning subgraph F such
$\deg_F v = f(v)$. Our intuition tells us that we should be able to find an f-barrier. Let
B = (∅,{v},VG \ {v}). From (*),

$$\delta(B) = h(B) + f(v) - \deg v > 0;$$

B is the required f-barrier.

As these three examples suggest we have the following characterization of an
f–factor.

Theorem 5.1 Let f: VG → N be a function. Then G has an f-factor or an
f–barrier but not both.   □

The original proof appears in [wT 52]. Tutte later [wT 54] deduced Theorem 5.1
from Theorem 2.1. As Theorem 2.1 is easy to deduce from Theorem 5.1 the two
theorems are equivalent. However this is a beautiful example of how little this
really means in a particular case, especially when we come to apply Theorem 5.1.
It is, if you like, the reformulations and further insights which Theorem 5.1
provides, that lead to so many powerful applications. It is for this reason, as well
as demands of space, that we omit the proof of Theorem 5.1 and concentrate on its
uses.

Firstly let us make clear when the problem of the existence of an f-factor becomes trivial in the same way that it is obvious that G has no 1-factor if |VG| is odd. From examples (ii) and (iii) and Theorem 5.1, G cannot have an f-factor if f(VG) is odd or if deg v < f(v) for some v ∈ VG. We shall in fact find Theorem 5.2, which is easily deduced from Theorem 5.1, more useful than the latter.

Theorem 5.2 Let f : VG → N be a function. Then G has an f-factor or a maximum f-barrier, but not both.   □

We have already noticed the complementary nature of 1-factors and 2-factors in cubic graphs. Clearly this complementary property carries over to f-factors. If G has an f-factor F, then the subgraph G - EF is an f'-factor of G where f'(u)=degu - f(u). It is obvious then, that G has an f-factor if and only if it has an f'-factor. We will always assume below that B = (S,T,U) is a graph triple for G and f : VG → N. Further, let B' = (T,S,U). We can now describe δ(B) in terms of δ(B').

Lemma 5.3
(i)   δ(B) = h(B) + q(S,T) - f(S) - f'(T).
(ii)   δ(B) - δ(B') = h(B) - h(B').

Proof: From (*)

$$\delta(B) = h(B) - f(S) + f(T) - \deg(T) + q(S,T).$$ (1)

By the definition of f',

$$f'(T) = \sum_{t \in T} f'(t) = \sum_{t \in T} (\deg t - f(t)) = \deg(T) - f(T).$$ (2)

So from (1) and (2) we have

$$\delta(B) = h(B) + q(S,T) - f(S) - f'(T).$$ (3)

From (3), with B' and f' replacing B and f, respectively, and S and T interchanged, we get

$$\delta(B') = h(B') + q(T,S) - f'(T) - f(S).$$ (4)

So, from (3) and (4),

$$\delta(B) - \delta(B') = h(B) - h(B').$$ □

Lemma 5.4 $B = (S,T,U)$ is a maximum f-barrier if and only if $B' = (T,S,U)$ is a maximum f'-barrier.

Proof: Suppose that $B = (S,T,U)$ is an f-barrier and C is a component of $<U>$. Then

$$J(B;C) = f(VC) + q(VC,T) \tag{1}$$

and

$$J(B';C) = f'(VC) + q(VC,S)$$

$$= (\sum_{u \in VC} \deg u) - f(VC) + q(VC,S). \tag{2}$$

But

$$\sum_{u \in VC} \deg u = q(VC, G \setminus VC) + q(VC)$$

$$= q(VC,S) + q(VC,T) + 2|EC|. \tag{3}$$

So, from (1), (2) and (3)

$$J(B';C) = q(VC,T) - f(VC) + 2(q(VC,S) + |EC|)$$

$$\equiv J(B;C) \quad (\mathrm{mod}\ 2).$$

Hence C is odd, relative to B' and f', if and only if C is odd, relative to B and f. So $h(B) = h(B')$ and, from Lemma 5.3, $\delta(B) = \delta(B')$. Hence since $\delta(B) > 0$, $\delta(B') > 0$ and B' is an f'-barrier. The converse is also true. Furthermore, B is a maximum f-barrier if and only if B' is a maximum f'-barrier. $\qquad\qquad\qquad\qquad\qquad\qquad\qquad\qquad\qquad\qquad\square$

The key to using Theorem 5.2 is the Transfer Theorem, Theorems 5.5 and 5.6. The idea behind the Transfer Theorem (and its variants) is roughly this. Given a maximum f-barrier $B = (S,T,U)$, we can examine what happens if we transfer elements between two of the sets S, T and U. We then obtain a new graph-triple $B_1 = (S_1,T_1,U_1)$. Since B is a maximum f-barrier we then have $\delta(B) \geq \delta(B_1)$. Notice here that even if B_1 is not an f-barrier, this is still true since in that case $\delta(B_1) \leq 0$ and $\delta(B) > 0$.

To make this precise we need some additional notation. Let $B = (S,T,U)$ be a graph-triple. Suppose that $x \in S \cup U$. We define $\alpha(x)$ as follows:-

(i)   if $x \in S$ then $\alpha(x)$, is the number of odd components C of $\langle U \rangle$ relative to B such that $q(x,VC) > 0$; see Figure 5.2(a);

(ii)   if $x \in U$ write $B_1 = (S \cup \{x\}, T, U \setminus \{x\})$. Now $\alpha(x)$ is the number of odd components C of $\langle U \setminus \{x\} \rangle$ relative to B_1 such that $q(x,VC) > 0$; see Figure 5.2(b).

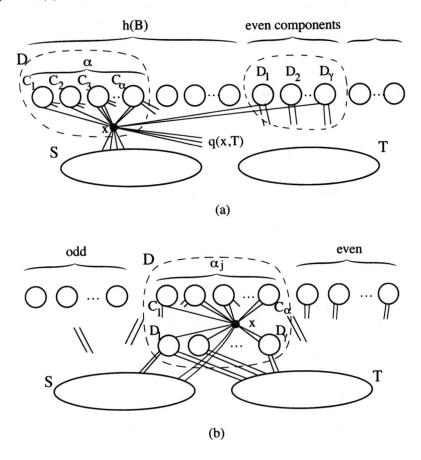

(a)

(b)

Figure 5.2

Finally define $\beta(x)$ as follows:

$$\beta(x) = \begin{cases} 0 & \text{if } \alpha(x) + f(x) + q(T,x) \text{ is even} \\ 1 & \text{if } \alpha(x) + f(x) + q(T,x) \text{ is odd.} \end{cases}$$

We then have the two following theorems which will be referred to jointly as the Transfer Theorem.

Theorem 5.5 Let $B = (S,T,U)$ be a maximum f-barrier. Then
(i) $f(x) \le \alpha(x) + q(T,x) - \beta(x)$ $(x \in S)$;
(ii) $f(x) \ge \alpha(x) + q(T,x) - \beta(x)$ $(x \in U)$;
(iii) if $f(x) = \alpha(x) + q(T,x) - \beta(x)$ for some $x \in S \cup U$, then x can be transferred between the sets S and U without affecting the maximality of B.

Proof: (i) Suppose that $x \in S$. We now 'transfer' x from S to U. Write $B_1 = (S \setminus \{x\}, T, U \cup \{x\})$. Then, see Figure 5.2(a), in the transfer, we lose $\alpha = \alpha(x)$ odd components of $<U>$ and gain possibly a new odd component D. This component D of $<U \cup \{x\}>$ is the component containing x. Hence

$$h(B_1) = h(B) - \alpha(x) + \theta(D) \tag{1}$$

where $\theta(D) = 1$ or 0 according as D is or is not an odd component of $<U \cup \{x\}>$.

Now consider Figure 5.2(a). Suppose that the odd components of $<U>$ to which x is joined by at least one edge, are $C_1, C_2,..., C_\alpha$ and the even components to which x is joined are $D_1, D_2,..., D_\gamma$. Then

$$J(B_1;D) = \sum_{i=1}^{\alpha} J(B;C_i) + \sum_{i=1}^{\gamma} J(B;D_i) + f(x) + q(x,T)$$

$$\equiv \alpha(x) + f(x) + q(x,T) \pmod{2}.$$

So $\theta(D) = 1$ if $\alpha(x) + f(x) + q(x,T)$ is odd and $\theta(D) = 0$ otherwise. Hence $\theta(D) = \beta(x)$. Therefore, from (1),

$$h(B_1) = h(B) - \alpha(x) + \beta(x). \tag{2}$$

Using (*) and (2) and remembering that B is a maximum f-barrier

$$\delta(B) = h(B) - f(S) + f(T) - \deg(T) + q(S,T)$$

$$\ge \delta(B_1)$$

$$= h(B_1) - f(S \setminus \{x\}) + f(T) - \deg(T) + q(S \setminus \{x\}, T)$$

$$= h(B_1) - f(S) + f(x) + f(T) - \deg(T) + q(S,T) - q(x,T)$$

$$= (h(B) - \alpha(x) + \beta(x)) - f(S) + f(x) + f(T) - \deg(T) + q(S,T) - q(x,T).$$

Hence $f(x) \leq \alpha(x) - \beta(x) + q(x,T)$.

(ii)   Suppose $x \in U$. We now 'transfer' x from U to S. The argument now exactly parallels that of (i). Write $B_1 = (S \cup \{x\},T,U \setminus \{x\})$. Then, see Figure 5.2(b), in the 'transfer' we gain $\alpha(x)$ odd components of $<U \setminus \{x\}>$ and lose possibly an odd component D of $<U>$. This component D is the component containing x. Hence

$$h(B_1) = h(B) + \alpha(x) - \theta(D) \tag{3}$$

where $\theta(D) = 1$ or 0 according as D is or is not an odd component of $<U>$.

Now consider Figure 5.2(b). Suppose that the odd components of $<U \setminus \{x\}>$ to which x is joined by at least one edge are $C_1, C_2,..., C_\alpha$ and the even components to which x is joined are $D_1, D_2,..., D_\gamma$. Then

$$J(B;D) = \sum_{i=1}^{\alpha} J(B_1;C_i) + \sum_{i=1}^{\gamma} J(B_1;D_i) + f(x) + q(x,T)$$
$$\equiv \alpha(x) + f(x) + q(x,T) \pmod 2.$$

Hence D is an odd component of $<U>$ if and only if $\alpha(x) + f(x) + q(x,T)$ is odd. Hence $\theta(D) = \beta(x)$. So (3) becomes

$$h(B_1) = h(B) + \alpha(x) - \beta(x). \tag{4}$$

Using (*) and (4) we have

$$\delta(B) = h(B) - f(S) + f(T) - \deg(T) + q(S,T)$$

$$\geq \delta(B_1)$$

$$= h(B_1) - f(S \cup \{x\}) + f(T) - \deg(T) + q(S \cup \{x\},T)$$

$$= h(B_1) - f(S) - f(x) + f(T) - \deg(T) + q(S,T) + q(x,T)$$

$$= h(B) + \alpha(x) - \beta(x) - f(S) - f(x) + f(T) - \deg(T) + q(S,T) + q(x,T)$$

Hence $f(x) \geq \alpha(x) - \beta(x) + q(x,T)$.

(iii)   If $f(x) = \alpha(x) - \beta(x) + q(x,T)$ then it easily follows that $\delta(B) = \delta(B_1)$. Hence x can be transferred between the sets S and U without affecting the maximality of B. $\square$

Theorem 5.6 Let $B' = (T,S,U)$ be a maximum f-barrier. Let f' be as defined on p.133. Then

(i) $f'(x) \leq \alpha(x) + q(T,x) - \beta(x)$ $(x \in S)$;

(ii) $f'(x) \geq \alpha(x) + q(T,x) - \beta(x)$ $(x \in U)$;

(iii)   if $f'(x) = \alpha(x) + q(T,x) - \beta(x)$, for some $x \in S \cup U$, then x can be transferred between the sets S and U without affecting the maximality of B'.

Proof: Suppose that $B' = (T,S,U)$ is a maximum f-barrier. Then, from Lemma 5.4, $B = (S,T,U)$ is a maximum f'-barrier. We now use Theorem 5.5 directly with f' replacing f.   □

Theorem 5.7 Suppose that G has an f-barrier. Then G contains a maximum f-barrier $B = (S,T,U)$ such that $f(t) \geq 2$ for all $t \in T$.

Proof: Let $B = (S,T,U)$ be a graph-triple. Let $T^* = \{t : t \in T, f(t) \leq 1\}$. Now choose a maximum f-barrier $B = (S,T,U)$ such that $|T^*|$ is a minimum. If $|T^*|=0$ then we are done. So assume that we can choose $x \in T^*$ i.e. $x \in T$ and $f(x) \leq 1$. We now 'transfer' x to S to see what information we get. The general situation is illustrated in Figure 5.3.

Let $B_1 = (S \cup \{x\}, T \setminus \{x\}, U)$. Suppose x is joined by at least one edge to the odd components $C_1, C_2,..., C_a$ of $<U>$ and to the even components $D_1, D_2,..., D_b$ of $<U>$ and to no other component of $<U>$. Then when x is transferred from T to S, the number of odd components of $<U>$ is reduced by at most 'a'. Therefore

$$h(B_1) \geq h(B) - a \geq h(B) - q(x,U). \tag{1}$$

Since B is maximum

$$\delta(B) \geq \delta(B_1) \tag{2}$$

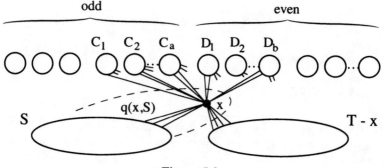

Figure 5.3

i.e. using (*),

$$h(B) - f(S) + f(T) - \deg(T) + q(S,T)$$

$$\geq h(B_1) - f(S \cup \{x\}) + f(T \setminus \{x\}) - \deg(T \setminus \{x\}) + q(S \cup \{x\}, T\setminus\{x\}). \quad (3)$$

Now since

$$q(S \cup \{x\}, T \setminus \{x\}) = q(S,T) - q(x,S) + q(x,T), \quad (4)$$

we obtain from (1), (3) and (4), after a little manipulation,

$$f(x) \geq q(x,T). \quad (5)$$

Notice that equality in (5) implies equality in (2) with the new T^* relative to B_1 having one fewer element. Hence the inequality (5) must be strict by the minimality of $|T^*|$. Since $x \in T^*$ it follows that $q(x,T) = 0$ and $f(x) = 1$. This means that $\delta(B) = \delta(B_1) + 1$. But $\delta(B) \equiv \delta(B_1)$ (mod 2), which is a contradiction; see Exercise 14. □

Now Tutte's 1-factor theorem is an easy consequence of Theorem 5.7.

Corollary 5.8. G either has a 1-factor or a 1-barrier but not both.

Proof: Suppose that f: $VG \rightarrow N$ satisfies $f(u) = 1$ for all $u \in VG$ and that G does not have a 1-factor. Then, by Theorem 5.1, G has an f-barrier. Choose, using Theorem 5.7, a maximum f-barrier $B = (S,T,U)$ such that $f(t) \geq 2$ for all $t \in T$. But $f(u) = 1$ for all $u \in VG$. Hence $T = \varnothing$ and, see example (i), S is a 1-barrier, in the earlier sense of Theorem 2.1. □

6. Applications
We give four applications of Theorem 5.2 simply to show its might and power. The reader should refer to [A-K 85] and [L-P 86] for further evidence of its fecundity.

We note here that all of the results in Section 5 also apply to pseudographs. We shall need this in the proof of the next theorem.

Theorem 6.1 Let G be an r-regular graph. If k is an integer $(0 \leq k < r)$, then there exists a spanning subgraph F of G such that $\deg_F u \in \{k, k+1\}$ for all $u \in VG$.

Proof: The theorem is trivially true if $k \in \{0, r-1, r\}$. So suppose that $0 < k \leq r-2$. If the theorem holds for even graphs then it holds also for odd graphs, for if VG is odd we can form $G \cup G$ and apply the result to give the appropriate spanning subgraph of $G \cup G$. Restricting this spanning subgraph to G gives the result for odd graphs. Hence we may suppose $|VG| = 2q$ ($q \geq 1$).

Form a pseudograph G' from G by adding a new vertex x with q loops and joining x to each vertex of G by a single edge. In G' the degree of x is $4q$ and the degree of all other vertices is $r + 1$. Define $f: VG' \to N$ by

$$f(u) = \begin{cases} 2q & u = x, \\ k + 1 & u \neq x. \end{cases}$$

Clearly any f-factor of G' when restricted to G gives a spanning subgraph satisfying the theorem. On the other hand, suppose F is a spanning subgraph of G with $\deg_F(u) \in \{k, k + 1\}$ for all $u \in VF$. Let $L = \{u \in VF : \deg u = k\}$. Since $|VF|$ is even, $|L|$ is even. Write $|L| = 2p$ ($0 \leq 2p \leq |VF|$). Construct F' by adding to F the vertex x together with $2p$ edges joining x to each vertex of L and $q - p$ loops at x. F' is then an f-factor of G'. The problem therefore reduces to proving that G' has an f-factor.

Assume otherwise. Then, by Theorem 5.2, G' has a maximum f-barrier. Let $B = (S, T, U)$ be such a maximum f-barrier. First of all we show that we can choose B so that $x \in U$. If $x \in S$ then, by the Transfer Theorem,

$$2q = f(x) \leq \alpha(x) + q(T, x) - \beta(x). \tag{1}$$

But $\alpha(x)$ is the number of odd components of $<U>$ joined to x so certainly $\alpha(x) \leq q(U, x)$. From (1) therefore

$$2q \leq q(U, x) + q(T, x) = q(U \cup T, x) \leq 2q. \tag{2}$$

The final inequality follows since there are at most $2q$ edges (as opposed to loops) incident to x. Therefore (2) implies that (1) is an equality throughout. Therefore by Theorem 5.5(iii) we can transfer x to U without affecting the maximality of B.

Now suppose $x \in T$. We use Theorem 5.6, however we must be a little careful. $B' = (T, S, U)$ in the statement of Theorem 5.6 is now replaced in this context by $B = (S, T, U)$. Hence Theorem 5.6(i) gives

$$2q = f'(x) \leq \alpha(x) + q(S, x) - \beta(x) \leq q(U, x) + q(S, x) \leq 2q. \tag{3}$$

Therefore equality holds throughout and we can transfer x to U without affecting the maximality of B.

Hence we may now assume that B is a maximum f-barrier and that $x \in U$. Now B is an f-barrier and so from Lemma 5.3,

$$1 \leq \delta(B) = h(B) + q(S,T) - f(S) - f'(T). \tag{4}$$

Since $x \in U$, and recalling that x is joined to all the vertices in G, $h(B) \leq 1$. Therefore, from (4),

$$0 \leq q(S,T) - (k + 1)|S| - ((r + 1)|T| - (k + 1)|T|),$$

i.e. $(k + 1)|S| + (r - k)|T| \leq q(S,T).$ (5)

Thus $(r + 1) \min\{|S|,|T|\} \leq q(S,T).$

On the other hand, since G is r-regular we must have

$$r \min \{|S|,|T|\} \geq q(S,T).$$

Hence $\min\{|S|,|T|\} = 0$ which means $q(S,T) = 0$. Since $0 < k < r$, from (5), $|S| = |T| = 0$. Therefore, from (4), $\delta(B) = h(B) = 1$. Therefore, since $U = VG'$, $<U> = G'$ is the only odd component of $<U>$. Since $J(B;U) = f(VG') = 2q(k+1) + 2q \equiv 0 \pmod 2$, we arrive at a contradiction. □

A much simpler proof of Theorem 6.1 is given in [cT 81] but we used the longer method to illustrate the technique.

A second application leads to another generalization of Tutte's 1-factor theorem. The generalization is to defect-d matchings and was first proved by Berge [cB 58]. Obviously if $d > |VG|$ or if $|VG|-d$ is odd, there can be no defect-d matching. We do not prove the next result since a proof, using Theorem 5.2 and the Transfer Theorem, follows a pattern similar to that of Theorem 6.1.

Theorem 6.2 Let d $(0 \leq d \leq |VG|)$ be an integer such that $|VG| - d$ is even. Then G has a defect-d matching if and only if $h_G(S) \leq |S| + d$ for all $S \subseteq VG$. □

Another application of Tutte's work on f-factors is to a result of Erdös and Gallai in an area of Graph Theory which is seemingly unrelated to factors. This area is that of degree sequences.

Let $d = (d_1, d_2, ..., d_n)$ be a sequence of positive integers. We say that d is a **degree sequence** or d is **graphic** if there is a graph G with $VG = \{u_i : i = 1, 2, ..., n\}$ and $\deg u_i = d_i$ for all $i = 1, 2, ..., n$. Clearly for d to be a degree sequence, $\sum_{i=1}^{n} d_i \equiv 0 \pmod 2$. But this is not a sufficient condition. Necessary and sufficient conditions are given by Erdös and Gallai in [E-G 60]; see also [H-S 78] for a survey.

Theorem 6.3 The sequence $d = (d_1, d_2, ..., d_n)$ with $d_1 \geq d_2 \geq ... \geq d_n$ is a degree sequence if and only if

(i) $\sum_{i=1}^{n} d_i \equiv 0 \pmod 2$ and

(ii) $\sum_{i=1}^{r} d_i \leq r(r - 1) + \sum_{i=r+1}^{n} \min(r, d_i)$ for each r such that $1 \leq r \leq n - 1$.

Proof: Define $f(u_i) = d_i$ for $i = 1, 2, ..., n$. Then we see that d is a degree sequence if and only if K_n has an f-factor.

So suppose $\sum_{i=1}^{n} d_i \equiv 0 \pmod 2$ and d is not graphic. Hence K_n does not have an f–factor. Choose, see Theorem 5.2, a maximum f-barrier $B = (S, T, U)$. By the Transfer Theorem 5.5, remembering that the graph we are considering is complete, we have

> for $x \in S$ $f(x) \leq \alpha(x) + q(T, x) - \beta(x) \leq 1 + |T|$ (1)

> for $x \in U$ $f(x) \geq (\alpha(x) - \beta(x)) + q(T, x) \geq -1 + |T|.$ (2)

By (iii) of the Transfer Theorem we know that if equality holds in (1), then we can transfer x from S to U without affecting the maximality of B. We may repeat this procedure until $f(x) < 1 + |T|$ for all $x \in S$. Note that the vertices x which have been transferred from S to U in this process satisfy $f(x) = 1 + |T|$.

We can now perform a similar operation on the vertices of U for which equality holds in (2), transferring these vertices from U to S. This can be done until $f(x) > -1 + |T|$ for all $x \in U$. The vertices x that have been moved to S all have $f(x) = -1 + |T|$ and so no vertex has been moved from S to U and back again.

Hence without loss of generality we may assume that in (1) and (2), strict inequalities apply to all vertices of S and U.

Assume that $T = \varnothing$. Then, from Lemma 5.3, $\delta(B) = h(B) - f(S)$. Since $\delta(B) > 0$ and $h(B) \leq 1$ (K_n is complete) we have $S = \varnothing$ and $h(B) = 1$. This in turn implies that $U = VK_n$ and $J(B;K_n)$ is odd. But

$$J(B;K_n) = f(VK_n) = \sum_{i=1}^{n} d_i.$$

Therefore $\sum_{i=1}^{n} d_i \equiv 1 \pmod{2}$, which is a contradiction of (i). Hence $T \neq \varnothing$.

Write $|T| = w$ ($1 \leq w < n$) ($w < n$ from Exercise 18). Since d is non-increasing

$$\sum_{i=1}^{w} d_i \geq f(T). \tag{3}$$

From Lemma 5.3, and because $\delta(B) > 0$,

$$f(T) > f(S) + \deg(T) - q(S,T) - h(B)$$

$$= f(S) + w(n - 1) - w|S| - h(B). \tag{4}$$

By definition

$$|S| + w + |U| = n \tag{5}$$

and so from (3), (4) and (5)

$$\sum_{i=1}^{w} d_i \geq f(T) > f(S) + w(w + |U| - 1) - h(B). \tag{6}$$

Write

$$\theta = \sum_{i=1}^{w} d_i - (f(S) + w(w + |U| - 1) - h(B)) \tag{7}$$

i.e. θ is the difference between the outside terms in (6).

If $\theta = 1$ then the first inequality in (6) must be an equality. Hence

$$f(T) = \sum_{i=1}^{w} d_i. \tag{8}$$

But

$$f(S) + f(T) + f(U) = \sum_{i=1}^{n} d_i = 2e. \tag{9}$$

Therefore, from (7), (8) and (9)

$$\theta = (2e - f(S) - f(U)) - (f(S) + w(w + |U| - 1) - h(B))$$

$$\equiv f(U) + w|U| - h(B) \quad (\text{mod } 2)$$

$$= J(B;<U>) - h(B)$$

$$\equiv 0 \quad (\text{mod } 2).$$

This is a contradiction since $\theta = 1$. Therefore $\theta \geq 2$ and from (6), since $h(B) \leq 1$,

$$\sum_{i=1}^{w} d_i > f(S) + w(w - 1) + w|U|. \tag{10}$$

From equation (1) and the subsequent comment, $f(x) < 1 + |T| = 1 + w$, i.e. $f(x) \leq w$, for all $x \in S$. Similarly from equation (2) and the subsequent comment, $f(x) \geq w$ for all $x \in U$.

Suppose that $x \in S$. Then $f(x) = d_{\sigma(x)}$ where $\sigma(x) \in \{1,2,...,n\}$. Since $f(x) \leq w$, $f(x) = \min\{w, d_{\sigma(x)}\}$. Hence, because d is a non-decreasing sequence, writing $|S| = a$,

$$f(S) = \sum_{x \in S} \min\{w, d_{\sigma(x)}\}$$

$$\geq \sum_{i=n-a+1}^{n} \min\{w, d_i\}. \tag{11}$$

Since $|U| = n - w - a$,

$$w |U| = w \left(\sum_{i=w+1}^{n-a} 1 \right) \geq \sum_{i=w+1}^{n-a} \min\{w, d_i\}. \tag{12}$$

Finally, from (10), (11) and (12)

$$\sum_{i=1}^{w} d_i > w(w - 1) + \sum_{i=w+1}^{n} \min\{w, d_i\}, \quad (1 \leq w \leq n - 1) \tag{13}$$

which violates condition (ii).

Suppose now that d is graphic. Then clearly (i) of the statement of the theorem is true. On the other hand suppose that, for some w $(1 \leq w \leq n - 1)$, the inequality in (13) holds. Define the graph triple $B = (S,T,U)$ on K_n as follows. $T = \{u_i : i = 1,2,...,w\}$, $S = \{x: x \in VG\backslash T, f(x) < w\}$ and $U = \{x : x \in VG\backslash T, f(x) \geq w\}$. We choose S and T in this way so that they accord with the characterization of S and T given by equations (1) and (2). Now inequality (13) implies inequality (10) and so on, working backwards until $\delta(B) > 0$. So B is an f-barrier and K_n has no f-factor which is a contradiction since d is graphic. $\Box$

Tutte's f-factor theorem is particularly easy to use with regular graphs. Given $r \geq 3$ and $1 \leq \lambda \leq r$ the next theorem determines all values of k for which every r-regular graph with edge-connectivity λ has a k-factor. For example Theorem 2.4 says that when $\lambda = r - 1$ then any r-regular graph with r odd, has a 1-factor. This theorem and Petersen's Theorem, Theorem 3.3, are both corollaries of the next theorem. We will write $\lambda \equiv \lambda(G)$ for the edge-connectivity of G. Also we write $\lambda^* = 2\lfloor \frac{1}{2}\lambda \rfloor + 1$.

Theorem 6.4 Let G be an r-regular multigraph with edge-connectivity λ, $r \geq 3$, $1 \leq \lambda \leq r$. Then G has a k-factor for the following values of k:
(i)   if r is even, all even numbers at least 2 and at most $r - 2$, together with, if $|VG|$ is also even, all odd numbers at least r / λ and at most $r - r / \lambda$;
(ii)   if r is odd and $\lambda \geq 2$, all positive even integers not more than $(\lambda^* - 1)r/\lambda^*$ and all odd integers at least r / λ^* and at most $r - 2$.
For no other value of k, $1 \leq k \leq r - 1$, can a graph G be guaranteed to have a k-factor.

Proof: See [B-S-W 85]. Suppose that f: $VG \rightarrow N$ satisfies $f(u) = k$ for all $u \in VG$ i.e. an f-factor is a k-factor. Tutte's f-factor theorem states that G has a k-factor if and only if G has no f-barrier. Hence G has a k-factor if and only if $\delta(B) \leq 0$, for all graph triples $B = (S,T,U)$. Recall that

$$\delta(B) = h(B) - f(S) + f(T) - \deg(T) + q(S,T). \qquad (1)$$

Abbreviate by writing $|S| = s$, $|T| = t$, $h = h(B)$ and $q = q(S,T)$. We shall abbreviate $q(S,T)$ **only** in this context. Then (1) becomes, remembering that G is regular,

$$\delta(B) = h - ks + kt - rt + q. \qquad (2)$$

So, from (2), G has a k-factor if and only if for all choices of $B = (S,T,U)$

$$h \le ks + (r - k)t - q. \tag{3}$$

The situation is illustrated in Figure 6.1.

If $s = t = 0$ then (3) must hold, since $<U> = G$ and $J(B;G) = k|VG| \equiv 0$ (mod2). Hence $h = 0$ and (3) is satisfied for all the values of $k|VG|$ under consideration. Henceforth we may assume that $s + t \ge 1$.

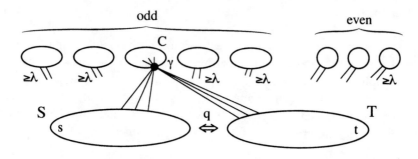

Figure 6.1

Clearly, see Figure 6.1,

$$q \le r \min\{s,t\}. \tag{4}$$

Let E be the set of edges joining U to $S \cup T$. Then, since each component of $<U>$ is joined by at least λ edges to $S \cup T$, we have $|E| \ge \lambda h$. So, since G is r-regular and there are q edges between S and T,

$$\lambda h \le |E| \le (rs - q) + (rt - q)$$

i.e.

$$\lambda h \le r(s + t) - 2q. \tag{5}$$

Amazingly, with just one exception, the theorem follows almost directly by manipulating equations (3), (4) and (5). There are five cases to consider and in only one case $(r$ odd, k even, λ even, $\frac{1}{2}r \le k \le (\lambda r) / (\lambda + 1))$ do we need to dig a little deeper. We consider just two cases below to provide the flavour of the proof. We start with Petersen's Theorem.

Suppose that r and k are even. Consider any odd component C of $<U>$. Then since C is odd and k is even

$$J(B;C) = f(VC) + q(VC,T)$$

$$= k|VC| + q(VC,T)$$

$$\equiv q(VC,T) \quad (\text{mod } 2).$$

Hence $q(VC,T)$ is odd and in particular $q(VC,T) > 0$. Also since r is even, $q(VC,S \cup T)$ is even. Therefore $q(VC,S)$ is odd and in particular, $q(VC,S) > 0$. It follows easily that

$$h \leq q(S,U) \leq rs - q \tag{6}$$

and $\quad h \leq q(T,U) \leq rt - q.$ $\tag{7}$

So multiplying (6) by k / r and (7) by $(r - k) / r$ and then adding we obtain (3). Since $\delta(B) < 0$ always, G has a k-factor. This is Petersen's Theorem.

Now suppose that r and $|VG|$ are even, k is odd and $r / \lambda \leq k \leq r - r / \lambda$. Since r is even, λ is even and so $\lambda \geq 2$. Suppose $a \geq 0$, $b \geq 0$ then from (4)

$$(a + b)q \leq r(as + bt). \tag{8}$$

From (5) and (8)

$$r\lambda h - r(as + bt) \leq r(r(s + t) - 2q) - (a + b)q,$$

i.e.

$$h - \frac{as}{\lambda} - \frac{bt}{\lambda} \leq \frac{rs}{\lambda} + \frac{rt}{\lambda} - \frac{2q}{\lambda} - \frac{aq}{r\lambda} - \frac{bq}{r\lambda}.$$

So

$$h \leq \left(\frac{r + a}{\lambda}\right)s + \left(\frac{r + b}{\lambda}\right)t - \frac{1}{\lambda r}(2r + a + b)q. \tag{9}$$

Now set $a = k\lambda - r \geq 0$ and $b = (r - k)\lambda - r \geq 0$ in (9) to obtain (3). $\qquad\square$

Finally we briefly mention two theorems which are essentially applications of Tutte's f-factor theorem although very subtle variations in the Transfer Theorems are required in their proofs.

A connected graph G is called t-**tough** if $t.\omega(G - S) \leq |S|$ for all subsets S of VG such that $\omega(G - S) > 1$ where $\omega(G - S)$ is the number of components of

G- S. Enomoto, Jackson, Katerinis and Saito [E-J-K-S 85] proved the following theorem - a result which was first conjectured by Chvátal [vC 73].

Theorem 6.5 Let $k \geq 1$ be an integer. Every k-tough graph has a k-factor if $k|VG|$ is even and $|VG| \geq k + 1$.   □

For further work on toughness and factors, see [pK 90].

Let G be a graph. If $X \subseteq VG$ let $N(X) = \bigcup_{x \in X} N(x)$ where $N(x)$ is the neighbourhood of x. The **binding number** b(G) [dW 73] of G is the minimum value of $|N(X)| / |X|$ taken over all non-empty subsets X of VG such that $N(X) \neq VG$.

Anderson [iA 71] proved that if $|VG|$ is even and $b(G) \geq \frac{4}{3}$, then G has a 1–factor. Woodall [dW 73] proved that if $b(G) \geq \frac{3}{2}$, then G has a Hamiltonian cycle. Woodall also discussed the intimate relationship of the toughness of a graph and its binding number. Finally the next result is proved by Katerinis and Woodall [K-W 87].

Theorem 6.6 Let $k \geq 2$ be an integer and let G be a graph with $p \geq 4k - 6$ vertices and binding number b(G) such that kp is even and $b(G) > (2k-1)(p-1) / (k(p - 2) + 3)$. Then G has a k-factor.   □

We have roamed far from the Petersen graph in this chapter. However although it appears only peripherally in the broad outline of the theory of factors it did play its role at a crucial time in the development of the theory. Hence it motivated what was to follow.

Exercises
1.   Let G be a (2k + 1)-regular graph. Suppose that $S \subseteq VG$ and U is an odd component of $G \setminus S$. Prove that U is joined to S by an odd number of edges.

2.   Let $|VG| \equiv 0$ (mod 2). If $S \subseteq VG$, prove that $h_G(S) \equiv |S|$ (mod 2).

3.   Find the smallest cubic graph which contains a bridge **and** a 1-factor.

4.   Find the smallest cubic graph which does not contain a 1-factor.

5. Let G be a bipartite graph with parts X, Y where |X| = m and let G
   contain a matching which covers every vertex of X. Suppose deg x ≥ d
   for all x ∈ X. Show that G contains at least d! matchings covering X if
   d ≤ m and at least d! / (d-m)! such matchings if d ≥ m.

6. Let G be a (k - 1)-edge-connected k-regular graph with an even number of
   vertices. Produce G' from G by removing k - 1 edges. Prove that G'
   contains a 1-factor.

7. Prove that every cubic bridgeless graph contains a 2-factor.

8. Prove that $\chi(\overline{P}) = 6$ (or that $\overline{P}$ is 1-factorable).

9. Show directly that P is factor-covered.

10. Let d be an integer with $0 \leq d \leq |VG|$ such that $|VG| + d$ is even.

    (a) Prove that G has a defect-d matching if and only if $h_G(S) \leq |S| + d$ for
        all $S \subseteq VG$ [cB 58].

    (b) Prove that G is d-covered if and only if
            (i) $h_G(S) \leq |S| + d$ for all $S \subseteq VG$,
        and (ii) $h_G(S) = |S| + d$ implies that $<S> = \overline{K}_{|S|}$ [L-G-H 75].

11. Prove that if G is (r - 1)-connected r-regular for $r \geq 2$ and $|VG| + d$ is
    even, then G is d-covered [L-G-H 75].

12. Prove that every even connected vertex-transitive graph is 0-covered and
    every odd connected vertex-transitive graph is 1-covered [L-G-H 75].

13. Find (r - 2)-connected r-regular graphs G for $r \geq 3$ with $|VG| + d$ even
    which do not have a defect-d matching.

14. Suppose that G is a graph and B = (S,T,U) is a graph triple in G,
    relative to a fixed f-factor. Prove that $\delta(B) \equiv 0$ (mod 2).

    [Hints (i) verify that $\deg(T) = 2|ET| + q(T,S) + q(T,U)$;
            (ii) verify that $q(T,U) = \sum_{C \text{ odd}} J(B;C) + \sum_{C \text{ even}} J(B;D) - f(U)$.
           (iii) $f(VG) \equiv 0$ (mod 2).]

15. Give an example of a bipartite graph with parts X and Y satisfying Hall's
    condition which does not contain a matching covering all the vertices of X.

16.   Give an example of a bipartite graph with parts X and Y such that (i) it satisfies Hall's conditions; (ii) it does not contain a matching covering all of X; (iii) subject to (i) and (ii) it is edge-maximal.

17.   By examining the examples given in Exercises 15 and 16, or otherwise, try to describe possible "barriers".

      [Hint: Consider a finite subset $S \subseteq X$ such that $|S| = |N(S)|$.]

18.   In the proof of Theorem 6.3, why is $w < n$?

      [Hint: prove that $w = n$ implies $\delta(B) = 0$.]

19.   (a) Prove the following.
            (i) K_n has a 1-factor if and only if n is even.
            (ii) K_n has a defect-1 matching if and only if n is odd.
            (iii) K_n is 1-factor-covered if and only if n is even.
            (iv) K_n is defect-1 covered if and only if n is odd.
            (v) K_n is 1-factorable if and only if n is even.

      (b) Generalize the above statements (and definitions, where necessary) to r-factors for r a positive integer.

References

[A-K 85]   J. Akiyama and M. Keino, Factors and Factorizations of Graphs - A survey, *J. Graph Th.*, 9, 1985, 1-42.

[iA 71]   I. Anderson, Perfect matchings of a graph, *J. Comb. Th.* B, 10, 1971, 183-186.

[iA 89]   I. Anderson, *A First Course in Combinatorial Theory*, Second Edition, Clarendon Press, Oxford, 1989.

[A-N-S 83]   R. Aharoni, C.St.J.A. Nash-Williams and S. Shelah, A general criterion for the existence of transversals, *Proc. London Math. Soc.*, 47, 1983, 43-68.

[rA 84]   R. Aharoni, Matchings in graphs of size $\aleph_1$, *J. Comb. Th.*, B, 36, 1984, 113-117.

[rA 88]   R. Aharoni, Matchings in infinite graphs, *J. Comb. Th.* B, 44, 1988, 87-125.

[cB 58]   C. Berge, Sur le couplage maximum d'un graphe, *C.R. Acad. Sci. Paris*, 247, 1958, 258-259.

[B-S-W 85]   B. Bollobás, A. Saito and N.C. Wormald, Regular factors of regular graphs, *J. Graph Th.*, 9, 1985, 97-103.

[vC 73]   V. Chvátal, Tough graphs and hamiltonian circuits, *Disc.Math.* 5, 1973, 215-228.

[D-M 74]   R.M. Damerell and E.C. Milner, Necessary and sufficient conditions for transversals of countable set systems. *J. Comb. Th.* A, 17, 1974, 350-374.

[eE 31]   E. Egerváry, Matrixok kombinatoriustulajdonságainó, *Mat. Fiz. Lapok*, 38, 1931, 16-28. (Hungarian with German summary.)

[E-J-K-S 85] H. Enomoto, B. Jackson, P. Katerinis and A. Saito, Toughness and the existence of k-factors, *J. Graph Th.*, 9, 1985, 87-95.

[E-0-K 88]   H. Enomoto, K. Ota and M. Kano, A sufficient condition for a bipartite graph to have a k-factor, *J. Graph Th.*, 12, 1988, 141–151.

[E-G 60]   P. Erdös and T. Gallai, Graphs with prescribed degrees of vertices (in Hungarian), *Mat. Lapok*, 11, 1960, 264-274.

[hF 92]   H. Fleischner, private communication.

[H-S 78]   S.L. Hakimi and E.F. Schmeichel, Graphs and their degree sequences: A survey, in *Theory and Applications of Graphs*, Lecture Notes in Mathematics No.642, Springer-Verlag, Berlin, 1978, 225-235.

[mH 48]   M. Hall, Distinct representatives of subsets, *Bull. Amer. Math. Soc.*, 54, 1948, 922-926.

[pH 35]   P. Hall, On representatives of subsets, *J. London Math. Soc.*, 10, 1935, 26-30.

[H-V 50]   P.R. Halmos and H.E. Vaughan, The marriage problem, *Amer. J. Math*, 72, 214-215, 1950.

[K-W 87]   P. Katerinis and D. Woodall, Binding numbers of graphs and the existence of k-factors. *Quart. J. Math. Oxford*, 38, 1987, 221–228.

[pK 90]   P. Katerinis, Toughness of graphs and the existence of factors, *Disc. Math.*, 80, 1990, 81-82.

[aK 86]   A.B. Kempe, A memoir on the theory of mathematical form, *Phil. Trans. Roy. Soc. London*, 177, 1886, 1-70.

[dK 31]   D. König, Graphok és matrixok, *Mat. Fiz. Lapok*, 38, 1931, 116–119, (Hungarian with German summary).

[L-G-H 75]   C.H.C. Little, D.D. Grant, and D.A. Holton, On defect-d matchings in graphs, *Disc. Math.*, 13, 1975, 41-54.

[lL 72]   L. Lovász, On the structure of factorizable graphs, *Acta Math. Acad. Sci. Hungar.*, 23, 1972, 179-195.

[lL 75]   L. Lovász, Three short proofs in graph theory, *J. Comb. Th. B*, 19, 1975, 269-271.

[L-P 86]   L. Lovász and M. Plummer, Matching Theory, *Akadèmiai Kiadò*, Budapest, 1986.

[cN 78]   C.St.J.A. Nash-Williams, Another criterion for marriage in denumerable societies, Advances in Graph Theory, (Cambridge Combinatorial Conference, Cambridge, 1973), *Ann. Disc. Math.*, 3, 1978, 165-179.

[cN 82]   C.St.J.A. Nash-Williams, A glance at graph theory - part 1, *Bull. London Math. Soc.*, 14, 1982, 177-212.

[jP 91]   J. Petersen, Die Theorie der regulären Graphen, *Acta Mathematica*, 15, 1891, 193-220.

[jP 98]   J. Petersen, Sur le théorème de Tait, *L'Intermédiaire des Mathematiciens*, 5, 1898, 225-227.

[P-S 76]   K.P. Podewski and K. Steffens, Injective choice functions for countable families, *J. Comb. Th. B*, 21, 1976, 40-46.

[cT 81]   C. Thomassen, A remark on the factor theorems of Lovász and
             Tutte, *J. Graph. Th.*, 5, 1981, 441-442.

[pT 80]   P.G. Tait, Note on a theorem in the geometry of position, *Trans.
             Roy. Soc. Edinburgh*, 29, 1880, 657-660.

[wT 47]   W.T. Tutte, The factorization of linear graphs, *J. London Math.
             Soc.*, 22, 1947, 107-111.

[wT 52]   W.T. Tutte, The factors of graphs, *Can. J. Math.*, 4, 1952,
             314–328.

[wT 54]   W.T. Tutte, A short proof of the factor theorem for finite graphs.
             Can.J. Math., 6, 1954, 347-352.

[mW 70]   M.E. Watkins, Connectivity of transitive graphs, *J. Comb. Th.*, 8,
             1970, 23-29.

[dW 73]   D. Woodall, The binding number of a graph and its Anderson
             number, *J. Comb. Th.*, B, 15, 1973, 225-255.

5

Beyond the Four Colour Theorem

0. Prologue

"The Four Colour Problem has been solved by K. Appel, W. Haken and J. Koch. But what about the other mathematicians who have been working on the problem? I imagine one of them outgribing in despair, crying 'What shall I do now?' To which the proper answer is 'Be of good cheer. You can continue in the same general line of research. You can study the Hajós and Hadwiger Conjectures. You can attack the problem of 5-flows and you can try to classify the tangential 2–blocks.' "

In this optimistic vein, Tutte [wT 78], rallied possibly disheartened 'Mapmen'. We have already given some space to Hajós and to Hadwiger in Section 2.5. Now we turn our attention to the last two problems mentioned by Tutte above, namely the problem of 5-flows and the classification of tangential 2-blocks.

Some real progress has been made, not only on the questions themselves in [bD76], [bD 81], [fJ 76] and [pS 81b], but also on the intimate relationship between them - [dW 79], [dW 80], [pS 81a] .

In fact P.D. Seymour [pS 81b] has shown that every bridgeless graph has a nowhere-zero 6-flow.

1. Flows

Let G be a finite pseudograph. Orient G by putting arrows on each edge $e \in EG$, so that one end of e is distinguished as the tail $t(e)$ of e and the other as the head $h(e)$ of e. Hence $t(e) = h(e)$ if and only if e is a loop.

If $v \in VG$ and $e \in EG$, define

$$\eta(e,v) = \begin{cases} 1 & \text{if } v = h(e), \\ 0 & \text{if } e \text{ is a loop or } v \text{ is not an end of } e, \\ -1 & \text{if } v = t(e). \end{cases}$$

Let H be a finite additive abelian group. Then an **H-flow on G** is a mapping $f: EG \to H$ such that for each vertex v of G,

$$\sum_{e \in EG} \eta(e,v) \, f(e) = 0. \tag{1}$$

If we regard G, when oriented, as an electrical network and think of f(e) as the current in each edge e of the network, then equation (1) is simply the Kirchhoff law which guarantees conservation of current through a node of a network.

In Figure 1.1 we show a $(\mathbf{Z}_2 \times \mathbf{Z}_2)$-flow on K_4.

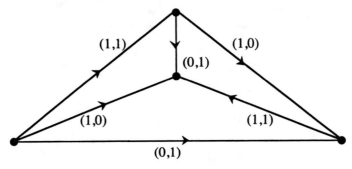

Figure 1.1

Actually in this example, the edges may be oriented at will without upsetting the $(\mathbf{Z}_2 \times \mathbf{Z}_2)$-flow. This is because all elements in $\mathbf{Z}_2 \times \mathbf{Z}_2$ are self-inverse. In other words, $(a,b) = -(a,b)$ for all $(a,b) \in \mathbf{Z}_2 \times \mathbf{Z}_2$. Hence the Kirchhoff law (1) holding at each vertex is independent of the direction of the arrows.

In our definition of an H-flow on G we assumed that H was finite. We shall however consider **Z**-flows on G where **Z** is the additive group of integers. In this connection Tutte [wT 49], proved the following.

Theorem 1.1 If f′ is a **Z**-flow on G and $k > 0$, then there is a **Z**-flow f on G such that for each $e \in EG$,

$$f(e) \equiv f'(e) \pmod{k} \text{ and } 1 - k \le f(e) \le k - 1. \qquad \square$$

We illustrate the theorem using the graph of Figure 1.2, where $k = 6$.

The **Z**-flow f may clearly be regarded in an obvious manner as a $\mathbf{Z}_k$-flow where $\mathbf{Z}_k$ is the set of integers modulo k under the obvious addition. As a result, we do not need to extend our definition to include non-finite groups. Where we are tempted to use **Z**-flows we shall restrain ourselves and deal instead with the associated $\mathbf{Z}_k$-flow for some $k > 0$.

We define the **support of f,** Sup f, to be the subset of EG on which the flow f is non-zero. Hence Sup f = $\{e \in EG : f(e) \ne 0\}$.

Then an H-flow f, is a **nowhere-zero H-flow,** if Sup f = EG.

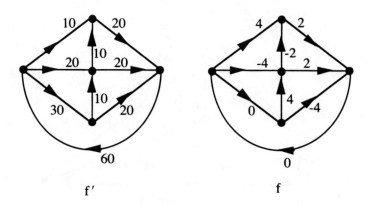

f'   f

Figure 1.2

In Figure 1.2, Sup f' = EG while Sup f ⊂ EG but Sup f ≠ EG. Hence f' is a nowhere-zero **Z**–flow and f is not.

We now make four important observations concerning nowhere-zero H-flows.

Lemma 1.2 If G has a nowhere-zero H-flow, then G is bridgeless.

Proof: Let e_0 be a bridge of G and suppose G has an H-flow f. Let v_0 be an endvertex of e_0 and let K be the component of G - e_0 which contains v_0; see Figure 1.3.

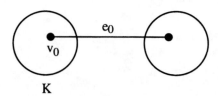

K

Figure 1.3

For each v ∈ VK, we have $\sum_{e \in EG} \eta(e,v) f(e) = 0$.

Hence $\sum_{v \in VK} \sum_{e \in EG} \eta(e,v) f(e) = 0$.

But in this summation, for $e \neq e_0$, $\eta(e,h(e))$ $f(e)$ + $\eta(e,t(e))$ $f(e) = 0$ and $\eta(e,v) = 0$ for all v other than $h(e)$ and $t(e)$. Hence the double summation reduces to

$$\eta(e_0,v_0) \, f(e_0) = 0.$$

Hence $f(e_0) = 0$ and f is not a nowhere-zero flow.   □

Lemma 1.3 If G has a nowhere-zero H-flow associated with some orientation of G, then it has a nowhere-zero H-flow associated with any orientation of G.

Proof: Suppose G has an orientation D with which is associated a nowhere–zero H-flow f. Now consider another orientation D'. If the orientation of the edge e is different in D' from that in D, then replace $f(e)$ by $-f(e)$ as the flow on the edge e. Otherwise retain the same edge flow.

Hence we have a nowhere-zero H-flow on G under the orientation D'.   □

As a result of Lemma 1.3 we shall henceforth speak of G as **supporting a nowhere-zero H-flow** to mean that in any orientation of G there exists a nowhere-zero H-flow.

Lemma 1.4 The number of nowhere-zero H-flows on G depends only on the order of the group H.

Proof: A proof of this result may be found in [dW 79]. The lemma is originally due to Tutte [wT 54].   □

As a result of Lemmas 1.3 and 1.4 we know that if G supports a nowhere-zero H-flow, then it will support a nowhere-zero H'-flow, where the order of H and H' are the same. If this order is k, we take G has a **nowhere-zero k–flow**, to mean that for **any** orientation of G and **any** finite group H of order k, there exists an H-flow on G.

Lemma 1.5 Let G support a nowhere-zero k-flow. Then G supports a nowhere-zero s-flow for any $s \geq k$.

Proof: Suppose that G supports a nowhere-zero k-flow. Then, see [fJ 79], p.207, G supports a nowhere-zero Z-flow with values in $[1-k, k-1]$ i.e. the values on edges are non-zero and lie between $1-k$ and $k-1$. To prove this matroidal techniques are required which are beyond the scope of this book; see [dW 76]. Now the required s-flow can easily be constructed from the given Z-flow by replacing negative numbers by their inverses modulo s.   □

We can now define $F(G)$ to be the least integer $k > 0$, such that G supports a nowhere-zero k-flow. If G has a bridge, then we know from Lemma 1.2 that no such value of k exists. In this case we let $F(G) = \infty$.

Tutte [wT 54] has conjectured that if G is 2-connected, then $F(G)$ is defined and finite. This conjecture was settled by Jaeger [fJ 76].

Theorem 1.6 Every 2-connected pseudograph supports a nowhere-zero 8-flow.

This result has now been pushed further by Seymour [pS 81a].

Theorem 1.7 Every 2-connected pseudograph supports a nowhere-zero 6-flow.

We give a proof of this result in the next section.

To date in this chapter, the Petersen graph has been noticeable by its absence.

Proposition 1.8 $F(P) > 4$.

Proof: Suppose $F(P) \leq 4$. In view of Lemmas 1.4 and 1.5, we may suppose that there is a nowhere-zero Z_4-flow f on P.

Now since P is cubic and $\sum_{e \in EG} \eta(e,v)f(e) = 0$, then at any given vertex v of P, the values of $\eta(e,v) f(e)$ at edges incident on v are either $1,1,2$ or $-1,-1,2$. If $1,1,2$ are the flow values associated with v, call it a **positive vertex**. Otherwise call it a **negative vertex**.

Now the edges with $f(e) = \pm 1$ form a 2-factor of P. Further in any cycle of this 2-factor, the positive and negative vertices alternate showing that such cycles have an even number of edges. Since the edges with $f(e) = 2$ form a 1-factor of P, then P must be 1-factorable, in contradiction of Theorem 4.3.1.   □

In fact Tutte was aware of the situation regarding the Petersen graph and had actually proposed the following conjecture in [wT 54].

Conjecture 1.9 Every bridgeless pseudograph has a nowhere-zero 5-flow.

It is striking that although this conjecture was made in 1954 it was not until 1975 that it was even known whether 5 could be replaced by some larger number k (independent of the graph). Jaeger first replaced 5 by 8 and then Seymour replaced 5 by 6. Unfortunately the arguments of both Jaeger and Seymour depend heavily on the fact that $8 = 2 \times 2 \times 2$ and $6 = 3 \times 2$ and so do not immediately suggest a method of proof for Conjecture 1.9.

On the other hand, Steinberg [rS 84] has shown that the conjecture is true for the projective plane. Möller, Carstens and Brinkmann [M-C-B 88] extend this result to orientable surfaces of genus at most two and nonorientable surfaces of genus at most four.

Now it turns out that we can characterize in terms of edge colouring, those bridgeless cubic pseudographs which have a nowhere-zero Z_4-flow.

Theorem 1.10 Let G be a bridgeless cubic pseudograph. Then $F(G) \leq 4$ if and only if G is 3-edge-colourable.

Proof: The argument of the proof of Proposition 1.8 shows that if $F(G) \leq 4$, then G is 3-edge-colourable.

On the other hand, if G is 3-edge-colourable, then colour the edges with the three 'colours' $(0,1), (1,0)$ and $(1,1)$. If these colours are considered as elements of $Z_2 \times Z_2$, then, because all elements are self-inverse, the colouring defines a nowhere-zero $(Z_2 \times Z_2)$-flow on G for any given orientation of G. □

As a consequence of this result Tutte's conjecture in section 2.6 concerning snarks can be rephrased in the language of flows; see [wT 69].

Conjecture 1.11 Let G be a bridgeless cubic pseudograph. If G does not support a nowhere-zero 4-flow, then G is subcontractible to P.

Actually an even stronger version of this conjecture is implicitly made in Tutte [wT66]; see [dW 80].

Conjecture 1.12 Let G be a bridgeless pseudograph. If G does not support a nowhere-zero 4-flow, then G is subcontractible to P.

We note in passing that if either of the last two conjectures can be proved to be true, then they would provide an alternative proof of the Four Colour Theorem, based on the non-planarity of P.

Suppose we restrict our attention to bridgeless planar graphs. Let $\chi^*(G)$ denote the face chromatic number of G, that is the smallest number of colours necessary to colour the faces so that no two adjacent faces have the same colour.

Theorem 1.13 Let G be a bridgeless planar graph. Then $F(G) = \chi^*(G)$.   □

We do not give a proof here. A rigorous proof can be based on a duality proof of the Tutte polynomials; see [dW 79]. We are content to prove that if G is a bridgeless planar graph then $F(G) \leq 4$.

Obviously all we need do is to prove this for a bridgeless plane graph G. By the Four Colour Theorem, the map in the Euclidean plane determined by G is 4-colourable and therefore the regions into which G divides the plane can be coloured with four colours c_1, c_2, c_3, c_4. A 4-flow in G can now be obtained by letting each edge e which lies between two regions with colours c_i and c_j carry a flow value $i - j$ in the direction in which we would travel along e so as to keep the region with colour c_i on our left and the region with colour c_j on our right. We give an example in Figure 1.4.

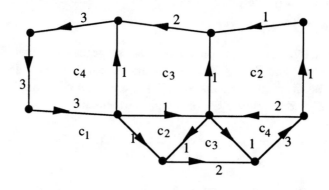

Figure 1.4

So Conjecture 1.9 becomes a theorem for planar graphs.

Theorem 1.14 Every bridgeless planar graph has a nowhere-zero 5-flow.

Proof: From the Five Colour Theorem, Theorem 2.2.5, if G is a bridgeless planar graph, then $\chi^*(G) \leq 5$. Hence from Theorem 1.13, $F(G) \leq 5$.   □

2. A Result of Seymour

Having established the importance of $F(G)$ in the last section, we present a result due to P.D. Seymour [pS 81a], which not only considerably advances our knowledge, but can also be presented in an elementary way. We need three lemmas before we can prove that every bridgeless pseudograph supports a nowhere-zero 6-flow.

Lemma 2.1 For $k > 0$, if there exists a pseudograph G with $k < F(G) < \infty$, then there exists a (simple) 3-connected cubic graph G' with $k < F(G') < \infty$.

Proof: Let $\mathbf{F} = \{G : k < F(G) < \infty\}$. Suppose that $\mathbf{F} \neq \emptyset$. Hence there exists $G \in \mathbf{F}$. We may assume that G is connected, for if not, then some component of G belongs to $\mathbf{F}$.

Since $F(G) < \infty$, then G is 2-edge-connected. If it has edge-connectivity two, let $\{e_1, e_2\}$ be an edge cut of G. Now contract the edge e_1. Then the pseudograph G_1 so obtained remains 2-edge-connected and $F(G) = F(G_1)$.

We continue in this way until we obtain a 3-edge-connected pseudograph $G_2 \in \mathbf{F}$. From G_2, we delete any loops without affecting the flow status of G_2. Hence we have a loopless pseudograph $G_3 \in \mathbf{F}$.

The minimum degree of any vertex of G_3 is at least three since G_3 is 3–edge–connected. If $v \in VG_3$ and deg $v \geq 4$, then we form a new graph from G_3 by replacing v with a cycle containing deg v vertices such that each new vertex is adjacent to one vertex in $N_{G_3}(v)$. We show this construction in Figure 2.1. By replacing all vertices of G_3 by cycles in this way we obtain the graph G_4. Now it is easy to verify, see Exercise 3, that $F(G) \leq F(G_4)$ by showing that $F(G_3) \leq F(G_4)$, $F(G_2) \leq F(G_3)$ and $F(G) \leq F(G_2)$.

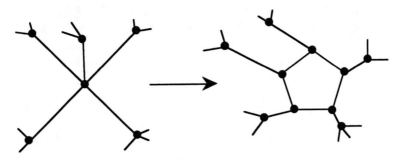

Figure 2.1

The graph G_4 so constructed is simple, cubic, and 3-edge-connected and hence, Theorem 1.2.2, 3-connected.   □

Before we can prove the next lemma we need a definition. Let $X \subseteq EG$. Then $\langle X \rangle_2$ is defined to be the smallest set $Y \subseteq EG$ such that

(i) $X \subseteq Y$, and

(ii)   there is no cycle C of G with $0 < |EC \setminus Y| \leq 2$.

Obviously $<X>_2$ is well-defined since if Y_1 and Y_2 satisfy (i) and (ii), then so does $Y_1 \cap Y_2$.

We may think of $<X>_2$ as being the smallest set containing X with the property that if any cycle C protrudes from the set, then there are at least three protruding edges.

Suppose $X = \{12,23,34,45,15\}$ in the standard labelling of P. Then $<X>_2=X$. However if $Y = X \cup \{11'\}$, then $<Y>_2 = EP$. Indeed if $e \in EP \backslash X$, then $<X \cup \{e\}>_2 = EP$.

Those readers familiar with matroid theory [dW 76], will recognize $<X>_2$ as a generalization of the closure operator.

Lemma 2.2 Let $X \subseteq EG$. If $<X>_2 = EG$, then G has a **Z**-flow f with $EG\backslash X \subseteq$ Sup f and such that $-2 \leq f(e) \leq 2$ for each $e \in EG$.

Proof: We use induction on $|EG \backslash X|$. The result is trivially true if this number is zero. So we suppose $|EG \backslash X| > 0$ and $<X>_2 = EG$.

Now there must exist a cycle C in G with $0 < |EC \backslash X| \leq 2$, otherwise $<X>_2=X \neq EG$. But $<X>_2 \subseteq <X \cup EC>_2$, so $<X \cup EC>_2 = EG$. Hence, by induction, there exists a **Z**-flow g on G with $EG \backslash (X \cup EC) \subseteq$ Sup g such that $-2 \leq g(e) \leq 2$ for each $e \in EG$. Associated with g is an orientation of G. Let h be a **Z**-flow on G associated with this orientation of G, such that Sup$h= C$ and $-1 \leq h(e) \leq 1$ for each $e \in EG$. This can be done by choosing an arbitrary direction around C as positive. Any edge oriented in this direction is given a flow of $+1$, while any edge oriented oppositely has a flow of -1.

Since $0 < |EC \backslash X| \leq 2$, we may choose an integer n such that $n \neq -\dfrac{g(e)}{h(e)}$ (mod3) for any $e \in EC \backslash X$. Then let $f' = g + nh$. Clearly f' is a **Z**-flow on G. Further, if $e \in EC \backslash X$, then $f'(e) \neq 0$ (mod 3) by the choice of n, while if $e \in EG \backslash (X \cup EC)$, then $f'(e) = g(e) \neq 0$ (mod 3) by the choice of g. Hence $E(G) \backslash X \subseteq$ Sup f'.

Now by Theorem 1.1 we know there exists a **Z**-flow f on G such that for each $e \in EG$, $f(e) \equiv f'(e)$ (mod 3) and $-2 \leq f(e) \leq 2$. Obviously $EG \backslash X \subseteq$ Sup f and the result follows.   □

Lemma 2.3 If G is a 3-connected graph, then there are vertex-disjoint cycles $C_1, C_2, \ldots, C_r$ such that $\langle \bigcup_{i=1}^{r} EC_i \rangle_2 = EG$ for some $r \geq 1$.

Proof: This follows by induction. See Exercise 5.   □

We note that in the Petersen graph, if $C_1 = (1,2,3,4,5)$ and $C_2 = (1',2',3',4',5')$ then $<EC_1 \cup EC_2>_2 = EG$.

The main result of this section now follows; see [pS 81a].

Theorem 2.4 Every bridgeless pseudograph supports a nowhere-zero 6-flow.

Proof: By Lemma 2.1 we need only consider simple 3-connected cubic graphs. Let G be such a graph. From Lemma 2.3 we know that there exist vertex-disjoint cycles $C_1, C_2, \ldots, C_r$ such that $\langle \bigcup_{i=1}^{r} EC_i \rangle_2 = EG$ Let $X = \bigcup_{i=1}^{r} EC_i$.

Now Lemma 2.2 shows that there exists a **Z**-flow g on G with $EG \backslash X \subseteq Supg$ and $-2 \leq g(e) \leq 2$ for each $e \in EG$. Since X is a union of disjoint cycles, then there exists a **Z**-flow g' with Sup g' = X and such that $-1 \leq g'(e) \leq 1$ for each $e \in EG$. Define $f = g + 3g'$. Then f is a **Z**-flow.

If $e \in EG \setminus X$, then $f(e) = g(e) \in \{\pm 1, \pm 2\}$, while if $e \in X$, then $f(e) = g(e) \pm 3$, where $-2 \leq g(e) \leq 2$. Hence $f(e) \neq 0$ and $-5 \leq f(e) \leq 5$ for all $e \in EG$. Thus G supports a nowhere-zero 6-flow.   □

Based on this theorem, Younger [dY 83] provides an algorithm for finding a 6–flow in any bridgeless graph.

In view of the preceding theorem Conjecture 1.12 can be restated.

Conjecture 2.5 If $F(G) \in \{5,6\}$, then G is subcontractible to P.

For a nice synthesis of some known and related results, see [fJ 79]. There are many variations of these flow problems; see, for example, [aB 83], [aK 87].

3. Projective Geometry

A graph can be represented as a set of points in a finite projective geometry over the Galois field GF(2). By so representing a graph it is hoped that an examination of the known properties of the more general system will enable theorems about graphs to be obtained.

In [wT 66], Tutte used this line of attack. One of the most striking conjectures that he makes is

Conjecture 3.1 The only tangential 2-blocks are the Fano, Desargues and Petersen blocks.

We will not be able to present here all the details necessary to understand fully the reasons for and the implications of this conjecture. However, we hope to present the reader with sufficient details so that both its essentially geometric nature and also its intrinsic beauty can be appreciated.

Before we even explain the concepts of the conjecture let us explore its content.

The conjecture implies as special cases the Four Colour Theorem and the conjecture that every snark is subcontractible to P, Conjecture 3.1.9. Further it is equivalent to the conjecture that any bridgeless pseudograph which does not support a 4–flow is subcontractible to P, Conjecture 1.12. This equivalence was proved directly by Seymour [pS 81a]. We shall see that his result means that a seemingly intractible geometrical problem is reduced to the conceptually simpler problem of characterizing which pseudographs do not have 4-flows. Unfortunately the equivalence between these two conjectures is best established in the context of matroid theory, which we have no room to develop here.

Of course it is natural to ask at this stage, why, if Conjectures 3.1 and 1.12 are equivalent and the latter conjecture is conceptually easier, should we concern ourselves with the former?

Well for one thing, a solution of the problem in either of its forms would lend insight in the other area. But the fact that each conjecture implies the other is, on the surface of it, quite remarkable and we cannot marvel at this phenomenon until we have some appreciation of the conjecture stated for the first time in this section.

Again we note that we cannot give the full story in the text. However extra details may be gleaned from the exercises at the end of the chapter. Before we go any further though we need to establish some background.

A **projective space** (**P,L,e**) is a triple consisting of a set of **points P**, a set of **lines L** and an incidence relation **e**. We say p is on the line L instead of peL and so on. The rules which this incidence relation satisfy are:

(i)   any two distinct points are on exactly one line;
(ii)   if x, y, z and w are four distinct points, no three of which are collinear, and if the line xy intersects the line zw, then xz intersects yw;
(iii)   each line contains at least 3 points.

A **projective plane** is a triple **(P,L,e)** such that:
(i)   any two distinct points are on exactly one line;
(ii)   any two distinct lines have exactly one point in common;
(iii)   there exist four points, no three of which are collinear.

Example
(i)   The Fano configuration (or Fano block - the only blocks we consider are 2–blocks, see p.172, and so we use the terms interchangeably) shown in Figure 3.1 is a projective plane. It consists of the seven points and the seven lines of Figure 3.1 (one of which is represented by a closed curve). (The coordinates are explained in the next example.)

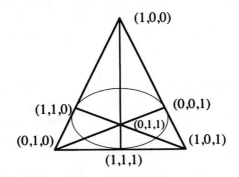

Fano Configuration

Figure 3.1

(ii)   Suppose that **(P,L,e)** is a projective space. Choose $x \in P$ and $L \in L$ so that $x \notin L$. The **plane** (x,L) in **(P,L,e)** is the ordered triple (P',L',e') with

$P' = \{y \in P : y = x$ or yx intersects $L\}$,
$L' = \{L' \in L : L'$ passes through two distinct points of $P'\}$, and
e' is the restriction of **e**.

It is easy to check that a plane in a projective space is a projective plane.

Corresponding to each projective plane there is a cardinal number n called its **order** such that

(i)   each line contains exactly $n + 1$ points,
(ii)   each point is incident with exactly $n + 1$ lines,
(iii) the plane contains $n^2 + n + 1$ points and $n^2 + n + 1$ lines.

The most concrete and familiar examples of finite projective spaces are the **projective geometries**. Indeed the projective geometries are the only finite projective spaces of dimension at least three. Before we define these objects we need some definitions.

The finite field with q elements will be denoted by $GF(q)$.

For any field F, consider the vector space V of all vectors $(a_0,..., a_n)$, $a_i \in F$. If $u, v \in V \setminus \{0\}$ we write $u \sim v$ if there exists some non-zero member λ of F such that $u = \lambda v$. It is easy to check that $\sim$ is an equivalence relation on $V \setminus \{0\}$. The equivalence classes under this relation are the points of the **projective geometry** of dimension n over F. When F is the finite field $GF(q)$ we denote this projective geometry by $PG(n,q)$.

Example
(i)   It is usual to choose one point in each equivalence class and to identify the equivalence class with this point. The point which is chosen has its first non-zero coordinate equal to one. This is the usual coordinatization. Thus $PG(2,3)$ contains the points $(0, 0, 1)$, $(0, 1, 0)$, $(0, 1, 1)$, $(0, 1, 2)$, $(1, 0, 0)$, $(1, 0, 1)$, $(1,0, 2)$, $(1, 1, 0)$, $(1,1, 1)$, $(1, 1, 2)$, $(1, 2, 0)$, $(1, 2, 1)$, $(1, 2, 2)$.

(ii)   The Fano block $PG(2,2)$ has points $(0,0,1)$, $(0,1,0)$, $(0,1,1)$, $(1,0,0)$, $(1,0,1)$, $(1,1,0)$, $(1,1,1)$. It is represented by the Fano configuration in Figure 3.1. $PG(2,2)$ is a projective plane of order 3. Thus 3 points lie on each line and through each point pass three lines.

A **projective space of dimension** 2 is simply a projective plane. We know that since $PG(2,q)$ is a projective plane with $q + 1$ points on each line, there exists a projective plane of order q for each prime power q. On the other hand there exist projective planes of prime power order q which are not isomorphic to $PG(2,q)$. It is not known whether there exists a plane of order n when n is not a prime power order. However, McKay, Lam, Thiel and Swiercz, [M-L-T-S 88], see also [rH 89], [L-T-S 88], [L-T-S 89], now claim that, after an exhaustive computer search, there is no projective plane of order 10. The famous Bruck–Ryser theorem does assert however that if $n \equiv 1$ or $2 \pmod 4$ and n is not the sum of two squares then there is no projective plane of order n.

A **0-chain** f on a graph G is a mapping f: $VG \to Z_2$ and a **1-chain** g on G is a mapping g: $EG \to Z_2$. We denote by $L_i(G)$ the set of i-chains on G for i= 0,1. It is straightforward to show that if we define $f(u_1 + u_2) = f(u_1) + f(u_2)$ and $g(e_1 + e_2) = g(e_1) + g(e_2)$, and use the usual scalar multiplication, then $L_0(G)$ and $L_1(G)$ are finite real vector spaces.

We define ∂g, the **boundary of a 1-chain** g, to be the 0-chain f such that for $u \in VG$,

$$f(u) = \sum g(e),$$

where the summation is over all edges e incident with u. Analogously, we define δh, the **coboundary of a 0-chain h,** to be the 1-chain k such that for $e = uv \in EG$,

$$k(e) = h(u) + h(v).$$

In Figure 3.2 we show a 1-chain f and its boundary ∂f for the graph of the cube. The values of f and ∂f are represented by numbers on the appropriate edge and vertex.

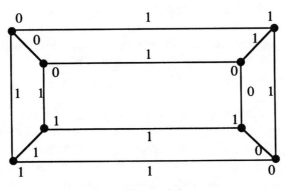

Figure 3.2

Let $\Delta(G)$ be the set of coboundaries of 0-chains of G. Then it is easy to see that if g is a 1-chain, then $g = \delta f \in \Delta(G)$ if and only if VG can be divided into two sets X, Y, such that Sup g is the set of edges in G which join vertices in X to vertices in Y. In other words, Sup g induces a bipartite subgraph of G. This follows since we may regard a 0-chain f as a colouring of the vertices in the two colours 0 and 1, and an edge has value 1 in δf if and only if it has one end of each colour.

A **1-cycle of G** is a 1-chain whose boundary is the 0-chain which is zero at every vertex. Let $\Gamma(G)$ be the set of 1-cycles of G. We note that the 1-chain f is in $\Gamma(G)$ if and only if Sup f is an Eulerian subgraph of G.

The set $\Delta(G)$ and $\Gamma(G)$ are both abelian groups, called, respectively, the **coboundary** and **cycle groups** of G. Naturally they are both subspaces of $L_1(G)$.

Two 1-chains f and g are said to be **orthogonal** if

$$\sum_{e \in EG} f(e) \, g(e) = 0.$$

It is an interesting (and classical) exercise in graph theory to show that a 1-chain is a 1-cycle if and only if it is orthogonal to every coboundary. Similarly a 1-chain is a coboundary if and only if it is orthogonal to every 1-cycle; see Exercise 8.

We can now consider representations of a graph in a projective geometry.

Let $\mathbf{P} = PG(d,2)$ denote a d-dimensional projective geometry over $\mathbf{Z}_2$. We shall regard $\mathbf{P}$ as being coordinatized so that the points of $\mathbf{P}$ are all the vectors $(x_1, x_2, ..., x_{d+1})$ with $x_i \in \mathbf{Z}_2$ and not all of the x_i's are zero.

Since we are dealing with $\mathbf{Z}_2$, the difference between the geometry $\mathbf{P}$ and the corresponding vector space is very slight. The only difference is in some of the nomenclature used and the fact that $\mathbf{P}$ contains no zero vector.

A k-**subspace Q** of the projective geometry $\mathbf{P}, k \geq 0$, (i.e. a subspace of $\mathbf{P}$ with (projective) dimension k) corresponds to a $(k + 1)$-dimensional subspace of the underlying vector space. In other words, $\mathbf{Q}$ contains a set of $k + 1$ linearly independent points together with all **non-zero** points dependent on $\mathbf{Q}$. Thus a 0-subspace consists of a single point, a 1-subspace has exactly three points and a 2-subspace has exactly seven. We call 1-subspaces, **lines** and 2-subspaces, **planes**.

We notice that there are 7 points and 7 lines in the Fano configuration. In general in $PG(d,2)$ there are as many 0-subspaces as there are $(d - 1)$-subspaces. The $(d-1)$-subspaces are usually referred to as **hyperplanes**.

Consider the graph G with $EG = \{e_i: i = 1, 2, ..., n\}$. Let $\Delta(G)$, considered as a subspace of the vector space, $L_1(G)$, have dimension r. To avoid trivialities we may suppose that $r \geq 1$. Let $\{f_1, f_2, ..., f_r\}$ be a basis for $\Delta(G)$. We identify f_i with the n-tuple such that its j-th coordinate is $f_i(e_j)$. We shall describe below how to select a basis for $\Delta(G)$ and we shall prove that $r = |VG| - \omega(G)$, where $\omega(G)$ is the number of components of G.

Let M be the matrix $[f_i(e_j)]$ and let s_j be the j-th column of M. Then $s_j \in PG(d,2)$, where $d = r - 1$. If $S = \{s_j : j = 1,2,...,n\}$, then S generates $PG(d,2)$ (remember row rank equals column rank).

The mapping $f: EG \rightarrow S$ defined by $f(e_j) = s_j$ is called the **direct representation** of G in $PG(d,2)$.

Analogously, if $\{g_1,g_2,...,g_s\}$ is a basis for $\Gamma(G)$, let $M^* = [g_i(e_j)]$. Again let s^*_j be the j-th column of M^* and let S^* be the set of all n columns of M^*. Now S^* generates $PG(d^*,2)$, where $d^* = s - 1$. Then the mapping $g: EG \rightarrow S^*$ defined by $g(e_j) = s^*_j$, is called the **dual representation** of G in $PG(d^*,2)$.

We now present algorithms for finding bases for $\Delta(G)$ and $\Gamma(G)$. First $\Delta(G)$.

Let F be an (edge) maximal spanning forest of G. Let $a_i = u_iv_i \in EF$. Now define F_i to be those vertices of F which are in the same component of $F \setminus a_i$ as u_i and define $\overline{F}_i$ to be those vertices of F which are in the same component of $F \setminus a_i$ as v_i. Write $E[F_i, \overline{F}_i]$ for the set of edges of G with one end in F_i and the other in $\overline{F}_i$. If finally we choose f_i to be the 1-chain with $\text{Sup} f_i = E[F_i, \overline{F}_i], i = 1, 2,..., |EF|$, then $\{f_i\}$ is a basis for $\Delta(G)$.

This algorithm is justified in Exercises 9 and 10. We simply note here that $r = |EF| = |VG| - \omega(G)$.

Naturally there will be a number of direct representations of G depending on the choice of basis. However, since the properties under discussion are invariant under a change of basis for $\Delta(G)$, we shall simply refer to **the** direct representation of G.

We now consider the direct representation of K_5 over **P**.

Let K_5 and F be as shown in Figure 3.3, and let $a_i = 1i$ for $i = 2,3,4,5$. Now $F_j = \{j\}$ and $\overline{F}_j = \{1,2,3,4,5\} \setminus \{j\}$ hence $E[F_j,\overline{F}_j] = \{ji: i \in \{1,2,3,4,5\} \setminus \{j\}\}$. Hence $\text{Sup } f_j = \{ji: i \in \{1,2,3,4,5\} \setminus \{j\}\}$.

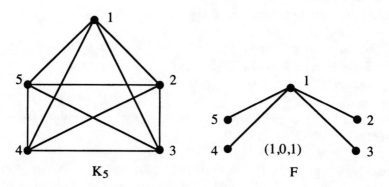

Figure 3.3

The matrix M is then

	12	13	14	15	23	24	25	34	35	45
f_1	1	1	1	1	0	0	0	0	0	0
f_2	1	0	0	0	1	1	1	0	0	0
f_3	0	1	0	0	1	0	0	1	1	0
f_4	0	0	1	0	0	1	0	1	0	1
S	s_1	s_2	s_3	s_4	s_5	s_6	s_7	s_8	s_9	s_{10}

Hence $r = 4$ and $d = 3$.

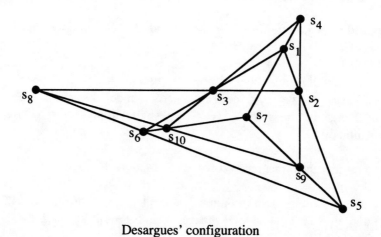

Desargues' configuration

Figure 3.4

The set S is called the **Desargues block.** This is because the ten points of S
define a 3-dimensional Desargues' configuration; see Figure 3.4. This

configuration has the triangles s_1, s_2, s_3 and s_7, s_9, s_{10}, perspective at the point s_4 and on the line s_5, s_6, s_8.

Now we show how to find a basis for $\Gamma(G)$, where G has no bridges.

As before let F be an (edge) maximal spanning forest of G. Let $a_i \in EG \setminus EF$. Then $F + a_i$ has a unique cycle C_i of G. If g_i is the 1-cycle with $\operatorname{Sup} g_i = EC_i$, then $\{g_i : i = 1, 2, ..., s\}$ is a basis for $\Gamma(G)$, where $s = |EG \setminus EF| = |EG| - |VG| + \omega(G)$.

The algorithm is proved in Exercises 11 and 12.

We illustrate the algorithm by constructing the dual representation of P over $\mathbf{P}$. As usual take the standard labelling of P and choose F as in Figure 3.5.

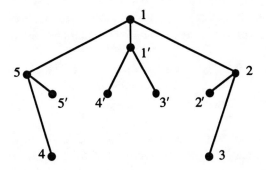

Figure 3.5

If we let $a_1 = 34$, $a_2 = 33'$, $a_3 = 44'$, $a_4 = 2'4'$, $a_5 = 2'5'$ and $a_6 = 3'5'$ then the matrix M^* is shown below.

	12	15	11'	23	22'	34	33'	45	44'	55'	1'3'	1'4'	2'4'	2'5'	3'5'
g_1	1	1	0	1	0	1	0	1	0	0	0	0	0	0	0
g_2	1	0	1	1	0	0	1	0	0	0	1	0	0	0	0
g_3	0	1	1	0	0	0	0	1	1	0	0	1	0	0	0
g_4	1	0	1	0	1	0	0	0	0	0	0	1	1	0	0
g_5	1	1	0	0	1	0	0	0	0	1	0	0	0	1	0
g_6	0	1	1	0	0	0	0	0	0	1	1	0	0	0	1
S^*	s_1^*	s_2^*	s_3^*	s_4^*	s_5^*	s_6^*	s_7^*	s_8^*	s_9^*	s_{10}^*	s_{11}^*	s_{12}^*	s_{13}^*	s_{14}^*	s_{15}^*

Hence $s = 6$ and $d^* = 5$. This is the dual representation of P in $PG(5,2)$. The set S^* is called the **Petersen block**.

To conclude this section we note that in the direct representation of G, distinct edges of G are represented by distinct elements of S; see Exercise 13. In both the direct and dual (where G is bridgeless) representations of G, no element of S and S^*, respectively, is zero; see Exercises 13 and 14.

4. Colourings and Projective Spaces

So far we have not shown how to interpret colourings of graphs in projective spaces. Obviously if we are to understand Conjecture 3.1, especially in the sense that it generalizes the Four Colour Theorem and Conjecture 1.11, then we must show how colourings come in to the picture. We rectify this omission in the current section and note again that we are only aiming to provide the flavour and not the substance of the story. For more details the reader should consult [wT 66].

Suppose that $\mathbf{P} = PG(d,2)$ and that $\mathbf{S} \subseteq \mathbf{P}$. Suppose also that $\mathbf{S}$ generates $\mathbf{P}$. Then $(\mathbf{P}, \mathbf{S})$ is said to be **graphic** if it corresponds to the direct representation of a graph and **cographic** if it corresponds to its dual representation.

A **colour-space** of $(\mathbf{P},\mathbf{S})$ is a $(d-2)$-subspace of $\mathbf{P}$ which does not contain any point of $\mathbf{S}$. If $(\mathbf{P},\mathbf{S})$ has no colour-space we call it a **2-block**. The Fano, Desargues and Petersen blocks are all 2-blocks.

Now Q is a **closed subset** of $\mathbf{S}$ if no point of $\mathbf{S} \setminus Q$ is linearly dependent on Q. Finally, a 2-block $(\mathbf{P},\mathbf{S})$ is **tangential** if for any closed subset Q of $\mathbf{S}$ there is a $(d-2)$-subspace of $\mathbf{P}$ that contains all the points of Q but no point of $\mathbf{S} \setminus Q$. (The above mentioned three 2-blocks are all tangential, see Exercise 15.)

Before proceeding further, let us say very crudely what all this means. Suppose we restrict ourselves in the first place to a graphic pair $(\mathbf{P},\mathbf{S})$ which directly represents a graph G. Then $(\mathbf{P},\mathbf{S})$ is a 2-block if and only if G is not 4–colourable. Furthermore it is a tangential 2-block if and only if G is a **subcontraction-minimal** non-4-colourable graph. In other words, G is not 4-colourable, but any subcontraction of G is.

Now recall that a special case of Hadwiger's conjecture ($k = 5$, see Section 2.5) states that the only subcontraction-minimal non-4-colourable graph is K_5. Conjecture 3.1 surmises that the only graphic tangential 2-block is the Desargues block - the direct representation of K_5. Since the special case of Hadwiger's conjecture is equivalent to the Four Colour Theorem we start to see how Tutte's tangential 2-block conjecture, Conjecture 3.1, does indeed generalize this theorem. Analogously Conjecture 3.1 suggests that the Petersen block is the only cographic tangential 2-block. This block is, of course, the dual representation of the Petersen graph. Now suppose we restrict ourselves to cographic pairs $(\mathbf{P},\mathbf{S})$ which dually represent a graph G. Then some at least of the tangential 2-blocks (P,S) represent cubic graphs and there must be subcontraction-minimal non–3–edge–colourable graphs. Conjecture 3.1 would claim that the Petersen block is the only such tangential 2-block. Hence the conjecture generalizes the snark Conjecture 3.19 - every snark is subcontractible to P.

In the above we have been tacitly assuming that the Fano block is neither graphic nor cographic. We have also been assuming that the Desargues block is not cographic and the Petersen block is not graphic. The problem of determining whether a given pair (P,S) is graphic or cographic can be solved by matroid theory. We shall not discuss this subject in the present book. The reader is urged to consult [wT 65] or [dW 76].

Although all of the above is handwaving it does at least begin to give a feel for the ideas. We now fill in a few details.

Suppose that (P,S) directly represents a graph G. A **colouring of** $\Delta(G)$ is an ordered pair (h_1,h_2) of coboundaries of G such that no edge of G has a zero coefficient in both h_1 and h_2.

Lemma 4.1 $\Delta(G)$ has a colouring if and only if G is 4-colourable.

Proof: Suppose that G is 4-colourable. Then 4-colour G with colours from $Z_2 \times Z_2$ so that vertex v_i has colour (a_i,b_i). Define the 0-chains f_1 and f_2 so that $f_1(v_i) = a_i$ and $f_2(v_i) = b_i$ for $i = 1, 2,..., |VG|$. Now suppose $e=v_iv_j \in EG$ and $\delta f_1(e) = \delta f_2(e) = 0$. Then $f_1(v_i) + f_1(v_j) = f_2(v_i) + f_2(v_j) = 0$ and so $(a_i,b_i) = (a_j,b_j)$. This contradiction shows that $(\delta f_1, \delta f_2)$ is a pair of coboundaries such that no edge of G has a zero coefficient in both.

Suppose that $\Delta(G)$ has a colouring $(\delta f_1, \delta f_2)$, $f_1, f_2 \in L_0(G)$. Now colour the vertex u, $(f_1(u),f_2(u))$. Consider an edge e = uv. Suppose u and v have the same colour. Then $f_1(u) = f_1(v)$ and $f_2(u) = f_2(v)$. Hence $\delta f_1(e) = \delta f_2(e) = 0$ which is a contradiction. Hence u and v have distinct colours and since G is assumed (trivially) to be connected, it follows that this is a 4-colouring of G.   □

A more intuitive way of looking at Lemma 4.1 is as follows. A colouring of $\Delta(G)$ is an ordered pair of coboundaries $(\delta f_1, \delta f_2)$. The supports of these coboundaries are the edges respectively, of a pair of spanning bipartite subgraphs H_1 and H_2, with $E(H_1) \cup E(H_2) = G$. Now if we 2-colour the vertices of H_1, red and white, and 2-colour the vertices of H_2, blue and green, then the vertices of G are 4–coloured with the colours (R,B), (R,G), (W,G) and (W,B).

Lemma 4.2 $\Delta(G)$ has a colouring if and only if a colour-space exists in $(\mathbf{P},\mathbf{S})$.

Proof: Suppose that $\mathbf{P}$ is d-dimensional. Recall from the last section, that a basis of $\Delta(G)$ can be chosen which produces a matrix M whose set of columns $\mathbf{S}$ generates $\mathbf{P}$.

Now there is a bijection between the non-zero elements of $\Delta(G)$ and the hyperplanes of $\mathbf{P}$; see Exercise 19. Furthermore, under this bijection, a non–zero 1-chain $f \in \Delta(G)$ is mapped onto a hyperplane V with the property that if $s_i \in \mathbf{S} \cap V$, then $f(e_i) = 0$.

Let (h_1,h_2) be a colouring of $\Delta(G)$. Then, under the above bijection, h_1 and h_2 are mapped onto distinct hyperplanes, U_1 and U_2 such that $U_1 \cap U_2$ contains no element of $\mathbf{S}$. For otherwise if $s_i \in \mathbf{S} \cap (U_1 \cap U_2)$, then $h_1(e_i) = h_2(e_i) = 0$ and yet (h_1,h_2) is a colouring of $\Delta(G)$.

So let $U = U_1 \cap U_2$. Then U is a (d - 2)-subspace which contains no point of $\mathbf{S}$. In other words, U is a colour-space of $(\mathbf{P},\mathbf{S})$. Hence given a colouring of $\Delta(G)$, there exists a colour-space of $(\mathbf{P},\mathbf{S})$.

On the other hand, given any colour-space U of $(\mathbf{P},\mathbf{S})$ it can be represented, not necessarily uniquely, as the intersection of two (d - 1)-subspaces of $\mathbf{P}$ and hence corresponds to a colouring of $\Delta(G)$ via the above bijection. □

Using the previous two lemmas we have shown that G is 4-colourable if and only if a colour-space exists in $(\mathbf{P},\mathbf{S})$. So considering $(\mathbf{P},\mathbf{S})$ as directly representing $G, (\mathbf{P},\mathbf{S})$ is a 2-block if and only if G is not 4-colourable.

Analogously a **colouring of** $\Gamma(G)$ is an ordered pair of cycles, (g_1,g_2), which are not both zero on any edge. Again, $\Gamma(G)$ has a colouring if and only if a colour-space exists in $(\mathbf{P},\mathbf{S})$, where $(\mathbf{P},\mathbf{S})$ represents G dually (see Exercise 20). Now if G is cubic, it is easy to see (Exercise 21) that G has a Tait colouring if and only if $\Gamma(G)$ has a colouring. So $(\mathbf{P},\mathbf{S})$, considered as dually representing G when G is cubic, is a 2-block if and only if G is not 3–edge–colourable.

Finally, although we do not attempt to prove them here, we present the following two results.

Lemma 4.3 If $(\mathbf{P},\mathbf{S})$ is a direct representation of a graph G and is also a tangential 2-block, then G is a contraction-minimal non 4-colourable graph. The converse is also true. □

Lemma 4.4 Let G be a cubic bridgeless graph. If **(P ,S)** is a dual representation of G and it is also a tangential 2-block, then G is a subcontraction-minimal non-Tait colourable graph. The converse is also true. □

By the Hadwiger conjecture for k = 5 (which in view of Lemma 2.5.2 and the Four Colour Theorem is known to be true) there exists only one sub-contraction-minimal non-4-colourable graph and that graph is K_5. Directly represented this gives the Desargues block. Tutte conjectures, Conjecture 3.1.9, that if G is cubic there exists only one subcontraction-minimal non-Tait colourable graph and that graph is P. Dually represented this gives the Petersen block.

Hence by the Four Colour Theorem, the Desargues block is the only graphic tangential 2-block. Tutte's Conjecture 1.11 implies that the only cographic tangential 2-block is the Petersen block. In fact, Seymour [pS 81b] shows that any new tangential 2-block must be cographic and hence that Conjectures 1.11 and 3.1 are equivalent.

5 . Epilogue
The reader should be struck by the prominent role played by P in this chapter. In fact herein lies the major claim for the centrality of P in the theory of graphs.

We note though, that P plays a dual role (sometimes precisely that) to K_5 in all the above discussions. For example, the Hadwiger conjecture states that

$$\chi(G) > 4 \text{ implies } G >_s K_5 \tag{i}$$

while Tutte's snark conjecture (Conjecture 3.1.9) states that

$$G \text{ is a snark implies } G >_s P. \tag{ii}$$

Corresponding statements may be found in both the contexts of nowhere-zero flows and the representation of graphs in projective spaces.

Suppose we call statements of the type of (i) above, 'vertex' statements while 'edge' statements will refer to those of the type of (ii). The classical way of transferring vertex statements to edge statements is by way of the line graph of a graph. Of course this connection between the right hand sides of (i) and (ii) holds since $P = L(K_5)$. Because of this very connection, P and K_5 share many properties. For instance, they have many of the same symmetry and spectral properties.

Now in Kuratowski's characterization of planarity, Theorem 1.6.3, the forbidden subgraphs are both K_5 and $K_{3,3}$. It is clear that $K_{3,3}$ is irrelevant to the Hadwiger conjecture since $\chi(K_{3,3}) = 2$. Is this the only reason why $L(K_{3,3}) \cong \overline{L(K_{3,3})}$ is relegated to a rather minor role in comparison to P in the theory of graphs? After all, $L(K_{3,3})$ and P also share many properties. Indeed $L(K_{3,3})$ is in some senses even more symmetrical than P; see [jS 74].

Finally we note that a **cycle cover** of a graph G is a set of cycles such that each edge of G belongs to at least one cycle of G. There is an intimate relationship between cycle covers and flows. This connection is explored in more detail in Section 9.10.

Exercises

1.   If $e \in EK_4$, then $<\{e\}>_2 = EK_4$.

2.   Let H be a non-null simple graph with $\delta(H) \geq 2$. Prove that there exists a subgraph B of H with at least 3 vertices so that B is 2-connected and at most one vertex of B is adjacent in H to vertices of H not in B.

   [Hint: A **block** is any maximal 2-connected induced subgraph. Define T to be the **block-tree** of H, where $VT = \mathbf{B} \cup \mathbf{C}$ and $\mathbf{B}$ is the set of blocks of H and $\mathbf{C}$ is the set of vertices of H which belong to at least two blocks. Two vertices of T are adjacent if and only if one vertex is an element $B \in \mathbf{B}$ and the other is an element of $\mathbf{C}$ which belongs to B in H. Consider any endvertex of T.]

3.   Using the notation of Lemma 2.1 prove that $F(G) \leq F(G_4)$.

4.   Prove that if G is a 3-connected graph, then G has a cycle C. Furthermore, prove that $<EC>_2$ is connected.

5.   Prove that if G is a 3-connected graph, then there exist vertex-disjoint cycles $C_1, C_2, \ldots, C_r$ such that $<\bigcup_{i=1}^{r} EC_i>_2 = EG$.
   [Hint: Choose $r \geq 1$ and as large as possible so that there exist vertex–disjoint cycles $C_1, C_2, \ldots, C_r$ with $<\bigcup_{i=1}^{r} EC_i>_2$ connected.

Put $X = \bigcup_{i=1}^{r} EC_i$ and $<X>_2 = Y$. Let U be the set of vertices of G incident with edges in Y and let $H = G \setminus U$. Suppose $VH \neq \emptyset$ and prove that if $v \in VH$, then v is adjacent in G to at most one vertex in U. Deduce that $\delta(H) \geq 2$.

Select a subgraph B of H satisfying the conditions of Exercise 2. Use the 3-connectedness of G to deduce that there exist distinct vertices $b_1, b_2 \in VB$, which are both adjacent in G to vertices in U.

Since B is a block, b_1 and b_2 lie on a cycle C_{r+1} entirely contained in B.]

6. Let $v \in VG$ and $e \in EG$. Define

$$\eta(e,v) = \begin{cases} 1 & \text{if } v \text{ is incident to } e, \\ 0 & \text{otherwise.} \end{cases}$$

Let $f \in L_1(G)$. Then the boundary $\partial f \in L_0(G)$ is defined by $\partial f(v) = \sum_{e \in EG} \eta(e,v) f(e)$. Let $g \in L_0(G)$. Then the coboundary $\delta g \in L_1(G)$ is defined by $\delta g(e) = \sum_{v \in VG} \eta(e,v) g(v)$.

Prove that (i) $\partial(f_1 + f_2) = \partial f_1 + \partial f_2$;
 (ii) $\delta(g_1 + g_2) = \delta g_1 + \delta g_2$.

Deduce that $\Delta(G)$ and $\Gamma(G)$ are abelian groups. Prove also that $\Delta(G)$ and $\Gamma(G)$ are subspaces of $L_1(G)$.

7. Let f be a non-zero 1-chain.

(a) Prove that $f \in \Gamma(G)$ if and only if Sup f induces an Eulerian subgraph of G.

(b) Is it true that $f \in \Delta(G)$ if and only if Sup f induces a bipartite subgraph of G?

8. Prove that (i) a 1-chain is a 1-cycle if and only if it is orthogonal to every coboundary;

(ii)   a 1-chain is a coboundary if and only if it is orthogonal to every 1-cycle.

[Hint: In an Eulerian graph an edge-cut has an even number of edges.]

9.   Let $f : EG \rightarrow PG(d,2)$ be a direct representation of G. Let F be an edge–maximal forest of G. Let f_i be the 1-chains defined in the text such that $Sup f_i = E[F_i, \overline{F_i}]$, i = 1, 2,..., $|EF|$ = r.

Prove that   (i) $|EF| = |VG| - \omega(G)$;
(ii) $f_i \in \Delta(G)$, i = 1, 2 ,...., r;
(iii) $f_1, f_2,...,f_r$ are linearly independent.

10.   Let G be a bridgeless graph. Let g: $EG \rightarrow PG(d,2)$ be a dual representation of G. Let F be an edge-maximal forest of G. Let g_i be the 1-chains defined in the text such that $Sup g_i = EC_i$, i = 1,2,...,$|EG\setminus EF|$ = s.

Prove that   (i) $|EG \setminus EF| = |EG| - |VG| + \omega(G)$;
(ii) $g_i \in \Gamma(G)$, i = 1, 2,..., s;
(iii) $g_1, g_2,..., g_s$ are linearly independent.

11.   Let G a bridgeless graph. Let f: $EG \rightarrow PG(d,2)$ be a direct representation of G. Let M be its associated matrix and $X = (x_1, x_2,...,x_n) \in PG(n-1,2)$. Using Exercise 8, deduce that X is a solution of MX=0 if and only if $X \in \Gamma(G)$.

12.   Let G be a bridgeless graph. Consider, see Exercise 6, $\Delta(G)$ and $\Gamma(G)$ as subspaces of the vector space $L_1(G)$. Prove that the dimension of $\Delta(G)$ is r and the dimension of $\Gamma(G)$ is s. Further prove that $\{f_1, f_2,...,f_r\}$ and $\{g_1, g_2,...,g_s\}$ are bases of $\Delta(G)$ and $\Gamma(G)$, respectively. Deduce that $\Delta(G)$ and $\Gamma(G)$ have (projective) dimension r - 1 and s - 1 respectively.

[Hint: Use Exercises 9, 10 and 11.]

13.   Let f: $EG \rightarrow PG(d,2)$ be a direct representation of G.

Prove that   (i)   no element $s \in S$ is the zero vector;
(ii) $s_i \neq s_j$ for $i \neq j$.

[Hint: (i) Recall that G is loopless. (ii) Otherwise, e_i and e_j must have equal coefficients in every coboundary and so both edges must have the same endvertices.]

14. Let g: EG → PG(d,2) be a dual representation of G.

    Prove that no element of S* is the zero vector.

    [Hint: Recall that in the dual representation we always assume that G is bridgeless.]

    [Remark: The analogue to Exercise 13(ii) is false. However, for all distinct i, j, $s_i^* \neq s_j^*$ if and only if no coboundary exists with exactly two non-zero coefficients. No difficulty arises in the subsequent theory even if this condition does not hold. We can always contract one edge of a 2-edge-cut as in Lemma 2.1.]

15. Prove that the Fano, Desargues and Petersen blocks are 2-blocks.

16. Prove that the Fano, Desargues and Petersen blocks are tangential.

17. Prove that the direct representation of P and the dual representation of K_5 are not tangential 2-blocks.

18. Prove that if G is connected then there are exactly two 0-chains with a given coboundary.

19. Suppose (P,S) directly represents a graph G. Prove that there exists a bijection from the set of non-zero elements of Δ(G) to the set of hyperplanes of P.

    [Hint: Let U be a (d-1)-subspace of P. Let f be a 1-chain defined by $f(e_i) = 0$ if and only if $s_i \in U$. Prove that $f \in \Delta(G)$. Remember Exercise 8(ii) and use a dimensionality argument. Notice finally that $|\Delta(G)| = 2^r - 1$, the number of hyperplanes.]

20. Suppose (P,S) dually represents a graph G. Prove that Γ(G) has a colouring if and only if a colour-space exists in (P,S).

21. Let G be a cubic graph. Prove that G has a Tait colouring if and only if Γ(G) has a colouring.

22. Suppose that G is a connected graph. Prove that F(G) = 2 if and only if G is Eulerian.

23. Suppose that G is a bridgeless cubic graph. Prove that F(G) = 3 if and only if G is bipartite.

24. Let G be a graph. An **Eulerian subgraph** of G is a spanning subgraph in which each vertex has even degree. Prove that G has a nowhere-zero 4–flow if and only if G is the union of two Eulerian subgraphs.

     [Hint: Suppose that φ is a nowhere-zero 4-flow for G. Let φ_1 be a nowhere-zero 2-flow on the subgraph induced by $E_1 = \{e \in EG : \varphi(e)$ is odd$\}$. Now put $E_2 = \{e \in EG : \varphi(e) + \varphi_1(e) = \pm 2\}$.]

25. Prove that Conjecture 1.12 is equivalent to the following conjecture. If G is a bridgeless graph which has no subgraph contractible to P, then G is the union of two Eulerian subgraphs; see [pS 81b].

26. Grötzsch [hG 59] proved that if G is a planar bridgeless graph which has no edge-cut of size three then $\chi^*(G) \leq 3$. Explain why it is natural to conjecture that "every bridgeless graph which has no edge-cut of size three has a nowhere-zero 3-flow".

References

[aB 83]   A. Bouchet, Nowhere-zero integral flows on a bidirected graph, *J. Comb. Th.*, B, 34, 1983, 279-292.

[bD 76]   B.T. Datta, Non-existence of six-dimensional tangential 2-blocks. *J. Comb. Th.*, B, 21, 1976, 171-193.

[bD 81]   B.T. Datta, Non-existence of seven-dimensional tangential 2–blocks, *Disc. Math.*, 36, 1981, 1-32.

[hG 59]   H. Grötzsch, Zur theorie der diskreten Gebilde VII, Ein Dreifarbensatz für dreikreisfreie Netze auf der Kugel, *Wiss. Z. Martin-Luther-Univ. Halle-Wittenberg Math. Natur. Reihe*, 8, 1958/1959, 109-120.

[rH 89]   R. Häggvist, Decompositions of complete bipartite graphs, *Surveys in Combinatorics* (ed. J. Siemens), London Math. Soc. Lecture Note Series 141, C.U.P., Cambridge 1989, 115-146.

[fJ 76]   F. Jaeger, On nowhere-zero flows in multigraphs, Proc. Fifth
              British Comb. Conf., *Utilitas Math.*, Winnipeg, 1976, 373-379.

[fJ 79]   F. Jaeger, Flows and generalized coloring theorems in graphs, *J.
              Comb. Th.* B, 26, 1979, 205-216.

[aK 87]   A. Khelladi, Nowhere-zero integral chains and flows in bidirected
              graphs, *J. Comb. Th.* B, 43, 1987, 95-115.

[L-T-S 88]   C. Lam, L. Thiel and S. Swiercz, A computer search for a
              projective plane of order 10, *Algebraic, Extremal and Metric
              Combinatorics* 1986 (ed. by M. Deza, P. Frankl and
              I.Rosenberg), London Math. Soc. Lecture Note Series 131,
              C.U.P., Cambridge 1988, 155–165.

[L-T-S 89]   C. Lam, L. Thiel and S. Swiercz, The non-existence of finite
              projective planes of order 10, *Can. J. Math.*, 41, 1989,
              1117–1123.

[M-C-B 88]   M. Möller, H. Carstens and G. Brinkmann, Nowhere-zero flows in
              low genus graphs. *J. Graph Th.*, 12, 1988, 183-190.

[M-L-T-S 88] J. McKay, C. Lam, L. Thiel and S. Swiercz, The projective plane
              of order 10 does not exist, submitted.

[pS 81a]   P.D. Seymour, Nowhere-zero 6-flows, *J. Comb. Th.*, B, 30,
              1981, 130-135.

[pS 81b]   P.D. Seymour, On Tutte's extension of the Four Colour Theorem,
              J. Comb. Th., B, 31, 1981, 82-94.

[jS 74]   J. Sheehan, Smoothly embeddable subgraphs, *J. Lond. Math.
              Soc.*, 9, 1974, 212-218.

[rS 84]   R. Steinberg, Tutte's 5-flow conjecture for the projective plane,
              J.Graph Th., 8, 1984, 277-285.

[wT 49]   W.T. Tutte, On the imbeddings of linear graphs in surfaces, *Proc.
              Lond. Math. Soc.*, 51, 1949, 474-483.

[wT 54]   W.T. Tutte, A contribution to the theory of chromatic polynomials,
              Can J. Math., 6, 1954, 80-91.

[wT 65]   W.T. Tutte, Lectures on matroids, *J. Res. Nat. Bur. Stands.*, 69B, 1965, 1-47.

[wT 66]   W.T. Tutte, On the algebraic theory of graph colorings, *J. Comb. Th.*, 1, 1966, 15-50.

[wT 69]   W.T. Tutte, A geometric version of the four colour problem with discussion, *Combinatorial Maths and Its Applications*, Univ. Nth. Carolina Press, Chapel Hill, 1969, 553-561.

[wT 78]   W.T. Tutte, Colouring problems, *Math. Intelligencer*, 1, 1978, 72-75.

[dW 76]   D.J.A. Welsh, *Matroid Theory*, Academic Press, London, 1976.

[dW 79]   D.J.A. Welsh, Colouring problems and matroids, *Surveys in Combinatorics*, (Proc. Seventh British Comb. Conf, Ed. B. Bollobás), London Math. Soc. Lecture Note Series 38, C.U.P., Cambridge, 1979, 229-257.

[dW 80]   D.J.A. Welsh, Colourings, flows and projective geometry, *Nieuw Archief Voor Wiskunde*, 28, 1980, 159-176.

[dY 83]   D. Younger, Integer flows, *J. Graph Th.*, 7, 1983, 349-357.

6
Cages

0 . Prologue

Now Albert had heard about Lions,
How they was ferocious and wild -
To see Wallace lying so peaceful,
Well, it didn't seem right to the child.

So straightway the brave little feller,
Not showing a morsel of fear,
Took his stick with its 'orse's 'ead 'andle
And pushed it in Wallace's ear.

You could see that the Lion didn't like it,
For giving a kind of a roll,
He pulled Albert inside the cage with 'im,
And swallowed the little lad 'ole.

(From "The lion and Albert" by Marriott Edgar (1932))

Cubic graphs are a class of graphs which have attracted considerable interest, not least because of the intimate relationship between planar cubic graphs and the Four Colour Theorem. One particular field of interest has always been the symmetries of cubic graphs. Another has been the existence of cubic graphs with specified constraints on their diameter or girth. These two fields of interest coincide in the study of cages i.e. the smallest possible cubic graphs with given constraints on their girth. In a certain sense some cages turn out to be the most symmetric graphs of all. Of course, P is a cage.

1 . Cages
A regular graph of degree 1 has no cycles. A regular graph of degree 2 has arbitrary girth. So we define a graph which is regular of degree r and of girth g to be an **(r,g)-graph** only for $r \geq 3$. Clearly g is also greater than or equal to 3. By the arguments of Section 1.5 the Petersen graph is a (3,5)-graph.

We first prove a result which links the number of vertices of an (r,g)-graph with r and g.

Theorem 1.1 Let G be an (r,g)-graph. Then $|V(G)| \geq f_0(r,g)$ where

$$f_0(r,g) = \begin{cases} 1 + r[(r-1)^{(g-1)/2} - 1]/(r-2), & g \text{ odd}, \\ 2[(r-1)^{g/2} - 1]/(r-2), & g \text{ even}. \end{cases}$$

Proof: (i) Suppose that g is odd. Let g = 2d + 1 (d≥1). Then we consider the vertices at most d from a fixed vertex v; see Figure 1.1(a) with r = 4, g = 5. Since there are no cycles smaller than g, it follows that at the first 'level' there is one vertex, the vertex v; at the second level there are r vertices; at the third, r(r-1); at the fourth, $r(r-1)^2$ and so on up to the d-th level. In total therefore

$$|VG| \geq 1 + r + r(r-1) + r(r-1)^2 + ... + r(r-1)^{d-1}$$

$$= 1 + r[(r-1)^d - 1]/(r-2).$$

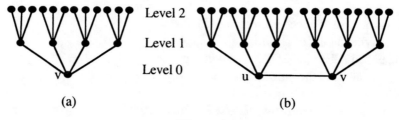

Level 2
Level 1
Level 0

(a)   (b)

Figure 1.1

(ii)   Suppose g is even. Let g = 2d (d≥2). Choose an edge uv. Then we consider the vertices at most distance d-1 from either of u and v, see Figure 1.1 (b), and repeat the argument of (i).   □

The first question that naturally arises is whether for each r and g, an (r,g)–graph exists. The answer is yes and it is fairly easy to give constructions for such graphs; see [E-S 63], [wT 66], [nS 67] [gW 73]. We will give only one of these constructions here, see Section 2, because in general they are not very revealing. The next result, although not addressing the question of existence directly, provides some insight into the question. To be honest though, the simplicity and beauty of its proof demands its inclusion; see [bB 78, p.104].

Theorem 1.2 Suppose r ≥ 4, g ≥ 3. Then there exists a graph G with δ(G)≥r and γ(G) ≥ g. Futhermore $|VG| \leq (2r)^g$.

Proof: Let $N = (2r)^g$. Let S(N) be the set of graphs (labelled) with vertex set {1,2,...,N} and with $|EG| = rN$. Then

$$|S(N)| = \left(\!\! \binom{N}{2} \atop rN \!\! \right). \tag{1}$$

Suppose $N \geq m \geq 3$. Then on these vertices, there are $\binom{N}{m}$ sets of m-subsets of vertices and on each of these subsets there are $((m-1)!)/2$ possible cycles of length m. So in total there are

$$\frac{1}{2}((m-1)!)\binom{N}{m} < \frac{1}{2m}N^m \tag{2}$$

possible choices for an m-cycle. Furthermore if we consider a fixed m-cycle then there are exactly

$$\left(\!\! \binom{N}{2} - m \atop rN - m \!\! \right) \tag{3}$$

graphs in S(N) which contain this cycle. This is because, having selected a fixed m edges, there are $rN - m$ edges which have to be chosen from $\binom{N}{2} - m$. Now let X denote the set of all possible m-cycles and let $Y = S(N)$. Consider the bipartite graph $\mathbf{B} = (X,Y)$ where $xy \in EB$ ($x \in X, y \in Y$) if and only if y contains x as a subgraph. Then, from (2) and (3)

$$|E(Y,X)| = |E(X,Y)| < \frac{N^m}{2m}\left(\!\! \binom{N}{2} - m \atop rN - m \!\! \right). \tag{4}$$

Hence the average degree α of $y \in Y$ satisfies, from (1) and (4)

$$\alpha < \frac{N^m}{2m}\left(\!\! \binom{N}{2} - m \atop rN - m \!\! \right) \Big/ \left(\!\! \binom{N}{2} \atop rN \!\! \right). \tag{5}$$

Thus (5) gives an upper bound for the average number of cycles of length m across all graphs in S(N). Now, see Exercise 4,

$$\sum_{m=3}^{g-1} \left(\frac{N^m}{2m}\left(\!\! \binom{N}{2} - m \atop rN - m \!\! \right) \Big/ \left(\!\! \binom{N}{2} \atop rN \!\! \right) \right) < N. \tag{6}$$

From (6) we deduce that there exists $G \in S(N)$ which contains at most $N - 1$ cycles whose length is at most $g - 1$. So in G we can ensure that $\gamma(G) \geq g$ by deleting no more than $N - 1$ edges. Hence there exists a subgraph H of G with at least $rN - (N-1) = (r-1)N + 1$ edges. Moreover, H has N vertices, more than $(r-1)N$ edges and $\gamma(H) \geq g$. Let

$H = \{G : |VG| \geq r - 1, |EG| \geq (r-1)|V(G)|\}$.

Then $H \in \mathbf{H}$. Also obviously if $K \in \mathbf{H}$ then $|VK| > r - 1$ for, if $|VK| = r - 1$, then $\binom{r-1}{2} \geq |EK| \geq (r-1)(r-1)$ which is a contradiction. Now suppose $\delta(H) \leq r - 1$. Select a vertex $v \in V(H)$ with $\deg v \leq r - 1$. Then $H - v$ has $N - 1$ vertices and more than $(r-1)(N-1)$ edges. Hence $H - v \in \mathbf{H}$. By repeating this process of deleting vertices of $\deg v \leq r - 1$, we must eventually obtain a subgraph $H^* \subseteq H \subseteq G$ such that $\delta(H^*) \geq r$, for we know that at each stage of the process we remain within $\mathbf{H}$ and yet each element of $\mathbf{H}$ has at least r vertices. Hence the process terminates when there remain no vertices of degree less than or equal to $r - 1$. Finally H^* has at most $N = (2r)^g$ vertices, $\delta(H^*) \geq r$ and $\gamma(H^*) \geq g$.   □

Let $f(r,g)$ be the number of vertices of a smallest (r,g)-graph i.e.$f(r,g)=\min\{|VG| : G$ is an (r,g)-graph$\}$. Bannai and Ito [B-I 81] have proved that $f(r,g) \neq f_0(r,g) +1$ for odd $g \geq 5$. An (r,g)-graph G is called an **(r,g)-cage** if $|VG| = f(r,g)$. Furthermore it is called a **Moore graph** ([H–S60]) if $|V(G)| = f(r,g) = f_0(r,g)$. Sometimes we will say a Moore graph is odd or even according as its girth g is odd or even. The most famous of all Moore graphs is the Petersen graph. To verify this we observe that P is a $(3,5)$–graph and from Theorem 1.1, $f_0(3,5) = 10 = |VP|$. Hence $f(3,5) = f_0(3,5)$ and so P is a $(3,5)$-cage and an (odd) Moore graph.

We shall see that there are suprisingly few Moore graphs. In fact, see Theorem 1.3, if a Moore graph of girth 5 exists then $r \in \{3,7,57\}$. When $r = 3$ such a graph exists as we have seen and P is that unique graph. The Hoffman-Singleton graph, Figure 3.11, is the unique $(7,5)$-Moore graph. It is unknown whether there exists a $(57,5)$-Moore graph. It is a considerable time since the question of the existence of the $(57,5)$-Moore graph was raised and it seems clear that the problem is not an easy one.

Sometimes in the literature (e.g. [H-S 60]) Moore graphs are considered from a different viewpoint. Suppose we consider an (r,g)-graph where $g=2d+1(d\geq1)$. Now if G is any regular graph of degree r and diameter d then $|VG| \leq f_0(r,g)$; see the proof of Theorem 1.1. Here the **diameter**, $\text{diam}(G)$, is defined by $d = \text{diam}(G) = \max \{d(u,v): u,v \in VG\}$. Therefore the maximum number of vertices in any such graph is at most $f_0(r,g)$. Hence such an odd Moore graph has the maximum number of vertices consistent with its degree and diameter constraints and the minimum number of vertices consistent with its degree and girth constraints. Similar remarks apply to even Moore graphs.

Suppose that $v_1, v_2,..., v_n$ is an ordering of the vertices of a graph G. Let M be the matrix whose (i,j)-th element is 1 if $v_i v_j \in EG$ and 0 otherwise. Then M is said to be an **adjacency matrix** of G.

Theorem 1.3 Suppose that $r \geq 3$. If a Moore graph of girth 5 and degree r exists then $r \in \{3,7,57\}$.

Proof: Let G be an (r,5)-graph. Suppose G is a Moore graph. Then, from Theorem 1.1, $|V(G)| = f_0(r,5) = r^2 + 1$. Write $n = r^2 + 1$. Let M be the adjacency matrix of G and let J be the $n \times n$ matrix whose elements are all equal to 1. We have

$$M^2 + M - (r-1)I = J. \tag{1}$$

Now to justify (1) we inspect the paths between the vertices in G. The (i,j)-th element in M and M^2 is the number of trails (i.e. paths with possibly repeated vertices and edges) of lengths 1 and 2 respectively, between the vertices v_i and v_j. So in M the (i,j)-th element is 0 or 1, according as $v_i v_j \notin EG$ or $v_i v_j \in EG$ and in M^2 it is 0 or r, according as $v_i v_j \in EG$ or $i = j$. Since $\gamma(G) = 5$ and diam G = 2 the (i,j)-th element of M^2 is 1 if $i \neq j$ and $v_i v_j \notin E(G)$. In all cases then, the (i,j)-th element of the left hand side of (1) is equal to 1.

Having established (1), the following is typical of all 'eigenvalue arguments' in this area. The graph-theoretic part of the argument is now completely over.

The eigenvalues λ of J, and their multiplicities $m(\lambda)$, are obtained by direct inspection. We find J has an eigenvalue n with $m(n) = 1$ and an eigenvalue 0 with $m(0) = n - 1$. On the other hand, if λ is an eigenvalue of M, then $\lambda^2 + \lambda - (r-1)$ is an eigenvalue of $M^2 + M - (r - 1)I$ i.e. $\lambda^2 + \lambda - (r-1)$ is an eigenvalue of J. Certainly, since G is r-regular, r is an eigenvalue of M; check, for example, that $MJ_0 = rJ_0$ where here J_0 is the $n \times 1$ matrix all of whose entries are 1. Since $r^2 + r - (r-1) = r^2 + 1 = n$, it follows that if $\lambda \neq r$ and λ is an eigenvalue of M, then

$$\lambda^2 + \lambda - (r-1) = 0. \tag{2}$$

From (2), $\lambda = (-1 \pm \sqrt{4r-3})/2$. Write $\lambda_1 = (-1 + \sqrt{4r-3})/2$ and $\lambda_2 = (-1 - \sqrt{4r-3})/2$. By direct inspection of M, $m(r) = 1$. Hence we obtain

$$1 + m_1 + m_2 = n = r^2 + 1, \tag{3}$$

where $m(\lambda_1) = m_1$ and $m(\lambda_2) = m_2$. Because the trace of M is 0,

$$r + m_1\lambda_1 + m_2\lambda_2 = 0. \tag{4}$$

Write $s = \sqrt{4r - 3}$. Then, from (3) and (4),

$$
\begin{aligned}
0 &= 2r + m_1(-1 + s) + m_2(-1 - s) = 2r - (m_1 + m_2) + s(m_1 - m_2) \\
&= 2r - r^2 + s(m_1 - m_2). \tag{5}
\end{aligned}
$$

From (5), if s is irrational, $r = 2$ which is a contradiction. Therefore s is rational. So

$$r = \tfrac{1}{4}(s^2 + 3) \tag{6}$$

and from (3), (5) and (6),

$$0 = 2r - r^2 + s(m_1 - (r^2 - m_1))$$

$$= \tfrac{1}{2}(s^2 + 3)/2 - \tfrac{1}{16}(s^2 + 3)^2 + 2m_1 s - \tfrac{1}{16}s(s^2 + 3)^2. \tag{7}$$

From (7),

$$s^5 + s^4 + 6s^3 - 2s^2 + (9 - 32m_1)s = 15. \tag{8}$$

Thus s divides the left hand side of (8) and so s divides 15. From (6), $s \neq 1$. Hence the only possibilites for s are 3, 5, 15 which means from (6) that $r \in \{3,7,57\}$.   □

Theorem 1.3 states that $f(r,5) \geq r^2 + 2$ unless possibly $r \in \{3,7,57\}$. Brown [wB 67b] showed that with these exceptions, $f(r,5) \geq r^2 + 3$. Robertson, Figure 3.7, showed that $f(4,5) = 19 = 4^2 + 3$ and proved thereby, that sometimes equality does hold. However, Kovacs [pK 81] proved that there are an infinite number of values of r for which $f(r,5) > r^2 + 3$.

Theorem 1.4 Suppose r is odd and $r \notin \{m^2 + m + 3, m^2 + m - 1, 7, 57\}$ for any $m \geq 0$. Then $f(r,5) \geq r^2 + 5$.

Proof: Again, like Brown's, the argument is an eigenvalue argument and exactly parallels the argument of Theorem 1.3. We simply sketch the proof. We know already that for these choices of r, using Theorem 1.3 and Brown's result, that $f(r,5) \geq r^2 + 3$. So assume that there exists an $(r,5)$-graph G, with $|VG| = r^2 + 3$ and r is constrained as in the statement. By an argument similar to that of Theorem 1.1, if $v \in VG$ there exist exactly two vertices at distance three from v. Let H be the graph whose vertex set is VG and where h_1 is adjacent to h_2 if and only if $d_G(h_1,h_2) = 3$. Then H is simply a union of cycles i.e. $\bigcup\limits_{i=1}^{b} C_{a_i}$ and $\sum\limits_{i=1}^{b} a_i = |VG|$. Let M, B be the adjacency matrices of G, H, respectively.

Repeating the argument of Theorem 1.3 which established equation (1) of that theorem, we have

$$M^2 + M - (r - 1)I = J - B.$$

The reason why the 'spectral' analysis of this equation is relatively easy is that B is a direct sum of circulant matrices and so the spectral properties of B are well-known; see [nB 74]. It turns out that M has (i) an eigenvalue r with $m(r) = 1$; (ii) $b - 1$ eigenvalues occurring as roots of $\lambda^2 + \lambda - (r-1) = -2$ and (iii) one eigenvalue satisfying each of the equations $\lambda^2 + \lambda - (r-1) = -2 \cos\left(\dfrac{2\pi j}{a_i}\right)$, $j = 1, 2,$ ... , $a_i - 1$; $i = 1, 2, ... , b$.

All of this was contained in [wB 67b]. The analysis now continues as in Theorem 1.3 although it is somewhat more complicated.   □

It is true to say that very little is known about the values of $f(r,g)$. The evidence, for what it is worth, seems to suggest that $f(r,g)$ is in general much closer to $f_0(r,g)$ than to the present upper bounds. These upper bounds will be discussed later. However, Theorem 1.2 gives some idea of the order of our present upper bounds. They are at worst the square of the lower bounds. This adds interest to Kovac's theorem since, although it yields only a constant increase over $f_0(r,g)$, there is some feeling that $f_0(r,g)$ is the correct order of magnitude. We can summarize the available evidence [rS 63], [F-H 64], [cB 66], [rS 66], [mA 71], [B-I 73], [rD 73] as follows.

1.   If there is an even (r,g)-Moore graph, then $g \in \{4,6,8,12\}$. Even Moore graphs exist when:
(a) $g = 4, r \geq 3$; $K_{r,r}$ (unique);
(b) $g = 6$; for all r such that there exists a projective plane, see Section 5.3, of order $r - 1$ (this construction is discussed below) and so in particular when $r - 1$ is a prime power;
(c) $g = 8$; for all r such that there exists a projective geometry of dimension 4 over $GF(r-1)$;
(d) $g = 12$; for all r such that there exists a projective geometry of dimension 6 over $GF(r - 1)$.

2.   If there is an odd (r,g)-Moore graph, then $g \in \{3,5\}$. Odd Moore graphs
exist when:
(a) $g = 3, r \geq 3$; K_{r+1} (unique);
(b) $g = 5, r \in \{3,7\}$; P (unique), Hoffman-Singleton (unique);
(c) $g = 5, r = 57$ undecided, although if it exists, it is not distance transitive; see
        Chapter 8.

The Moore graphs of girth 6 are interesting. Suppose **P**(r) is any projective
plane of order r - 1. Consider the bipartite graph (even Moore graphs are always
bipartite) with vertex set equal to the union of the sets of points and lines of **P**(r).
A point is joined (in the graph) to a line if and only if the point is incident (in the
projective plane) to the line. There are certainly non-isomorphic projective planes
of the same order (for example when the order is 9). Non-isomorphic projective
planes correspond to non-isomorphic graphs and so the constructions do not in
general lead to uniqueness. The Heawood graph is however the unique graph
corresponding to **P**(3) since **P**(3) is unique; see Figure 1.2(a) and Exercise 2.
We may regard **P**(3) as the 2-dimensional projective geometry over GF(2).
Coordinatized in the usual way, its points are $1 = (1,0,0)$, $2 = (1,0,1)$.
$3=(1,1,0)$, $4 = (1,1,1)$, $5 = (0,1,0)$, $6 = (0,1,1)$, $7 = (0,0,1)$ and its lines are
127, 135, 146, 236, 245, 347, 567. The point-line incidence graph of **P**(3) is
shown in Figure 1.2(b). The edges at level 2 are omitted but the rule for
incidence should be clear; e.g. 3 is joined to 136 because $3 \in \{1,3,6\}$. The
bipartition is indicated by solid and empty circles.

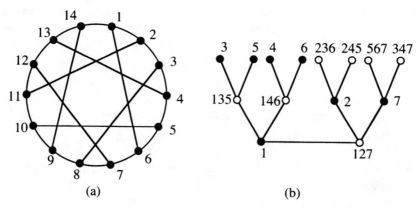

(a)   (b)

Figure 1.2

2.   Upper Bounds for f(r,g)
Realistically we know very little indeed about the values for $f(r,g)$. We suspect
that $f(r,g)$ is always quite close to $f_0(r,g)$. Table 1 indicates the very slight
evidence for this suspicion. Where two numbers are indicated those are the known
bounds, the second coordinate being approximate. More detail about this table is
included in Section 3.

As can be seen from even these small values, the distance between the bounds is ludicrously large. In fact the best general upper bound is when $r = 3$ where $f(3,g) \le 2^{g-1}$. Otherwise, the upper bound for $f(r,g)$ is at best almost of the order of $(f_0(r,g))^2$.

r \ g	3	4	5	6	7	8	9	10
3	4	6	10	14	24	30	58	70
4	5	8	19	26	(53,728)	80	$(161,2.3^7)$	$(242,4.3^7)$
5	6	10	30	42	(106,2730)	170	$(426,2.4^7)$	$(682,4^9)$
6	7	12	40	62	(187,7812)	312	$(937,2.5^7)$	$(1562,5^8)$
7	8	14	50	90	(302,18662)	(518,37324)	$(1814,2.6^7)$	$(3110,6^8)$

Values of f(r,g) for small r and g

Table 1

The next theorem is extremely useful since it allows us to sometimes replace 'having girth g' with 'having girth at least g'.

Theorem 2.1 Suppose $g \ge 3$. Let $C(r,g) = \{G: G$ is an (r,γ)-graph, $\gamma \ge g\}$. Then
(i) $f(r,g) = \min \{|VG|: G \in C(r,g)\}$;
(ii)   if $G \in C(r,g)$ and $|VG| = f(r,g)$ then G is an (r,g)-cage.

Proof: Suppose that $G \in C(r,g)$ and $|VG| = \min\{|VG| : G \in C(r,g)\}$. Then G is an (r,γ)-graph, $\gamma \ge g$. Suppose we can show that $\gamma = g$. Then G is an (r,g)–graph and, since all (r,g)-graphs are members of $C(r,g)$, it follows that $f(r,g) = |VG|$ and we are done.

If $g = 3$ then $K_{r+1} \in C(r,3)$ and $f(r,3) = r + 1$. Furthermore K_{r+1} is the unique $(r,3)$-cage. So now assume that $g \ge 4$. Firstly suppose that r is even. Let $u \in VG$. Let G^* be the graph with $VG^* = VG \setminus \{u\}$ and edges consisting of the union of $E(G \setminus \{u\})$ together with an arbitrary set of $(\deg_G u)/2$ independent edges joining pairs of vertices in $N_G(u)$. Notice that G^* contains no multiple edges, as G contains no triangles. Since G^* is r-regular and $|VG^*|<|VG|$, $\gamma(G^*) < g$. Let C be a cycle in G^* of order less than g. This cycle must contain some of the new edges introduced between the neighbours of u. Let $v_iw_i, i = 0, 1,..., t - 1, 0 \le t \le \frac{1}{2}r$, be the 'new' edges used in C and let A_i be the subpaths of C joining w_i to v_{i+1}, addition on the subscripts being modulo t. There is no loss of generality here in assuming the cycle runs from w_i through A_i to v_{i+1}, since otherwise we relabel. Now in G, A_i together with uw_i and uv_{i+1}, form a cycle which must be of length at least g. Hence $|EA_i|\ge g- 2$. So

$g > |EC| \geq t(g - 2) + t = t(g - 1)$

which means $t = 1$ and $|EA_0| = g - 2$. Hence uv_0, uw_0 together with A_0, form a cycle of order g and the result follows.

Suppose r ($r \geq 5$) is odd. Then our argument is similar to the even case. Choose an edge $uv \in EG$. Let G^* be the graph with $VG^* = VG \setminus \{u,v\}$ and edges consisting of the union of $E(G \setminus \{u,v\})$ together with an arbitrary set of $\frac{1}{2}$(deg u-1) independent edges joining the elements of $N_G(u) \setminus \{v\}$ in pairs, and an arbitrary set of $\frac{1}{2}$(deg v-1) independent edges joining the elements of $N_G(v) \setminus \{u\}$ in pairs. Again G^* must contain a cycle C of order less than g and this cycle must contain some of the new edges introduced. The only difference now is that the subpaths A_i of C may have first vertex $w_i \in N_G(u) \setminus \{v\}$ and final vertex $v_{i+1} \in N_G(v) \setminus \{u\}$. In this case we are only guaranteed that $|EA_i| \geq g - 3$. Even so we have

$g > |EC| \geq t(g-3) + t = t(g-2)$

which means $t = 1$ and $|EA_0| \in \{g-3, g-2\}$. Now A_0 is a $w_0 v_0$-path. So, without loss of generality, we may assume that $w_0, v_0 \in N_G(u) \setminus \{v\}$, in which case A_0, together with uw_0 and uv_0, forms a cycle of order g.   □

Corollary 2.2 $f(r,g) < f(r,g+1)$.

Proof: Suppose that G is an $(r,g+1)$-cage. Then $G \in C(r,g)$. Hence $f(r,g) \leq |VG| = f(r,g + 1)$. If $f(r,g + 1) = f(r,g)$, then $|VG| = f(r,g)$ and, by the theorem, G is an (r,g)-cage. This is impossible since $\gamma(G)$ is unique. Therefore $f(r,g) < f(r,g+1)$.   □

We do not know if $f(r,g) \leq f(r+1,g)$ but to provoke interest we conjecture:

Conjecture 2.3 $f(r,g) < f(r+1,g)$.   □

Theorem 2.4 Suppose g is odd. Then $f(r,g+1) \leq 2f(r,g)$.

Proof: Let G be an (r,g)-cage with $VG = \{u_i: i=1,2,...,n\}$. Form the bipartite graph B, with $VB = \{u_i,v_i: i=1,2,...,n\}$ and $EB = \{u_i v_j: u_i u_j \in E(G), 1 \leq i,j \leq n\}$. Then $|VB| = 2|VG|$ and B is r-regular. Further, for t odd, $(u_1,u_2,...,u_t)$ is a t–cycle in G if and only if $(u_1,v_2,u_3,...,u_t,v_1,u_2,...,v_t)$ is a 2t-cycle in B. For t even, $(v_1,v_2,...,v_t)$ is a t-cycle in G if and only if $(u_1,v_2,u_3,...,v_t)$ and $(v_1,u_2,v_3,...,u_t)$ are t-cycles in B. So if G contains a cycle of order $g + 1$, then, since g is odd, B contains a cycle of order $g + 1$ also. In this case, $\gamma(B) \geq g + 1$ since any cycle of G of order g corresponds to a cycle of order

2g in B and any cycle of order $s \geq g + 2$ corresponds to a cycle(s) of order at least s. So B is an (r,s)-graph for some $s \geq g + 1$. Therefore, using Corollary 2.2,

$$2f(r,g) = 2|VG| = |VB| \geq f(r,g) \geq f(r,g + 1). \qquad \square$$

Theorem 2.5

$$f(r,g) \leq \begin{cases} 2[(r-1)^{g-1} - 1] \, / \, (r-2), & g \text{ odd}, \\ 4[(r-1)^{g-2} - 1] \, / \, (r-2), & g \text{ even}. \end{cases}$$

Proof: Firstly suppose g is even. Then, from Theorem 2.4,

$$f(r,g) \leq 2f(r,g-1),$$

where g - 1 is odd. Hence the case g even, is a direct consequence of the case g odd. Now suppose g is odd. The strategy is then to construct an r-regular graph of girth at least g and to use Theorem 2.1. We do not give the details of Sauer's construction here; see [nS 67]. $\square$

Other improvements to these upper bounds include (i) $f(r,5) \leq 2r(r-1)$, r a prime [gW 73]; (ii) $f(r,6) < 4r^2$ [wB 67a]. We are content to give just one example of an upper bound construction which proves that if p is a prime ($p \geq 3$) then $f(p,5) \leq 2p(p-1)$.

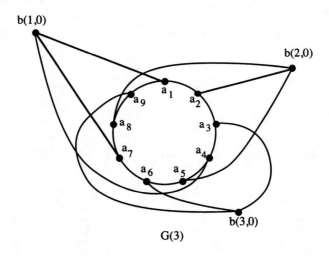

G(3)

Figure 2.1

Take p to be a prime ($p \geq 3$). Let $G = G(p)$ be the graph defined by (i) $VG = A \cup B$, A $= \{a_1, a_2, ..., a_{p2}\}$, B $= \{b(i,j) : i = 1, 2, ..., p; j = 0, 1, ... , p-3\}$

and $A \cap B = \emptyset$, (ii) $<A> \cong C_{p^2}$, $a_i a_{i+1} \in EG$, $i = 1, 2, ..., p^2$, where the indices are modulo p^2, (iii) $E<B> = \emptyset$, (iv) $a_k b(n + mj, j) \in E(G)$ if and only if $k = mp + n$ $(0 < n \leq p)$, $j = 0, 1, ..., p - 3$, and in the first b-coordinate addition is modulo p.

It is easy to verify that $G(p)$ is regular of degree p. It is quite simple but needs more effort to verify that $\gamma(G(p)) = 5$; see [gW 73]. In Figure 2.1. we illustrate $G(3)$.

3. Small Cages

Table 1 gives some values for $f(r,g)$, $3 \leq r \leq 7$, $3 \leq g \leq 10$. In Figures 3.2 - 3.11 we give some of the cages corresponding to these values. Of course, we know already that K_{r+1}, $K_{r,r}$, P and the Heawood graph are, respectively, the $(r,3)$–cage $(r \geq 3)$, the $(r,4)$-cage $(r \geq 3)$, the $(3,5)$-cage and the $(3,6)$-cage. Even at a first glance, the reader will notice how often a subdivision of P occurs as a subgraph of a cage with odd girth. Many of the 'bigger' small cages have been found [mS 78], [B-H 80], [O-W 79], [O-W 80], [O-W 81], [pW 82], [pW 83], [cE 84] as the result of long computer searches. Having found one of these small (r,g)-graphs it is often then a very difficult task to establish whether or not it is a cage, see Exercises 1, 2, 3, and whether or not it is unique. The sorts of argument that can be used are illustrated by the next example.

Example. We prove that $f(4,5) = 19$ ([nR 64]). The $(4,5)$-graph of Figure 3.7 has 19 vertices. The graph was small enough to be found without the use of a computer. We know therefore, using Theorem 1.1, that $17 \leq f(4,5) \leq 19$.

Suppose that $f(4,5) = 17$. Let G be a $(4,5)$-cage. Let T_1 be a spanning subtree of G as described in Theorem 1.1; see Figure 3.1(a).

Define m to be the number of 5-cycles which contain the edge e. Then m equals the number of paths of length 2 with first vertex v and not containing the vertex u (remember here that G has no cycles of length 3 and 4).

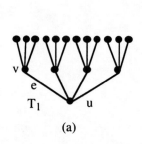

(a)

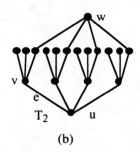

(b)

Figure 3.1

Hence $m = 3 \times 3 = 9$. There are 34 edges in G and so 34×9 counts the number of 5-cycles that arise by considering each edge of G. However, this means that each 5-cycle of G is counted 5 times. Therefore the number of 5–cycles in G is $(34 \times 9) / 5$ which is absurd. So $f(4,5) \geq 18$.

Suppose now that $f(4,5) = 18$. Let G be a (4,5)-cage. Let T_2 be a spanning subgraph of G; see Figure 3.1(b). We know G must contain such a subgraph since, using the same argument as in Theorem 1.1, G must contain the tree $T_2 \setminus \{w\}$. Then, since $\gamma(G) = 5$ and $r = 4$, the neighbourhood of w must, up to an isomorphism, be as shown in Figure 3.1(b). Consider the number m of 5-cycles which contain e. Then m is equal to the number of paths of length 2 with first vertex v, not containing the vertices u or w. Hence, again remembering that $\gamma(G) = 5$, $m = 3 \times 3 - 1 = 8$. So now G contains exactly $(34 \times 8) / 5$ 5–cycles which is absurd. So $f(4,5) \geq 19$. Therefore $f(4,5) = 19$.   □

In the Figures 3.2 - 3.11, we give some of the small (r,g)-graphs.

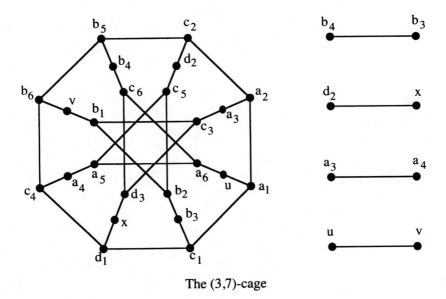

The (3,7)-cage

Figures 3.2

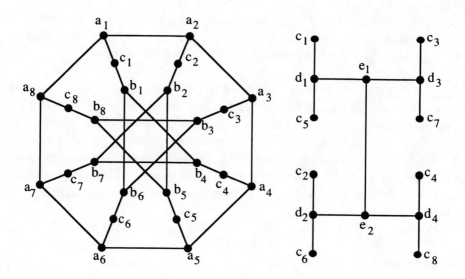

The (3,8)-cage

Figure 3.3

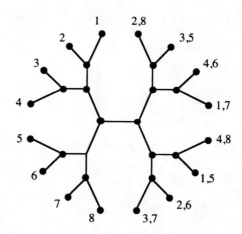

The (3,8)-cage again

Figure 3.4

In Figure 3.4, i is joined to j,k if $i \in \{j,k\}$.

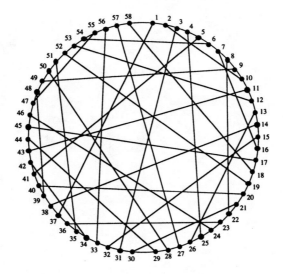

A (3,9)-cage

Figure 3.5

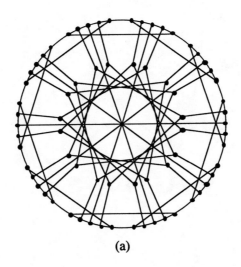

(a)

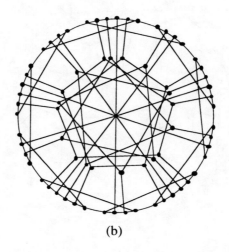

(b)

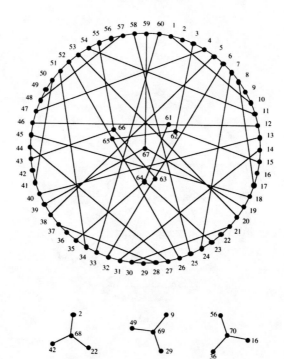

(c)

(3,10)-cages

Figure 3.6

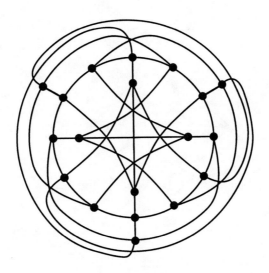

The (4,5)-cage

Figure 3.7

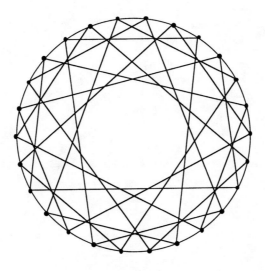

The (4,6)-cage

Figure 3.8

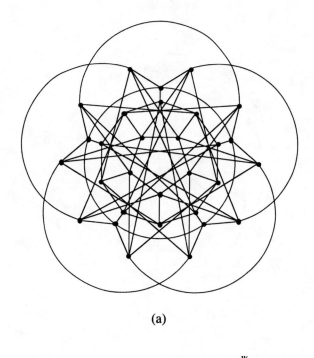

(a)

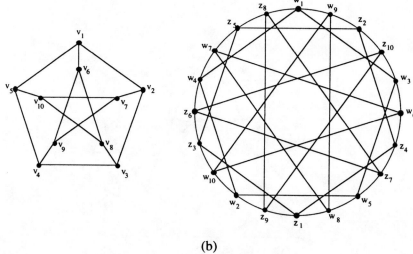

(b)

A (5,5)-cage

Figure 3.9

In Figures 3.9(a) and 3.9(b), we have two drawings of the same graph. In Figure 3.9(b), v_i is joined to w_i and z_i.

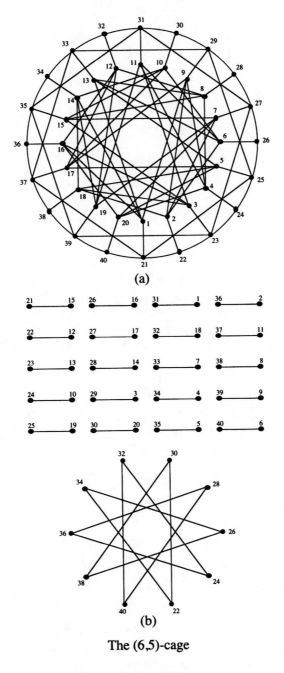

(a)

(b)

The (6,5)-cage

Figure 3.10

In Figure 3.10 we have two drawings of the same graph.

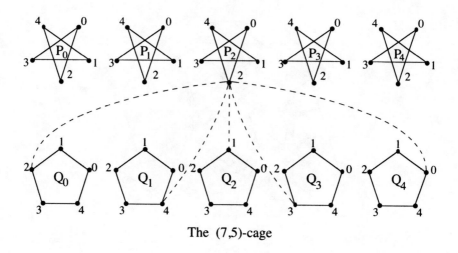

The (7,5)-cage

Figure 3.11

In Figure 3.11, $i \in P_j$ is joined to $i + jk \pmod 5$ of Q_k.

In Table 2 we give some references (by no means exhaustive) for Table 1.

r\g	3	4	5	6	7	8	9	10
3	[wT 66]	[wT 66]	[H-S 60]	[rS 63]	[wM 60]	[rS 63]	[bM 89]	[pW 83]
4	"	"	[nR 64]	[rS 66]	Theorems	"	Theorems	Theorems
5	"	"	[gW 73]	[rS 66]	1.1	"	1.1	1.1
6	"	"	[O-W 79]	[rS 66]	and	"	and	and
7	"	"	[H-S 60]	[O-W 81]	2.5		2.5	2.5

Table 2

Sometimes different authors have obtained the same results. Rather than give all the details here, we have given a somewhat larger list of references at the end of the chapter, to help the reader to progress further into the subject.

4   Symmetry and cages

Some Moore graphs are extraordinarily symmetric. For example the (3,5)-cage P has 120 paths of length 3 and each of these paths is the same, in the sense that there is an automorphism of the graph which will map any one of these paths onto any other. Again the (3,8)-cage (the Tutte-Coxeter graph) has 1440 paths of length 5 all of which are equivalent up to an automorphism. These particular Moore graphs, as well as the Heawood graph could, in a way which we shall now try to make clear, not be more symmetric. Indeed, all odd Moore graphs that are known have a high degree of symmetry. This adds some significance to the discovery that if the (57,5)-cage exists, then it lacks a certain type of symmetry [mA 71].

We need a few definitions before we can proceed. Suppose $s \geq 0$. By an **s–route** in a graph G, we mean a sequence of vertices $(v_0, v_1, ..., v_s)$ of G such that $v_i v_{i+1} \in E(G)$ for $i = 0, 1, ..., s - 1$ and $v_{i-1} \neq v_{i+1}$ for $i = 1, 2, ..., s - 1$. In Figure 4.1, $(1,2,4,1,3,4,5)$ is a 6-route, whereas $(1,2,4,5,4,2)$ is not a 5-route.

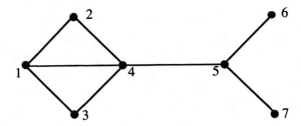

Figure 4.1

Notice that $1,2,4,5,6$ is different, as a 4-route, from $6,5,4,2,1$. Indeed, we call the latter the **inverse** of the former. In general, the **inverse** of the s-route $(v_0, v_1, ..., v_s)$ is the s-route $(v_s, v_{s-1}, ..., v_0)$. If π is an s-route, its inverse is denoted by π^{-1}. We will write $\pi = (v_0, v_1, ..., v_s)$ when we mean that π is the s-route $v_0, v_1, ..., v_s$. If $v_s v_{s+1} \in EG$ and $v_{s+1} \neq v_{s-1}$ then $\pi^* = (v_1, v_2, ..., v_{s+1})$, is called a **successor of** π. Hence in Figure 4.1, $(5,4)$ is a successor of $(6,5)$, and $(5,4,2), (5,4,1)$ and $(5,4,3)$ are all successors of $(7,5,4)$. Assuming $\gamma(G)$ is defined, if $0 \leq s < \gamma(G)$, an s-route is the same as an s-path (see Exercise 7).

Suppose G contains an s-route $(s \geq 0)$. G is **s-transitive** if the automorphism group of G acts transitively on all the s-routes. This means that if π and π^* are any two s-routes, then there exists an automorphism σ such that σ maps π onto π^* (we write $\sigma(\pi) = \pi^*$) where here we remember that π and π^* are sequences of vertices and so any automorphism of G has an induced action on the s-routes of G.

So vertex-transitivity, see Section 1.4, and 0-transitivity are the same. Recall that if the automorphism group acts transitively on the edges of G, then G is edge–transitive. So G is edge-transitive if for any two edges ab, cd there is an automorphism σ such that $\sigma\{a,b\} = \{c,d\}$. Edge-transitivity and 1–transitivity are not the same, e.g. $K_{1,n}$ is edge transitive but not 1-transitive.

The path P_{s+1} is s-transitive. For if the vertices of P_{s+1} are ordered so that $P = (v_0, v_1, v_2, ..., v_s)$ is an s-route, then π^{-1} is the only other s-route and the automorphism $\sigma = (v_0,v_s)(v_1,v_{s-1})\ldots\ldots$ satisfies $\sigma(\pi) = \pi^{-1}$. On the other hand, P_{s+1} is not edge transitive for $s \geq 3$, although it is for $s = 1,2$.

Suppose that $C_n = (v_1,v_2,\ldots,v_n)$ $(n \geq 3)$ is a cycle. Then C_n is s-transitive for all $s \geq 0$; see Exercise 6. In addition, it is clear that any number of disjoint copies of C_n $(n \geq 3)$ also gives rise to an s-transitive graph. Actually if G is any s–transitive graph, then the union of any number of disjoint copies of G is also s–transitive. So below we confine our attention to connected s-transitive graphs.

Funnily enough, although, as we have seen, the star $K_{1,n}$ $(n \geq 2)$ is edge-transitive and not 1-transitive, it **is** 2-transitive. Given the labelling of $K_{1,n}$ shown in Figure 4.2, we see that (u_i,v,u_j) $(i \neq j)$ are the only 2-routes. Since the automorphism group $A(K_{1,n})$ acts like the symmetric groups S_n on $\{u_i : i = 1,2,\ldots,n\}$, every 2-route must be sent into every other 2-route by $A(K_{1,n})$. Further $K_{1,n}$ is **not** 1-transitive since every automorphism fixes v. Hence the 1-route (v,u_i) can never be mapped by an automorphism into (u_i,v) for any i.

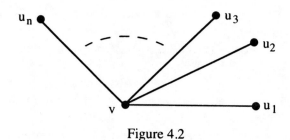

Figure 4.2

It would be nice if s-transitive graphs were also (s-1)-transitive graphs. We shall now see that this is the case except for trees. The $K_{1,n}$ example clearly fits into this category.

Lemma 4.1 Suppose that G is a non-trivial connected s-transitive graph. Then $\delta(G) = 1$ if and only if G is a tree.

Proof: Clearly if G is a non-trivial tree, then $\delta(G) = 1$. Now suppose G is connected, non-trivial, s-transitive and $\delta(G) = 1$. Let C be any cycle in G. Choose any s-route π whose vertex set is contained in C. We may always do this since we can 'run round' as much of the cycle, as often as we require. Now choose any s-route π^*, whose first vertex has degree 1. Again it is obvious since G is connected and contains a cycle, that we can always do this. There is no automorphism σ such that $\sigma(\pi) = \pi^*$, since the initial vertices of π and π^* have different degrees. This contradiction proves that G contains no cycles and so is a tree. ◻

In view of Lemma 4.1 we will usually assume that $\delta(G) \geq 2$.

Lemma 4.2 Suppose that G is a connected s-transitive graph $(s \geq 0)$. Then if $\delta(G) \geq 2$, G is k-transitive for all $k, 0 \leq k \leq s$.

Proof: Suppose that π_1 and π_2 are any two k-routes, $0 \leq k \leq s$. Since G is connected and $\delta(G) \geq 2$, we can extend π_1 and π_2 to s-routes π_1* and π_2*. We may choose $\sigma \in A(G)$, since G is s-transitive, so that $\sigma(\pi_1*) = \pi_2*$. Therefore σ maps π_1 onto π_2.   □

Notice that Lemma 4.2 implies that if G is s-transitive $(s \geq 0)$ and $\delta(G) \geq 2$, then G is transitive and so in particular G is regular. Also if $s \geq 1$ then G is 1-transitive which in turn implies that G is edge-transitive. The next result provides a useful test for s-transitivity.

Lemma 4.3 Suppose that G is connected and $\delta(G) \geq 2$. Suppose that G contains an s-route π $(s \geq 1)$ such that for each successor $\pi*$ of π, there exists $\sigma \in A(G)$ such that $\sigma(\pi) = \pi*$. Then G is s-transitive.

Proof: Let $\pi = (u_0, u_1, ..., u_s)$ be an s-route satisfying the conditions of the statement of the lemma. Let $M = \{\omega: \omega$ is an s-route and $\omega = \sigma(\pi)$ for some $\sigma \in A(G)\}$. Choose $\omega \in M$ where $\omega = (v_0, v_1, ..., v_s)$ and $\omega = \sigma(\pi)$, $\sigma \in A(G)$. Let $\omega* = (v_1, v_2, ..., v_{s+1})$ be a successor of ω. Then since $\sigma^{-1}(\omega)$ $= \pi$, $\sigma^{-1}(\omega*) = (u_1, u_2, ..., u_s, \sigma^{-1}(v_{s+1}))$. So $\sigma^{-1}(\omega*)$ is a successor of π. Hence there exists, by hypothesis, $\mu \in A(G)$ satisfying $\sigma^{-1}(\omega*) = \mu(\pi)$. Hence $\omega* = \sigma\mu(\pi) \in M$. So M contains all s-routes obtainable from π by repeatedly taking successors. Now provided $G \neq C_n (n \geq 3)$, it is at least intuitively clear that since G is connected and $\delta(G) \geq 2$, M contains all the s-routes of G. (The details of this argument are in [wT 66, pp 56-58].) So if M contains all the s–routes of G, G is s-transitive. If $G \approx C_n$ then again G is s-transitive and we have proved the lemma.   □

We now link the concepts of s-transitivity and girth.

Lemma 4.4 Let G be a connected s-transitive graph. Suppose that G is not a cycle and $\delta(G) \geq 2$. Then $s \leq \lfloor \frac{1}{2}(\gamma(G) + 2) \rfloor$.

Proof: Since $\delta(G) \geq 2$, $\gamma(G)$ is defined. Choose a cycle $C = (u_0, u_1, ..., u_t)$, with $t = \gamma(G) - 1$. Then, since G is connected and G is not a cycle, we may assume without loss of generality, that there exists $v \in VG$, $v \notin \{u_0, u_1, ..., u_t\}$ and $u_t v \in EG$.

Assume that $s \geq \gamma(G)$. Now $\pi = (u_0, u_1, ..., u_t, u_0)$ and $\omega = (u_0, u_1, ..., u_t, v)$ are $(t + 1)$-routes. Since $s \geq t + 1$, there must exist $\sigma \in A(G)$ with $\sigma(\pi) = \omega$. This is obviously not possible, since π has the same first and last element whereas ω does not. Hence $s < \gamma(G)$.

Since $s < \gamma(G)$ any s-route is a path of length s. Now let $\pi = (u_0, u_1, ..., u_s)$ be an s-route, where $u_i \in VC$ as before. Label the vertices of C so that there exists $v \in V(G)$, $v \notin VC$ and $u_{s-1}v \in E(G)$. Then $\omega = (u_0, u_1, ..., u_{s-1}, v)$ is an s-path. Using the s-transitivity there must exist $\sigma \in A(G)$ so that $\sigma(\pi) = \omega$. This implies that $\sigma(u_0) = u_0$, $\sigma(u_{s-1}) = u_{s-1}$ and $\sigma(u_s) = v$. But there exists a $u_0 u_{s-1}$-path of length $t - s + 2$ passing through the edge $u_s u_{s-1}$ and so there exists a $u_0 u_{s-1}$-path of length $t - s + 2$ passing through the edge $v u_{s-1}$. These two distinct (although not necessarily edge disjoint) paths, together form a cycle of length at most $2(t-s+2)$. Therefore $\gamma(G) \leq 2(t-s+2) = 2(\gamma(G)-s+1)$ and the result follows. $\square$

Suppose now that G is a connected graph and $\delta(G) \geq 2$. We say that G is **strictly s-transitive** $(s \geq 1)$ if G is s-transitive but not $(s+1)$-transitive. Tutte [wT 66], and many other authors, calls such a graph G, s-regular.

Example. The Petersen graph is strictly 3-transitive. We prove this as follows. Since $\gamma(P) = 5$, if P is s-transitive, then by Lemma 4.4, $s \leq 3$. Consider the 3-path $\pi = (1, 1', 3', 3)$. By inspection, $\sigma = (1 1' 3' 3 4 5)(2 4' 5')$ and $\mu = (1 1' 3' 3 2)(4 2' 5 4' 5')$ are automorphisms, and $\sigma(\pi) = (1', 3', 3, 4)$, $\mu(\pi) = (1', 3', 3, 2)$. Hence π is mapped in turn into both of its successors by an automorphism. Therefore, by Lemma 4.3, P is 3-transitive. Hence since P is not 4-transitive, P is strictly 3-transitive.

The study of s-transitivity for regular graphs of degree $k \geq 3$, is extremely difficult. We concentrate below on cubic graphs.

Lemma 4.5 Let G be a connected cubic strictly s-transitive graph $(s \geq 1)$. Then, for any two (not necessarily distinct) s-routes π and ω, there exists exactly one $\sigma \in A(G)$ such that $\sigma(\pi) = \omega$.

Proof: Suppose that there exist $\sigma_1, \sigma_2 \in A(G)$ such that $\sigma_1(\pi) = \omega = \sigma_2(\pi)$, where σ_1 and σ_2 are distinct. Then $\sigma_2^{-1}\sigma_1(\pi) = \pi$. So we may choose an s-route π such that there exists $\sigma \in A(G)$, σ not the identity, and $\sigma(\pi) = \pi$. Now it is at least intuitively clear (see [wT 66]) that, since σ is not the identity, we can choose π with $\sigma(\pi) = \pi$ and such that σ does not fix its two successors (there are exactly two successors since G is cubic). Suppose then that $\pi = (v_0, v_1, ..., v_s)$ and $N(v_s) = \{v_{s-1}, x, y\}$. Without loss of generality, suppose that $\sigma(x) = y$. Choose $z \in N(v_0) \setminus \{v_1\}$. Then by the s-transitivity of G, there exists $\mu \in A(G)$ such that $\mu(z, v_0, v_1, ..., v_{s-1}) = (v_0, v_1, ..., v_s)$. Hence

$\mu(v_s) \in \{x,y\}$. Relabelling if necessary, we may assume that $\mu(v_s) = x$. We now have $\mu(z,v_0,v_1,...,v_s) = (v_0,v_1,...,v_s,x)$ and $\sigma\mu(z,v_0,v_1,...,v_s) = (v_0,v_1,...,v_s,y)$.

Finally, by Lemma 4.3, this means that G is (s+1)-transitive. This contradiction completes the proof. □

Lemma 4.6 Let G be a connected cubic strictly s-transitive graph ($s \geq 1$). Then G contains exactly $|VG|.3.2^{s-1}$ paths of length s and $|A(G)| = |VG|.3.2^{s-1}$.

Proof: Let S be the set of s-routes. Since $s < \gamma(G)$ (by Lemma 4.4), an s-route is an s-path. Since G is s-transitive, the induced action of A(G) on S is transitive and since G is strictly s-transitive, the stabilizer of any element is trivial (by Lemma 4.5). Therefore A(G), when acting on S, has exactly one orbit and hence $|A(G)| = |S|$, via Lemma 1.4.5. Starting at a fixed vertex v, there exist exactly 3.2^{s-1} paths of length s, since G is cubic. Therefore $|S| = |VG|.3.2^{s-1}$. □

Suppose that G is a connected cubic strictly s-transitive graph ($s \geq 1$). Choose any fixed path π of length s in G. Let S be the set of paths of length s in G. Then the last two lemmas tell us that there exists a surjection f: S → A(G) defined by $f(\omega) = \sigma$, where $\sigma(\pi) = \omega, (\omega \in S, \sigma \in A(G))$.

The next theorem is certainly the most central, surprising and elegant in this area. Unfortunately its proof is too long to give here. The theorem is due to Tutte [wT47] and modified proofs using more group-theoretic techniques have been derived in the light of work by Sims [cS 67], Djokovič [dD 72], Gardiner [aG73,74,76] and Weiss [rW 81].

Theorem 4.7 Let G be a connected cubic s-transitive graph ($s \geq 1$). Then $s \leq 5$. □

Now we can make precise our statement that some cages are as symmetric as can be. Let G be a connected cubic strictly s-transitive graph. Then from Lemma 4.4, $s \leq \lfloor \frac{1}{2}(\gamma(G) + 2) \rfloor$. If equality holds, we call G a **Tutte graph**. Using Theorem 4.7, Tutte graphs exist only if $(s,\gamma(G)) \in \{(2,3),(3,4),(3,5),(4,6),(4,7),(5,8),(5,9)\}$. The (3,i)-cages for $i \in \{3,4,5,6,8\}$ are all Tutte graphs. Hence in the sense of s-transitivity these graphs are as symmetric as possible since s is as large as the girth will allow.

On the other hand the (3,7)-cage, the so-called McGee graph, is not a Tutte graph and is not even vertex-transitive. The reason for this lack of symmetry may well be that, since its order is 24, it is not a Moore graph. It is possibly not 'caged' sufficiently tightly to expect all the rough parts to be smoothed out.

We summarize this in Table 3 where $|A(G)|$ is found via Lemma 4.6.

Cage	Name	value of s so that G is strictly s-transitive	$\lfloor \frac{1}{2}(\gamma+2) \rfloor$	$|A(G)|$
(3,3)	K_4	2	2	24
(3,4)	$K_{3,3}$	3	3	72
(3,5)	P	3	3	120
(3,6)	Heawood	4	4	336
(3,7)	McGee	Not transitive	4	32
(3,8)	Tutte-Coxeter	5	5	1440

Table 3

We shall return to a discussion of the symmetries of these graphs in Chapter 8. In the meantime we note that in the sense of s-transitivity the Tutte-Coxeter graph is the most symmetric of all cubic graphs, since it has the largest value of s (five) among all such graphs.

Tutte's theorem has attracted a lot of attention in the literature. One of the most surprising extensions of the theorem is due to Weiss [rW 81] who proved the following theorem using the classification of finite simple groups.

Theorem 4.8 Suppose G is a connected regular graph of degree $r \geq 3$. If G is strictly s-transitive then $s \leq 5$ or $s = 7$.   ☐

Weiss's techniques are group-theoretic but it would be fair to say that the general strategy of his proof follows Tutte's own.

Strangely, Praeger [cP 89], shows that for any degree k and integer $(s \geq 1)$, there exist an infinite number of s-transitive digraphs of degree k (which are not (s+1)-transitive). Here we assume the obvious directed version of s-transitivity.

Ivanov, together with collaborators, has shown in a series of papers (see for example [aI 87a], [aI 87b] and [I-S 88]), how omnipresent the Petersen graph is when considering the structure of certain highly symmetric graphs. Unfortunately, the details are too lengthy to include here.

We shall return to this topic again in Chapter 8.

Exercises

1. Prove that P is the unique (3,5)-cage.

    [Hint: Consider the spanning subtree described in Theorem 1.1. and Figure 1.1(a).]

2. (a) Prove that the Heawood graph H is the unique (3,6)-cage.

    [Hint: as for Exercise 1 except now use Figure 1.1(b).]

    (b) Prove that for any $v \in VH, H - v$ is isomorphic to a subdivision of P.

3. Prove that $f(3,7) \geq 24$. Establish that the McGee graph (see Table 2) is the unique (3,7)-cage. (This requires considerably more work than Exercises 1 and 2.)

4. Verify inequality (6) in the proof of Theorem 1.2.

5. Let A be an adjacency matrix for P. Does there exist n such that $A^n = 0$? Why or why not?

6. Let $n \geq 3$. Prove that a cycle C_n of length n is s-transitive for each $s \geq 0$. Can C_n ever be strictly s-transitive?

7. Prove that if $0 \leq s < \gamma(G)$, then an s-route is an s-path.

8. Characterize those trees which are s-transitive.

9. Prove directly that the Heawood graph is 4-transitive.

10. Prove that P is not 4-transitive.

11. Show that M (Lemma 4.3) contains all s-routes unless $G \cong C_n$.

12. Give an example of a graph $G, \delta(G) \geq 2$, which is vertex-transitive but not edge-transitive.

13. Prove that if G is an edge-transitive graph with $\delta(G) \geq 1$, then either (i) G is transitive or (ii) G is bipartite; its bipartition corresponding to the two orbits of A(G).

14. A graph G is **symmetric** if it is transitive and edge-transitive. Prove that every connected symmetric graph of odd degree is 1-transitive.

[Hint: Choose a 1-route π. Let $M = \{\sigma(\pi): \sigma \in A(G)\}$. Verify that for any 1-route ω, either ω or ω^{-1} belongs to M. Using a counting argument, or otherwise, show that there exists an $\omega \in M$ such that $\omega^{-1} \in M$.]

15. Give examples of graphs which are connected and symmetric but not 1–transitive. (This is difficult; see [iB 70], [dH 81].)

16. A graph has property A_n if, for each sequence $u, v_1, v_2, \ldots, v_n$ of $n + 1$ vertices, there is another vertex adjacent to u but not to $v_1, v_2, \ldots, v_n$. Prove that P is the only graph with property A_2.

[Comments (i) This was first proved by Murty [uM 68]. (ii) Exoo and Harary [E-H 80] proved that for $1 \le n \le 6$, the $(n+1,5)$ - cage is a graph of smallest order with property A_n.]

References

[mA 71]   M. Aschbacher, The nonexistence of rank three permutation groups of degree 3250 and subdegree 57, *J. of Algebra*, 19, 1971, 538–540.

[aB 72]   A.T. Balaban, A trivalent graph of girth ten, *J. Comb. Th*, B, 12, 1972, 1-5.

[aB 73]   A.T. Balaban, Trivalent graphs of girth nine and eleven and relationships among cages, *Rev. Roumaine Math.*, 18, 1973, 1033-1043.

[B-I 73]   E. Bannai and T. Ito, On finite Moore graphs, *J. Fac. Sci. Univ. Tokyo Sect. I.A. Math.*, 20, 1973, 191-208.

[B-I 81]   E. Bannai and T. Ito, Regular graphs with excess one, *Discrete Math.*, 37, 1981, 147-158.

[B-C 71]   M. Behzad and G. Chartrand, *Introduction to the Theory of Graphs*, Allyn and Bacon, Boston, 1971.

[cB 66]   C.T. Benson, Minimal regular graphs of girths eight and twelve, *Can. J. Math.*, 18, 1966, 1091-1094.

[B-H 80]   N.L. Biggs and M.J. Hoare, A Trivalent graph with 58 vertices and girth 9, *Disc. Math.*, 30, 1980, 299-301.

[nB 74]   N. Biggs, *Algebraic Graph Theory*, Cambridge University Press, 1974.

[bB 78]   B. Bollobás, *Extremal Graph Theory*, Academic Press, London, 1978.

[iB 70]   I.Z. Bouwer, Vertex and edge-transitive, but not 1-transitive graphs, *Can. Math. Bull.*, 13, 1970, 231-237.

[wB 67a]   W.G. Brown, On hamiltonian regular graphs of girth six, *J. Lond. Math. Soc.*, 42, 1967, 514-520.

[wB 67b]   W.G. Brown, On the non-existence of a type of regular graphs of girth 5, *Can. J. Math.*, 19, 1967, 644-648.

[hC 50]   H.S.M. Coxeter, Self-dual configurations and regular graphs, *Bull. Amer. Math. Soc.*, 56, 1950, 413-455.

[rD 73]   R.M. Damerell, On Moore graphs, *Proc. Camb. Phil. Soc.*, 74, 1973, 227-236.

[dD 72]   D.Z. Djokovič, On regular graphs II., *J. Comb. Th. B*, 12, 1972, 252-259.

[cE 84]   C.W. Evans, A second trivalent graph with 58 vertices and girth 9, *J. Graph Th.*, 8, 1984, 97-99.

[E-S 63]   P. Erdös and H. Sachs, Reguläre Graphen gegebener Taillenweite mit minimaler Knotenzahl, *Wiss. Z. Univ. Halle Martin-Luther Univ. Halle-Wittenberg Math.-Natur. Reihe*, 12, 1963, 251-257.

[E-H 80]   G. Exoo and F. Harary, The smallest graphs with certain adjacency properties, *Discrete Math.*, 29, 1980, 25-32.

[F-H 64]   W. Feit and G. Higman, The nonexistence of certain generalized polygons, *J. Algebra*, 1, 1964, 114-131.

[aG 73]   A.D. Gardiner, Arc transitivity in graphs, *Quart. J. Math.* 24, 1973, 399-407.

[aG 74]   A.D. Gardiner, Arc transitivity in graphs II, *Quart. J. Math*, 25, 1974, 163-167.

[aG 76]   A.D. Gardiner, Arc transitivity in graphs III, *Quart. J. Math*, 27, 1976, 313-323.

[H-S 60]   A.J. Hoffman and R.R. Singleton, On Moore graphs with diameters 2 and 3, *IBM. J. Res. Develop*, 4, 1960, 497-504.

[dH 81]   D.F. Holt, A graph which is edge-transitive but not arc transitive, *J. Graph Th.*, 5, 1981, 201-204.

[tI 81]   T. Ito, On a graph of O'Keefe and Wong, *J. Graph. Th.*, 5, 1981, 87-94.

[aI 87a]   A. Ivanov, On 2-transitive graphs of girth 5, *Europ. J. Combin.*, 8, 1987, 393-420.

[aI 87b]   A. Ivanov, Graphs of girth 5 and diagram geometries connected with the Petersen graph, *Dokl. Akad., Nauk SSR*, 295, 1987, 529-533.

[I-S 88]   A. Ivanov and S. Shpektorov, Geometries for sporadic groups related to the Petersen graph I, *Comm. Alg.*, 16, 1988, 925-953.

[pK 81]   P. Kovacs, The non-existence of certain regular graphs of girth 5, *J. Comb. Th.* B, 30, 1981, 282-284.

[jL 70]   J.Q. Longyear, Regular d-valent graphs of girth 6 and $2(d^2-d+1)$ vertices, *J. Comb. Th.*, 9, 1970, 420-422.

[wM 60]   W.F. McGee, A minimal cubic graph of girth seven, *Can. Math. Bull.*, 3, 1960, 149-152.

[bM 89]   B. McKay, private communication.

[uM 68]   U.S.R. Murty, On critical graphs of diameter 2, *Math. Magazine*, 41, 1968, 138-140.

[O-W 79]   M. O'Keefe and P.K. Wong, A smallest graph of girth 5 and valency 6, *J. Comb. Th.* B, 26, 1979, 145-149.

[O-W 80]   M. O'Keefe and P.K. Wong, A smallest graph of girth 10 and valency 3, *J. Comb. Th.* B, 29, 1980, 91-105.

[O-W 81]   M. O'Keefe and P.K. Wong, The smallest graph of girth 6 and valency 7, *J. Graph Th.* 5, 1981, 79-85.

[cP 89]   C. Praeger, Highly arc-transitive digraphs, *Europ. J. Comb.*, 10, 1989, 281-292.

[nR 64]   N. Robertson, The smallest graph of girth 5 and valency 4, *Bull, Amer. Math. Soc.*, 70, 1964, 824-825.

[nS 67]   N. Sauer, Extremaleigenschaften regulärer Graphen gegebener Taillennweite, I and II, *Osterreich Akad. Wiss. Math.-Natur, Kl., S-B II*, 176, 1967, 9-25, ibid 176, 1967, 27-43.

[cS 67]   C.C. Sims, Graphs and finite permutation groups, *Math. Zeitschr.*, 95, 1967, 76-86.

[rS 63]   R. Singleton, Regular graphs of evengirth, Ph.D. thesis, Princeton University, 1963.

[rS 66]   R. Singleton, On minimal graphs of maximum even girth, *J. Comb. Th.*, 1, 1966, 306-332.

[mS 78]   M. Spill, Über minimale Graphen gegebener Taillenweite und Valenz, Diplomathesis, Technische Universität Berlin, 1978.

[wT 47]   W.T. Tutte, A family of cubical graphs, *Proc. Cambridge Phil. Soc.*, 43, 1947, 459-474.

[wT 66]   W.T. Tutte, *Connectivity in Graphs*, University of Toronto Press, 1966.

[gW 73]   G. Wegner, A smallest graph of girth 5 and valency 5 , *J. Comb. Th.* B, 14, 1973, 203-208.

[rW 81]   R.M. Weiss, The nonexistence of 8-transitive graphs, *Combinatorica* 1, 1981, 309-311.

[pW 82]   P.K. Wong, Cages: a survey, *J. Graph. Th.*, 6, 1982, 1-22.

[pW 83]   P.K. Wong, On the smallest graphs of girth 10 and valency 3, *Disc. Math.*, 43, 1983, 119-124.

7

Hypohamiltonian graphs

0. Prologue

It appears that there was a club and the president decided that it would be nice to hold a dinner for all the members. In order not to give any one member prominence, the president felt that they should be seated at a round table.

But at this stage he ran into some problems. It seems that the club was not all that amicable a little group. In fact each member only had a few friends within the club and positively detested all the rest. So the president thought it necessary to make sure that each member had a friend sitting on either side of him at the dinner.

Unfortunately, try as he might, he could not come up with such an arrangement. In desperation he turned to a mathematician. Not long afterwards, the mathematician came back with the following reply.

'It's absolutely impossible! However, if one member of the club can be persuaded not to turn up, then everyone can be seated next to a friend.'

'Which member must I ask to stay away?' the president queried.

'It doesn't matter', replied the mathematician. 'Anyone will do.'

'By the way, if you had fewer members in the club you wouldn't be faced with this strange combination of properties.'

So the president, on some pretext, excused himself from the dinner and was easily able to seat the members of the club so they all had a friend on either side.

How many club members were there? Who likes whom and who dislikes whom? Show that the solution is unique (to within the obvious symmetries).

1. Sousselier's problem

The Prologue is a loose translation of Problème No. 29: Le Cercle Des Irascibles, which appeared in *Revue Française de Recherche Opérationelle*, 1963. The problem was posed by René Sousselier [rS 63]. In the same journal in the following year, a solution was given by J.C. Herz, Th. Gaudin and Ph. Rossi [H-G-R 64].

Take the members of the club to be vertices and agree that two vertices are adjacent if the members are friends. The graph so obtained cannot be hamiltonian otherwise they could all be seated at the round table for the dinner. However, the graph has the property that when any vertex is deleted (along with all edges adjacent to the vertex) the resulting graph is hamiltonian.

Definition: If G is a non-hamiltonian graph and if G_v is hamiltonian for all $v \in VG$, then G is called a **hypohamiltonian graph**.

To solve the problem we need to find the smallest hypohamiltonian graph.

Lemma 1.1 If G is hypohamiltonian then $\delta(G) \geq 3$.

Proof: Let w be adjacent to v in G. Then, since G_w is hamiltonian, the degree of v in G_w is at least 2. $\square$

Lemma 1.2 If G is hypohamiltonian, then $\Delta(G) \leq \lfloor \frac{1}{2}(n - 1) \rfloor$.

Proof: Let $\deg v > \frac{1}{2}(n - 1)$. Now G_v contains a hamiltonian cycle and since $\deg v > \frac{1}{2}(n - 1)$, v is adjacent to two consecutive vertices on this cycle. Hence G itself contains a hamiltonian cycle. $\square$

Lemma 1.3 If G is hypohamiltonian, $|VG| \geq 10$.

Proof: By the degree arguments of Lemmas 1.1 and 1.2 it is immediate that $|VG| \geq 7$. But for $|VG| = 7$ we would require a cubic graph of odd order and so $|VG| \geq 8$.

Suppose that $|VG| = 8$ and $VG = \{0,1,2,3,4,5,6,7\}$. By the arguments above, G is cubic. Further, since G is hypohamiltonian we know that there exists a cycle through the vertices $\{1,2,3,4,5,6,7\}$. The vertex 0 is of degree 3. It clearly cannot be adjacent to both i and $i + 1$, $i = 1,2,3,4,5,6$ or to both 1 and 7. Hence, without loss of generality we may assume that 0 is adjacent to 1, 3 and 5. But then we either have 2 adjacent to 6 and 4 adjacent to 7 or 2 adjacent to 7 and 4 adjacent to 6. Both cases give rise to a hamiltonian graph.

Suppose then that $VG = \{0,1,2,3,4,5,6,7,8\}$. Again arguing on the degrees of the vertices, we see that $3 \leq \deg v \leq 4$ for all $v \in VG$. Hence it is clear that there exists at least one vertex of degree 4. Starting with the cycle (1,2,3,4,5,6,7,8) we see that, without loss of generality, we may assume that 0 is adjacent to 1,3,5,7. But we obtain a hamiltonian cycle in G if 2 is adjacent to 4, 6 or 8. Hence, without loss of generality, we assume 2 is adjacent to 5. Now if we repeat the arguments used on the vertex 2 for the vertex 8 we see that 8 is adjacent to 3 or 5. Since the edge 58 now contradicts Lemma 1.2, 8 must be adjacent to 3. Similarly 6 is adjacent to 1 and then 4 is adjacent to 7. But the graph we have now constructed is bipartite - the vertices labelled by even numbers being in one part and the odd vertices in the other. Hence the subgraph <0,2,3,4,5,6,7,8> is a subgraph of $K_{3,5}$ and cannot possibly be hamiltonian. □

Theorem 1.4 If G is hypohamiltonian of order 10, then it is isomorphic to the Petersen graph.

Proof: Suppose that G contains a vertex of degree 4. Then, without loss of generality, we may assume that this vertex is the vertex 0, which is adjacent to 1, 3, 5 and 7 on the cycle (1,2,3,4,5,6,7,8,9). Now the vertices 2,4,6,8,9 must be adjacent to at least one each of the vertices 1,3,5,7 otherwise G is hamiltonian. But then one of these four latter vertices has degree greater than 4. Hence G must be cubic.

Now we know that the Petersen graph is not hamiltonian. Since the Petersen graph is vertex-transitive, then all of its vertex-deleted subgraphs are isomorphic. It is readily seen, Lemma 1.5.2, that P_v is hamiltonian for all v ∈ VP. Hence P is hypohamiltonian.

Now any hypohamiltonian graph is 3-connected since G_v is 2-connected for all v ∈ VG. Hence if |VG| = 10, then G is the Petersen graph since by [B-C-C-S 76], P is the only non-hamiltonian 3-connected cubic graph on 10 vertices. □

So the answer to problem no. 29 is that the club has 10 members, all of whom have three friends. The friendships are given by the Petersen graph. Theorem 1.4 shows that this solution is unique.

But are there other hypohamiltonian graphs?

2. Petersen generalizations

The aim now is to find as many hypohamiltonian graphs as possible. We consider in this section those hypohamiltonian graphs which are generalizations of the Petersen graph. These results very quickly show that there are an infinite number of hypohamiltonian graphs.

Recall, Exercise 38 of Chapter 1, the definition of the 'standard' generalized Petersen graph $P(n,k)$. There are, of course, other generalizations. We consider a generalization of the Petersen graph which is due to Lindgren [wL 67] and Sousselier [rS 63].

For $k \geq 1$, $L(k)$ is the graph defined as follows. Let $VL(k) = \{0,1,...,6k+2\} \cup \{v\}$ and let $EL(k) = \{i\ i+1 : i = 0, 1,..., 6k+2\} \cup \{1+3j \sim 5+3j \text{ and } v \sim 3j : j=0, 1,..., 2k\}$, where addition is modulo $6k + 3$. Clearly, see Figure 1.1.7(c), the graph $L(1)$ is the Petersen graph and we know that this is hypohamiltonian from Theorem 1.4.

To show that $L(k)_u$ is hamiltonian for all $u \in VL(k), k \geq 2$, the appropriate cycles can be constructed. It is clear that $L(k)_v$ contains a hamiltonian cycle. By symmetry it only remains to show that this is true for $L(k)_0, L(k)_1$. We illustrate the general case using $k = 3$. In $L(3)_0$, a hamiltonian cycle is $(1,2,3,4,8,7,11,12,13,17,18,19,20,16,15,14,10,9,v,6,5)$, while in $L(3)_1$ we have $(2,3,...,18,v,0,20,19)$.

To prove that $L(k)$ does not contain a hamiltonian cycle is straightforward but tedious. We refer the reader to [wL 67]. We thus have the following result.

Proposition 2.1 $L(k)$ is hypohamiltonian.   □

Another generalization of the Petersen graph is given by Doyen and Van Diest [D–V 75]. Their construction is as follows. Let m, n be integers with $m \geq 2$, $n \geq 3$. We define $D(m,n)$ to have vertex set $VD(m,n) = \{u_i, v_j, w: i = 0,1,...,mn-1 \text{ and } j = 0,1,...,n-1\}$ and edge set $ED(m,n) = \{u_i u_{i+1}: i = 0,1,...,mn-1\} \cup \{v_j u_{j+kn}: j = 0,1,...,n-1 \text{ and } k = 0,1,..., m-1\} \cup \{wv_j: j = 0,1,...,n-1\}$, the addition on the subscripts of the vertices u_i being modulo mn.

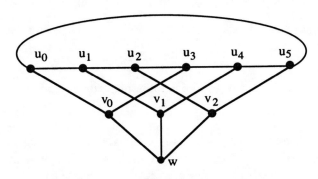

Figure 2.1

It is not difficult to see that $D(2,3) \cong P$, although the natural way of representing $D(2,3)$ is not a common presentation of P. Hence we show the drawing in Figure 2.1; compare this to Figure 1.1.7(a).

Once again it is straightforward to show that $D(m,3)$ is hypohamiltonian. The routine search for cycles is aided by the symmetries of the graphs. Thus we need only show that $D(m,3)_{u_0}$, $D(m,3)_{v_0}$ and $D(m,3)_w$ contain a hamiltonian cycle in order to be assured that all vertex-deleted subgraphs contain such a cycle. The symmetries can also be exploited to shorten the proof of the non-existence of a hamiltonian cycle in $D(m,3)$.

It is interesting to note that $D(m,n), n > 3$ is hamiltonian.

Proposition 2.2 $D(m,3)$ is hypohamiltonian for all $m \geq 2$.

$D(m,n)$ is hamiltonian for all $m \geq 2, n > 3$.   □

But Doyen and Van Diest [D-V 75] have also produced other families of hypohamiltonian graphs. Let $D(m,n,t)$ have vertex set $VD(m,n,t) = \{u_i, v_j, w_k: i = 0,1,..., mt - 1; j = 0,1,...,t-1; k = 0,1,...,nt-1\}$ and edge set $ED(m,n,t) = \{u_i u_{i+1}: i = 0,1,...,mt-1\} \cup \{w_k w_{k+1}: k = 0,1,...,nt-1\} \cup \{v_j u_{j+gt}, v_j w_{j+st}: j = 0,1,...,t-1; g = 0,1,2,...,m-1; s = 0,1,...,n-1\}$, the addition on the subscripts of u_i, w_k being, respectively, modulo mt and nt.

It is straightforward to find a hamiltonian cycle in $D(m,n,t)$ when t is even. By routine cycle chasing, it is not difficult to show that $D(m,n,3)$ and $D(m,n,5)$, where $m \geq 2$, $n \geq 2$, are two families of hypohamiltonian graphs. In addition, $D(m,n,t)$, for $t \geq 3$ and odd, have the property that all vertex-deleted subgraphs are hamiltonian unless $m = n = 1$. But there are problems associated with deciding whether or not $D(m,n,t)$, with m, n not both 1 and $t \geq 7$ and odd, is hypohamiltonian.

Simone Gutt [sG 77] has shown the following.

Proposition 2.3 $D(m,n,t)$ is hypohamiltonian for all odd $t \geq 3$ and for all $m \geq 2$.   □

We have already noted (Lemma 1.1), that every vertex in a hypohamiltonian graph has degree at least three. Thus it makes sense to ask whether or not the Petersen graph can be generalized to an infinite class of cubic hypohamiltonian graphs. Such a generalization is considered by Bondy [jB 72]; see Exercise 38 of Chapter

1. Using rather lengthy arguments, which in [jB 72] are simplified by recourse to diagrams, the following result can be proved.

Proposition 2.4 If $k \equiv 5$ (mod 6), then $P(k,2)$ is hypohamiltonian.   □

One of the more ingenious methods of construction of hypohamiltonian graphs is due to Chvátal [vC 73]. The graphs are produced by means of what he calls flip-flops.

We say that a pair of vertices (a,b) is **good** in a graph G if G contains a hamiltonian path with endvertices a and b. The pair $((a,b),(c,d))$ is **good** in G if G has a spanning subgraph consisting of two vertex disjoint paths whose endvertices are a, b and c, d, respectively.

Then a **flip-flop** is a quintuple (G,a,b,c,d), where G is a graph and a,b,c,d are distinct vertices in G with the properties

(i) $(a,d),(b,c),((a,d),(b,c))$ are good in G;
(ii)   none of $(a,b),(a,c),(b,d),(c,d),((a,b),(c,d)),((a,c),(b,d))$ are good in G;
(iii)   for each $v \in VG$, at least one of $(a,c),(b,d),((a,b),(c,d)),((a,c),(b,d))$ is good in G_v.

Figure 2.2 shows a flip-flop F_8, of order 8. The three defining properties are easily checked. For instance, the path a,u,b,v,x,c,w,d shows that (a,d) is good, while paths a,u,w,d and b,v,x,c show that $((a,d),(b,c))$ is good. Obviously (a,b) is not good in G. The other properties can all be easily checked.

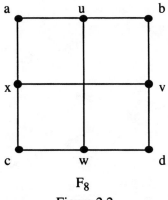

$$F_8$$
Figure 2.2

A flip-flop (G,a,b,c,d) is **cubic** when a,b,c,d are of degree 2 and all other vertices have degree 3. Hence F_8 is cubic.

Suppose that $f_i = (G_i, a_i, b_i, c_i, d_i)$, $i = 1, 2, 3$ are flip-flops. Define (f_1, f_2) to be the quintuple $(G_4, a_1, b_1, c_2, d_2)$, where G_4 is obtained from disjoint copies of G_1 and G_2 by joining c_1 to b_2 and d_1 to a_2. Further, define (f_1, f_2, f_3) to be the quintuple $(G_5, a_1, b_1, c_3, d_3)$ where G_5 is obtained from disjoint copies of G_1, G_2, G_3 by adding the new vertices u_1, v_1, u_2, v_2 and the new edges $u_1v_1, u_2v_2, c_1u_1, u_1a_2, d_1v_1, v_1b_2, u_2v_2, c_2u_2, u_2a_3, d_2v_2, v_2b_3$. These new graphs are shown in Figure 2.3.

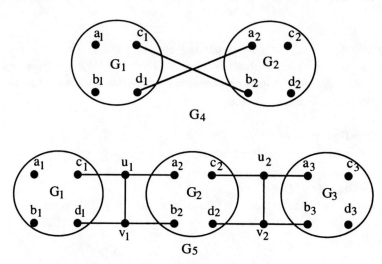

Figure 2.3

Lemma 2.5 If f_i, $i = 1, 2, 3$ are flip-flops, then so are (f_1, f_2) and (f_1, f_2, f_3). Moreover if the f_i are cubic, then so are (f_1, f_2) and (f_1, f_2, f_3).

Proof: It is straightforward to check all the properties involved. For instance, we note that since (a_1, d_1) and (a_2, d_2) are good, then there is a hamiltonian path of G_4 with endvertices a_1 and d_2. Hence (a_1, d_2) is good.   □

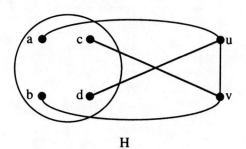

H

Figure 2.4

Finally if G is any graph and a,b,c,d are distinct vertices of G, then the graph **based on** (G,a,b,c,d) is the graph H obtained from G by adding the new vertices u, v and the new edges uv,ua,ud,vb,vc. This situation is shown in Figure 2.4.

We now show the relevance of flip-flops to hypohamiltonian graphs.

Lemma 2.6 If H is based on a flip-flop (G,a,b,c,d), then H is hypohamiltonian. Further, if H is based on a cubic flip-flop, then H is cubic.

Proof: Suppose that H contains a hamiltonian cycle in which u and v are consecutive. Then in G, at least one of (a,b),(a,c),(b,d) or (c,d) is good, contradicting the fact that (G,a,b,c,d) is a flip-flop.

If H contains a hamiltonian cycle in which u and v are not consecutive, then the cycle is (...,a,u,d,...,c,v,b,...) or (...,a,u,d,...,b,v,c,...). Hence ((a,b),(c,d)) or ((a,c),(b,d)) are good in G but this again contradicts a flip-flop property. Hence H is not hamiltonian.

In H_u, we have a hamiltonian cycle since (b,c) is good in G. Similarly for H_v. If w is any other vertex, then since at least one of (a,c),(b,d),((a,b),(c,d)), ((a,c,),(b,d)) is good in G_w, H_w contains a hamiltonian cycle.

The cubic property is obvious.   □

So we can produce hypohamiltonian graphs if we can produce suitable flip-flops. It should come as no surprise that the graph based on F_8 is the Petersen graph; see Figure 2.2. Four more flip-flops are shown in Figure 2.5.

It can be shown that L(2) [vC 73] and D(2,1,5) [C-S 77] are graphs based on flip-flops. Using the flip-flops F_8, F_{11} and F_{13} (see Figures 2.2 and 2.5), Chvátal [vC 73] produced three hypohamiltonian graphs of order 21, 23, 24, respectively, based on the flip-flops (F_8,F_{11}), (F_8,F_{13}), (F_{11},F_{11}). We denote the graph of order 21 by C(21).

Finally, the flower snarks J_k, k ≥ 5 and odd, which as we noticed in Section 3.2 can be regarded as a generalization of P, are hypohamiltonian. This is proved in [C-E 83]. The graph J_3 is not hypohamiltonian. However, since J_k is, perhaps surprisingly, isomorphic to D(2,1,k), for k odd, this gives us no new information; see Exercise 8.

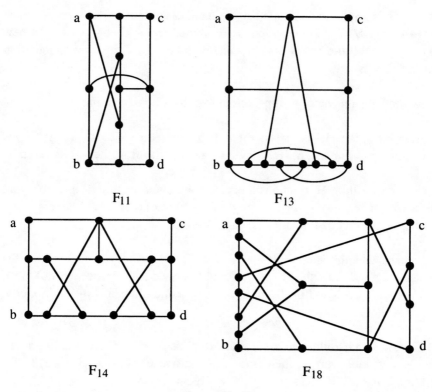

Some flip-flops

Figure 2.5

3. A graph for all vertices

Having discovered that there are an infinite number of hypohamiltonian graphs, the next question to be attacked is whether or not there exists a hypohamiltonian graph for each order n.

We show that there exists a hypohamiltonian graph for every order greater than or equal to 10 except for 11, 12, 14 and possibly 17. If we restrict ourselves to cubic graphs, we find that there exists a cubic hypohamiltonian graph of every even order from 18 on and for no other order apart from 10. The proofs of both these results rely on a simple construction, once the early orders have been established.

From Lemma 1.3 and Theorem 1.4 we know that there is no hypohamiltonian graph on fewer than 10 vertices and that the Petersen graph is the unique (cubic) hypohamiltonian graph of order 10.

n = 11. There are, in fact, no hypohamiltonian graphs of order 11 as was shown in [H-D-V 67]. To prove this we need the following lemmas.

Lemma 3.1 Let G be hypohamiltonian, $D \subset VG$ and $n - 1 \geq |D| \geq \frac{1}{2}n$. Then $<D>$ contains a set of at least $2|D| - n + 1$ edges such that the subgraph of $<D>$ induced by these edges is the union of pairwise (vertex) disjoint paths.

Proof: Let $u \in VG \setminus D$ and let H be any hamiltonian cycle of G_u. The number of edges of $H \cap <D>$ equals $|EH|$ minus the edges that have at least one vertex in $VG_u \setminus D$. Hence the number of edges of $H \cap <D>$ is minimized when the number of edges with at least one vertex in $VG_u \setminus D$ is maximized. Since $|D| \geq \frac{1}{2}n$, this maximum occurs when every vertex of $VG_u \setminus D$ is the end of two edges of H. Hence

$$|EH \cap E<D>| \geq |EH| - 2|VG_u \setminus D| = (n - 1) - 2(n - 1 - |D|) = 2|D| - (n - 1).$$

Since these edges are in H, they are the union of disjoint paths. □

Lemma 3.2 If G is hypohamiltonian, then $\Delta(G) \leq \lfloor \frac{1}{2}(n - 4) \rfloor$.

Proof: Suppose that v is a vertex with $\deg v \geq \lfloor \frac{1}{2}(n - 2) \rfloor$. Let $A = N(v)$, $B = VG \setminus N(v)$ and let H be a hamiltonian cycle in G_v. By Lemma 1.2 we have $\deg v \leq \lfloor \frac{1}{2}(n - 1) \rfloor$. Further, $<A> \cap H$ consists of isolated vertices since G is hypohamiltonian.

If n is even, then $\deg v = \frac{1}{2}(n-2) = |A|$. Since $|B| = |A| + 2$ (remember $v \in B$), it follows that $<B> \cap H$ consists exactly of $|A| - 1$ isolated vertices and one edge. If $<B>$ contains more than one edge, then H can be extended to a hamiltonian cycle for G. But by Lemma 3.1, $<B>$ contains at least $2|B| - n + 1 = 3$ edges. Hence we contradict the fact that G is hypohamiltonian.

So assume that n is odd. Then either $\deg v$ is $\frac{1}{2}(n-3)$ or it is $\frac{1}{2}(n-1)$. Suppose that $\deg v$ is $\frac{1}{2}(n-1)$. Because $<A> \cap H$ consists only of isolated vertices, $<B> \cap H$ has precisely 2 edges. If these two edges form a path u_1, u_2, u_3, then since G is not hamiltonian, every edge of $<B>$ (other than an edge joining u_1 to u_3) is adjacent to u_2. But by Lemma 3.1, $<B>$ contains at least 4 edges no three of which are incident with u_2. Hence we have a contradiction. On the other hand if the two edges of $<B> \cap H$ are independent, then the non-hamiltonicity of G forces each edge of $<B>$ to join pairs of the endvertices of these edges. Again this contradicts Lemma 3.1.

Hence we may suppose that n is odd and that deg $v = \frac{1}{2}(n - 1)$. In this case <B> must consist of $\frac{1}{2}(n-1)$ isolated vertices but this is contradicted by Lemma 3.1 which forces at least 2 edges in <B>.   □

So from Lemmas 1.1 and 3.2, an 11 vertex hypohamiltonian graph G must be cubic. Clearly no such graph exists.

n = 12. Herz, Duby and Vigué [H-D-V 67] also proved that there is no hypohamiltonian graph of order 12. They did this with the aid of a computer. Essentially this was done by assuming the existence of a cycle through 11 vertices and joining a new vertex of appropriate degree to this cycle taking due notice of all possible symmetries. Then edges were added between vertices on the initial cycle until the degree requirements were met. Of the possible graphs thus generated, none was hypohamiltonian.

n = 13. At this stage Herz et al found that they could not continue with an exhaustive search of the 13 vertex graphs because of time considerations. However, by restricting their attention to certain ways of joining a 13th vertex to a cycle through 12 vertices, they discovered the graph H(13) of Figure 3.1. Note the ever present Petersen graph. This graph is actually based on F_{11}, one of the flip-flops of Figure 2.5 ([vC 73]). There may be other 13 vertex hypohamiltonian graphs.

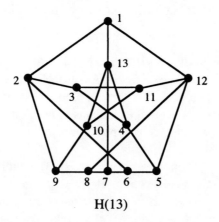

H(13)

Figure 3.1

n = 14. Once again this case was settled by recourse to the computer. Denniston ([D-V 75] p.236) showed that there is no cubic hypohamiltonian graph on 14 vertices. Later Collier and Schmeichel [C-S 78] used essentially the same approach as Herz et al and were able to show that there is no hypohamiltonian graph at all of order 14.

n = 15. In [H-D-V 67], the graph H(15) of Figure 3.2 was found. And Petersen strikes again! As do flip-flops, since H(15) is based on F_{13}; see Figure 2.5 and [vC 73].

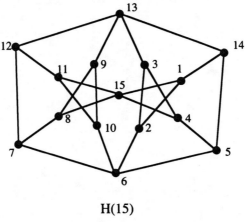

H(15)

Figure 3.2

n = 16. The graph L(2) shows that there are hypohamiltonian graphs of this order. However, in [C-S 78] a computer search failed to reveal any cubic hypohamiltonian graph of order 16.

n = 17. To date this remains the one open question. If a hypohamiltonian graph of this order exists, then a computer search has shown that its girth must be less than 5 [C-S 78].

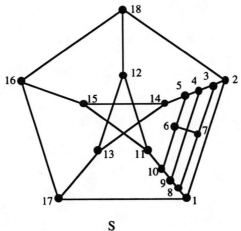

S

Figure 3.3

n = 18. The 18 vertex graph S of Figure 3.3 is due to Sousselier and was announced in [H-D-V 67]. The Petersen graph is again to the fore. We note that S is cubic and contains a subgraph isomorphic to F_8. Actually S is based on (F_8, F_8) [C-S 77]. In their computer search, Collier and Schmeichel [C-S 78] also produced the graph of Figure 3.4. This graph again contains a subgraph which is a subdivision of the Petersen graph.

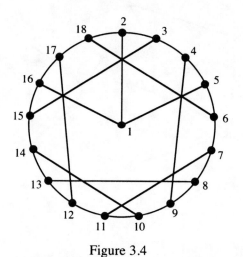

Figure 3.4

n = 19. One such graph is D(5,3), see [D-V 75].

n = 20. Here we have a cubic hypohamiltonian graph D(2,1,5), see [D-V 75]. This graph is based on the flip-flop F_{18}, of order 18 see [C-S 77] and Figure 2.5. The flower snark J_5 is seemingly another example but, in fact, it is isomorphic to D(2,1,5); see Exercise 8.

n = 21. Take C(21), the graph based on (F_8, F_{11}) [vC 73].

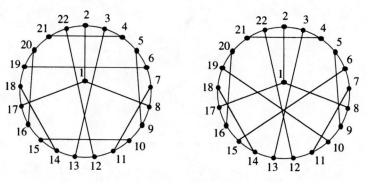

Figure 3.5

n = 22. L(3) gives a required hypohamiltonian graph here. In addition P(11,2),
see [jB 72], gives a cubic hypohamiltonian graph, as do the two graphs of Figure
3.5; see [C-S 78]. Both of these graphs contain a subgraph which is a subdivison
of the Petersen graph.

Having dispensed with the preliminaries we are now able to produce the first
theorem.

Theorem 3.3 There exists a hypohamiltonian graph for every order $n \geq 18$ and
for $n = 10, 13, 15, 16$.

Proof: We proceed in four steps.

(i)   If the hypohamiltonian graph G contains a vertex v of degree 3, then v
does not lie on a triangle.

Let $N(v) = \{u_0, u_1, u_2\}$. Since G_{u_2} is hamiltonian, it contains a hamiltonian cycle
which must be, in part, $(..., u_0, v, u_1, ...)$. But if (v, u_1, u_2) is a triangle in G,
this cycle can be modified to $(..., u_0, v, u_2, u_1, ...)$. Hence G is hamiltonian.

(ii)   Suppose that v and v' are vertices of degree three in graphs G, G' and
$N(v) = \{u_0, u_1, u_2\}$, $N(v') = \{u_0', u_1', u_2'\}$. We form a new graph H from G, G'
by removing v, v', respectively, and identifying the vertices u_i and u_i', i = 0,1,2.
H is simple by (i). We label these vertices w_0, w_1, w_2 and in the argument
below, according to context, regard them as vertices of G, G' or H. The
construction is shown in Figure 3.6. (Note that H(15) is constructed this way
from two copies of the Petersen graph.)

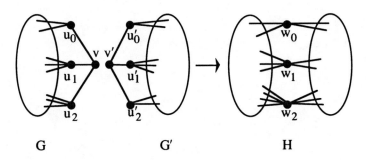

$$G \qquad\qquad G' \qquad\qquad H$$

Figure 3.6

(iii)   If G, G' are hypohamiltonian, then H formed by the construction in (ii), is
also hypohamiltonian.

Suppose that H is hamiltonian. Then w_0, w_1, w_2 are on some hamiltonian cycle in H which is therefore the union of three paths Q_i from w_{i+1} to w_{i+2} and avoiding w_i, $i = 0, 1, 2$, addition being modulo 3. But two of these paths must lie in one of G, G'. Suppose without loss of generality that Q_0, Q_1 lie in G. Then Q_0 followed by Q_1 is a hamiltonian path in G_v which joins u_0 to u_1. Inserting the vertex v and the edges vu_0, vu_1 gives rise to a hamiltonian cycle in G. This contradiction proves that H is not hamiltonian.

Now let x be any vertex of H and suppose without loss of generality that $x \in VG$. Since G_x contains a hamiltonian cycle, $G - \{x, v\}$ contains a hamiltonian path joining u_0 and u_1, say. Similarly $G' - \{u'_2, v'\}$ contains a hamiltonian path joining u'_0 to u'_1. Putting these two paths together in H_x gives a hamiltonian cycle. Hence H_x is hamiltonian for all $x \in VH$.

(iv)   We now complete the proof. As we have shown by examples prior to the theorem, there exist hypohamiltonian graphs of order $10, 13, 15, 16, 18, 19, 20, 21$ and 22. We note that all of these graphs have two non-adjacent vertices of degree 3. Since the construction of (ii) gives a hypohamiltonian graph of order $n_1 + n_2 - 5$ from two such graphs of order n_1 and n_2, we are thus able to construct higher order hypohamiltonian graphs. Because our small graphs contain two non-adjacent vertices of degree 3, each new graph will contain two such vertices and the construction can be continued indefinitely. We thus construct a hypohamiltonian graph for every order greater than 22.   □

The proof used here is that of Thomassen [cT 74a]. When he proved the corresponding theorem in 1974 the cases 14 and 19 were still open. We now deal with the cubic case.

Theorem 3.4 There exists a cubic hypohamiltonian graph of even order n for $n = 10$ and for all $n \geq 18$. Cubic hypohamiltonian graphs exist for no other orders.

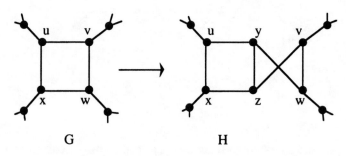

Figure 3.7

Proof: Suppose that G is a cubic hypohamiltonian graph that contains an induced cycle on the four vertices, u,v,w,x. Then remove the edges uv, wx and add two new vertices y,z and the edges yu,yw,yz,zv,zx to form the cubic graph H with |VH| = |VG| + 2. This construction is illustrated in Figure 3.7.

If H is hamiltonian, then it is straightforward to show that G is also. This is accomplished by considering the various possibilities for the hamiltonian cycle in H. For instance, if this cycle were such that in H we had (....,u,x,...,v,z,y,w,...), then (...,u,x,...,v,w,...) would be the corresponding hamiltonian cycle in G.

Similarly if $s \in VH \cap VG$, then the hamiltonian cycle in G_s is easily amended to a hamiltonian cycle in H_s. In H_y we simply take the hamiltonian cycle of G_v, remove the edge wx, and insert in its place the path w,v,z,x. Hence if G is hypohamiltonian, then so is H.

The examples prior to Theorem 3.3 give cubic hypohamiltonian graphs of order 10, 18, 20, 22 and show that such graphs do not exist for 12, 14, 16. The graph T(1) of Figure 3.8 gives a cubic hypohamiltonian graph of order 24 which has an induced cycle on four vertices. If we apply the construction above to this graph, the theorem follows.   □

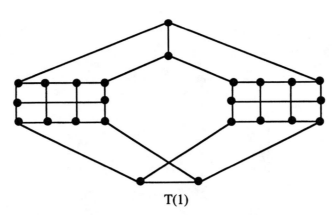

T(1)

Figure 3.8

The theorem above is due to Thomassen [cT 81]. Earlier versions of Theorems 3.3 and 3.4 were proved by Chvátal [vC 73] using the flip-flop concept.

4. Questions alive and dead
If f(n) is the number of non-isomorphic hypohamiltonian graphs of order n, what is f(n)?

This is a fairly natural question. Using flip-flops, Chvátal [vC 73] has shown that $f(n) \to \infty$ as $n \to \infty$ and Doyen and Van Diest [D-V 75] have found linear lower bounds for $n = 3k$ and $n = 5k$. However, the best result to date is by Collier and Schmeichel [C-S 77] who have shown that $f(n)$ is larger than a certain exponential function. Before we can prove this theorem we need a further construction.

A flip-flop (G,a,b,c,d) is said to be **strong** if: (a,c) is not good in $G - \{b,d\}$; (a,b) is not good in $G - \{c,d\}$; (b,d) is not good in $G - \{a,c\}$; and (c,d) is not good in $G - \{a,b\}$.

It is straightforward to check that the flip-flops $F_8, F_{11}, F_{13}, F_{14}$ and F_{18} are strong.

Suppose B is an arbitrary non-empty bipartite graph with no isolated vertices and with parts $U = \{u_1, u_2, ..., u_m\}$, $V = \{v_1, v_2, ..., v_n\}$ with $m, n \geq 2$ and suppose (G, a_0, b_0, c_0, d_0) is a flip-flop. Then we define the graph $[G,B]$ to be the disjoint union of G and B together with the new vertices $\{a_i, d_i: i = 1, 2, ..., m\} \cup \{b_j, c_j: j = 1, 2, ..., n\}$, cycles $J_1 = (a_1, u_1, d_1, a_2, u_2, d_2, ..., a_m, u_m, d_m)$ and $J_2 = (b_1, v_1, c_1, b_2, v_2, c_2, ..., b_n, v_n, c_n)$, and the edges $a_0 a_i, d_0 d_i, i = 1, 2, ..., m$ and $b_0 b_j, c_0 c_j, j = 1, 2, ..., n$.

The graph $[G, K_{3,3}]$ is illustrated in Figure 4.1, where (G, a_0, b_0, c_0, d_0) is an arbitrary flip-flop.

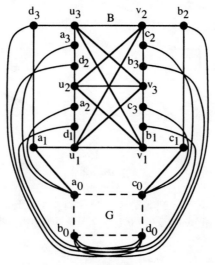

$[G, K_{3,3}]$

Figure 4.1

Theorem 4.1 If (G, a_0, b_0, c_0, d_0) is a strong flip-flop and B is bipartite with no isolated vertices and each part greater than one, then [G,B] is hypohamiltonian. □

The proof uses standard techniques so we do not include it here; see [C-S 77]. It is not difficult to see that [G,B] ≠ [G,B'] whenever B ≠ B' and so by considering bipartite graphs B of the appropriate orders we have the following result.

Theorem 4.2 There exist constants $\alpha, \beta > 0$ such that $f(n) \geq \exp(\alpha n + \beta)$ for all $n \geq 20$. □

Does every hypohamiltonian graph have girth greater than 4? Is there a hypohamiltonian graph containing a triangle?

These questions arose in [H-D-V 67] and [vC 73] in view of the fact that all of the hypohamiltonian graphs that were known at the time had girth 5 and because of Bondy's observation that no vertex of degree 3 is contained in a triangle.

The first question is dismissed by reference to Theorem 4.1 and T(1). If $B \cong K_{2,2}$, then [G,B] has girth at most 4. In fact there are hypohamiltonian graphs which contain large numbers of 4-cycles. Combining previous results we have the following theorem.

Theorem 4.3 Except for possibly $n = 21,22,27$, there exists a girth 4 hypohamiltonian graph for every order $n \geq 20$.

Moreover if $n = 3(m_1 + n_1) + w$, where $w \in \{8,11,18\}$ and $m_1, n_1 \geq 2$, then there exists a hypohamiltonian graph that contains at least $\binom{m_1}{2}\binom{n_1}{2}$ cycles of length 4.

Proof: Since the number of vertices in [G,B] is $3(m_1 + n_1) + |VG|$, where B is a spanning subgraph of K_{m_1,n_1} with $m_1, n_1 \geq 2$ and with no isolated vertices, we can show that there exists a [G,B] of order n for all $n \geq 30$, by using $G = F_8$, F_{11} or F_{18}. The other orders between 20 and 30 can be filled in using other strong flip-flops.

For the second part of the theorem we note that K_{m_1,n_1} contains $\binom{m_1}{2}\binom{n_1}{2}$ cycles of length 4. □

But hypohamiltonian graphs of girth 3 do exist. In [cT 74b], Thomassen constructed the graph T(2); see Figure 4.2. Moreover there are hypohamiltonian graphs of almost every order which contain triangles and further, contain large numbers of triangles.

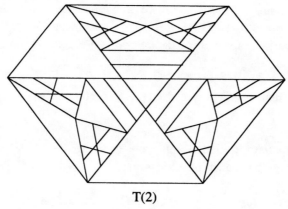

T(2)

Figure 4.2

Before proving this we need a construction. Let $f_i = (G_i, a_i, b_i, c_i, d_i)$ be flip-flops for $i = 1, 2, ..., n$. Then $F(n) = (...((f_1, f_2), f_3), ..., f_n)$ is a flip-flop defined by successive application of Chvátal's binary operation; see G_4 in Figure 2.3. Let $G(n)$ be the graph based on $F(n)$, see Figure 2.4, and let $G'(n)$ be the graph $G(n)$ together with the additional edges ud_j, $j = 1, 2, ..., n - 1$, ua_k, $k = 2, 3, ..., n$.

Theorem 4.4 $G'(n)$ is a hypohamiltonian graph with $n - 1$ triangles.

Proof. $G'(n)$ certainly has $n - 1$ triangles. To show that $G'(n)$ is not hamiltonian we assume that it is and obtain contradictions in a similar way to the proof of Lemma 2.6. Since $G(n)_w$ is hamiltonian for all $w \in VG(n)$ by Lemmas 2.5 and 2.6, then $G'(n)_w$ must be also. The addition of edges to a hamiltonian graph cannot destroy a hamiltonian cycle.   □

As we have noted earlier, the Sousselier cubic hypohamiltonian graph S on 18 vertices is based on (F_8, F_8). Using Theorem 4.4 we can use S to produce an 18 vertex hypohamiltonian graph with one triangle. This is, of course, a smaller graph than $T(2)$.

Now we can show that there are large numbers of girth 3 hypohamiltonian graphs. The theorem is due to Collier and Schmeichel [C-S 77].

Theorem 4.5 Except for possibly $n = 19, 20, 22, 25$, there exists a girth 3 hypohamiltonian graph for every order $n \geq 18$. Moreover, if $n = 8p + 10$, there exists a hypohamiltonian graph which contains at least p triangles.

Proof: As we have seen, flip-flops $F_8, F_{11}, F_{13}, F_{14}, F_{18}$ exist which are of order 8,11,13,14,18, respectively. Various combinations of these give flip-flops f of order 16,19,21,22,24,25,....,31 and thereafter we can get flip-flops of order |f| + 8 via (f,F_8). Hence we have hypohamiltonian graphs G(n) of the orders required by the theorem, formed from these flip-flops. The graphs G'(n) are then the required girth 3 graphs. If we construct G'(n) from n copies of F_8, then G'(n) has n - 1 triangles (see Theorem 4.4) and order 8n + 2. Now let n=p+1 and the proof is complete.   □

It was observed by Thomassen [cT 81] that no known cubic hypohamiltonian graph had girth **greater** than 5. Do such graphs exist?

Moving in a different direction, Chvátal [vC 73] asked if there exists a **planar** hypohamiltonian graph.

The following theorem is due to Thomassen; see [cT 76].

Theorem 4.6 There exist infinitely many planar hypohamiltonian graphs.

Proof: The graph in Figure 4.3 is a planar hypohamiltonian graph on 57 vertices. It was produced by Hatzel in [wH 79]. We note that this graph contains at least one pair of non-adjacent vertices of degree 3 so we can use the construction shown in Figure 3.6 to produce infinitely many planar hypohamiltonian graphs starting from Hatzel's graph.   □

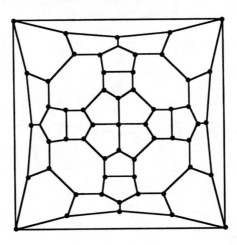

Hatzel's graph

Figure 4.3

We note that in [cT 76] Thomassen used a graph on 105 vertices as his starting point.

Is there a cubic planar hypohamiltonian graph? This question of Chvátal [vC 73] was again settled by Thomassen [cT 76].

Theorem 4.7 There exist infinitely many cubic planar hypohamiltonian graphs.

Proof: Suppose that G is a cubic planar hypohamiltonian graph with induced cycle u,v,w,x. Then we produce a new graph G′ from G by removing the edges uv,wx and adding the cycle (u′,v′,w′,x′), and the edges uu′,vv′,ww′,xx′ as shown in Figure 4.4.

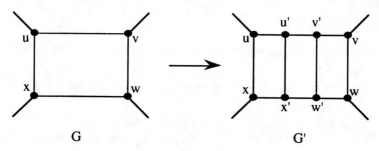

Figure 4.4

It is now not difficult to show, using techniques illustrated in previous theorems, that if G is cubic planar hypohamiltonian, then so is G′.

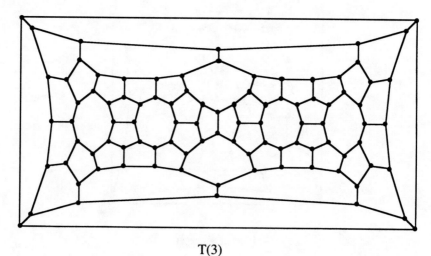

T(3)

Figure 4.5

Hence we require only a starting graph in order to prove the current result. This is provided by the graph T(3) with 94 vertices of Figure 4.5. The proof of its hypohamiltonicity is given in [cT 81].   □

This leaves open the following questions.

What is the smallest order N (respectively N_c) for which there exists a planar (respectively cubic planar) hypohamiltonian graph?

Does there exist a planar (respectively cubic planar) hypohamiltonian graph for every order N+ m (respectively N_c + 2m) for all positive integers m?

We have no information on these questions for non-cubic graphs, but the proof of Theorem 4.7 does give us a cubic planar hypohamiltonian graph for every order 94 + 4m. Is Hatzel's graph the smallest planar hypohamiltonian graph? Is T(2) the smallest cubic planar hypohamiltonian graph?

If the deletion of an edge e from a hypohamiltonian graph G does not create a vertex of degree 2, then is G - e hypohamiltonian? If the addition of a new edge to a hypohamiltonian graph of girth greater than 4 does not create a cycle of length less than 5, then is it true that it does not create a hamiltonian cycle [vC73]?

The answer to both these questions is no and is given by the graph T(4) of Figure4.6 [cT 74b]. This graph can be shown to be hypohamiltonian. The deletion of any edge gives a non-hypohamiltonian graph. If an edge is inserted between the vertices u and v, then the heavy edges together with uv, form a hamiltonian cycle. However no cycle of T(4) + {uv} is of length less than 5.

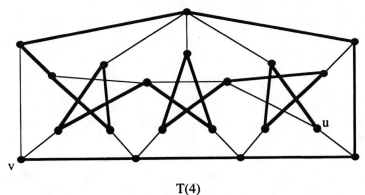

T(4)

Figure 4.6

Given a graph G is there a hypohamiltonian graph which has a subgraph isomorphic to G [vC 73]?

For G a bipartite graph with both parts of size greater than 1, the question is answered by Theorem 4.1. This is also proved in [cT 81] where it is also noted that there are many planar bipartite graphs which are subgraphs of planar hypohamiltonian graphs. Is this the case for any planar bipartite graph?

Now K_3 can be the subgraph of a hypohamiltonian graph by Theorem 4.4. What about K_4? In [cT 81] it is shown that K_4 is not contained in any planar hypohamiltonian graph.

Is there a hypohamiltonian graph with no vertex of degree 3 [C-S 77]? If so, does there exist a regular hypohamiltonian graph of degree k for all $k \geq 3$? For what k does there exist a hypohamiltonian graph of connectivity (minimum degree) k [cT 78]?

To illustrate how little is known in answer to the first of these questions, the best result in this direction is due to Häggkvist [rH 87] who has proved that every hypohamiltonian graph has a vertex of degree less than $\frac{8}{17}|VG|$.

Thomassen has proved the following result.

Theorem 4.8 Every planar hypohamiltonian graph contains a vertex of degree 3.

Proof: A nice proof of this result can be found in [cT 78].   □

As far as the other questions go, nothing is known for k > 3. Can a hypohamiltonian graph with n vertices have a vertex of degree greater than $\frac{1}{3}(n - 1)$ What is the maximum number of edges among hypohamiltonian graphs with n vertices [vC 73]?

Theorem 4.9 For every $n = 4k + 2, k \geq 23$, there exists a hypohamiltonian graph of order n, having $\frac{1}{4}(n - 90)^2 + 137$ edges, with maximum degree $\frac{1}{2}n - 44$. Such a graph contains a spanning cubic planar hypohamiltonian graph as well as $K_{\frac{1}{2}(n-90),\frac{1}{2}(n-90)}$.

Proof: Let G be the graph obtained from T(3) by the construction of Figure 4.4. Let S be the set of vertices added to VT(3) to produce VG, along with the four vertices on the cycle of length 4 in T(3). The graph induced by S is bipartite. Add edges to G so that the bipartite graph becomes a complete bipartite graph. Then we still have a hypohamiltonian graph. This graph has the properties stated in the theorem. The result now follows by iteration.   □

The above result is in [cT 81].

We note that by Lemma 1.2, no vertex in a hypohamiltonian graph can have degree greater than $\frac{1}{2}n$. Hence a hypohamiltonian graph can have at most $\frac{1}{4}n^2$ edges. The graph given in Theorem 4.9 then, is close to being best possible.

Do there exist infinitely many vertex-transitive hypohamiltonian graphs [cT 76]?

To date the only known vertex-transitive hypohamiltonian graphs are the Petersen graph and the Coxeter graph. Interestingly the Coxeter graph, see Figure 5.1 of Chapter 9, is based on the flip-flop F_{26} shown in Figure 4.7.

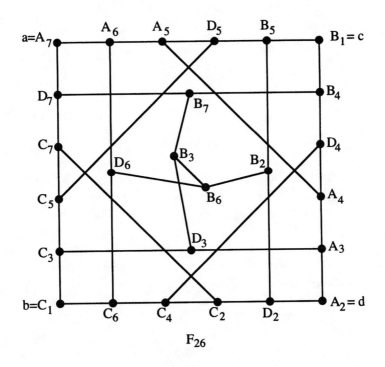

F_{26}

Figure 4.7

Do there exist infinite locally finite hypohamiltonian graphs [cT 76]?

Thomassen has shown the existence of infinite hypohamiltonian graphs but none of them are locally finite [cT 76].

5 . A nine point theorem

As P is hypohamiltonian, there is a cycle through any nine vertices of P. The rather surprising result to be proved in this section is that there is a cycle through any nine vertices in **every** cubic 3-connected graph. This theorem appears in [H-M-P-T 82].

We shall make use of the following three reductions of cubic 3-connected graphs.

Let G be a cubic 3-connected graph, e = uv ∈ EG, N(u) = {u_1,u_2,v} and N(v)= {v_1,v_2,u}. Then the **e-reduction** of G is the multigraph G - {u,v} with the edges u_1u_2 and v_1v_2 added. This reduction is shown in Figure 5.1.

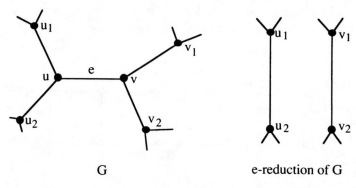

G   e-reduction of G

Figure 5.1

As usual we denote the cyclic edge-connectivity of a graph G by $\lambda_c(G)$; see page 85. If $\lambda_c(G) \geq 4$ and G ≠ K_4, then any e-reduction is 3-connected. This is because if G - {u,v} has a 2-edge cyclic cut S then S ∪ {e} is a 3-edge cyclic-cut of G.

We recall from p. 84 that a 3-cut-reduction is a graph obtained from G by contracting a component of G - E to a single vertex where E is an edge cutset of size 3. Corresponding to E there are two 3-cut-reductions H and J which are shown in Figure 5.2. We will use the notation of Figure 5.2 below.

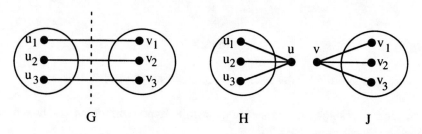

G   H   J

Figure 5.2

It is clear that any 3-cut-reduction of G is a simple graph which is both cubic and 3-connected. If G is bipartite, then so are any of its 3-cut reductions. For example suppose that (X,Y) is the induced bipartition of $H - \{u\}$; see Figure 5.2. Assume that u_1, u_2 and u_3 do not all belong to the same part. Then without loss of generality $u_1 \in X$ and $u_2, u_3 \in Y$. Hence $q(X,Y) \equiv 2 \pmod 3$ and $q(Y,X) \equiv 1 \pmod 3$ which is impossible. Therefore u_1, u_2 and u_3 all belong to the same part and H is bipartite.

Our third reduction involves cubic 3-connected graphs which contain 4-cycles. Let $C = (u_1, u_2, u_3, u_4)$ be a 4-cycle in G and let $v_i \in N(u_i) \setminus VC$ for $i=1,2,3,4$. Then the cubic multigraphs $G_1 = (G - VC) \cup \{v_1v_2, v_3v_4\}$ and $G_2 = (G - VC) \cup \{v_2v_3, v_1v_4\}$ are called **4-cycle reductions of G with respect to C.** These reductions are illustrated in Figure 5.3. Both G_1 and G_2 are connected since G is 3-connected.

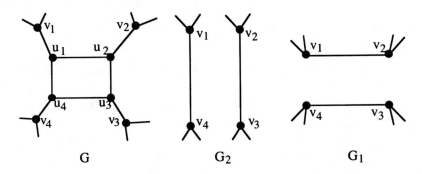

$$G \qquad\qquad G_2 \qquad\qquad G_1$$

Figure 5.3

Lemma 5.1 If G is a cyclically 4-edge-connected cubic graph of order greater than 8 and $C = (v_1, v_2, v_3, v_4)$ is a 4-cycle of G, then at least one of the 4-cycle reductions with respect to C is 3-connected.

Proof: Let $G_1 = (G - VC) \cup \{v_1v_2, v_3v_4\}$. Suppose that G_1 is not 3–connected. Then G_1 has a set E_1 of one or two edges such that $G_1 - E_1$ has two components H_1 and J_1, say. Since G is 3-connected, we may assume without loss of generality, that $v_1v_2 \in EH_1$ and $v_3v_4 \in EJ_1$. Any other possibilities (see Exercise 20) would imply that G is not cyclically 4–edge–connected. Again since $\lambda_c(G) \geq 4$, $|E_1| = 2$. Now consider the other 4–cycle reduction $G_2 = (G - VC) \cup \{v_2v_3, v_1v_4\}$. Then, by the same argument, G_2 contains a set E_2 consisting of two edges such that $G_2 - E_2$ has two components, H_2 and J_2, say, with $v_2v_3 \in EH_2$ and $v_1v_4 \in EJ_2$. Then $G^* = (G - VC) - (E_1 \cup E_2)$ has at least 4 components because v_1, v_2, v_3 and

v_4 must belong to distinct components M_1, M_2, M_3 and M_4, respectively. This in turn is because $v_1v_2 \in EH_1$, $v_3v_4 \in EJ_1$, $v_2v_3 \in EH_2$ and $v_1v_4 \in EJ_2$. Again, since G is 3-connected, $q(VM_i, VG \setminus VM_i) \geq 3$.

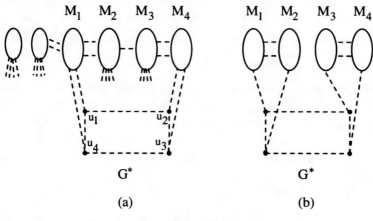

Figure 5.4

The situation so far is illustrated in Figure 5.4(a) with the elements of $EG \setminus EG^*$ shown by broken lines. Now since there are at least 3 edges 'leaving' each M_i and in total no more than 8 edges 'entering' all the M_i's together, it follows that there are precisely 4 components with precisely 3 edges 'leaving' each component; see Figure 5.4(b). Since $\lambda_c(G) \geq 4$ it follows that $|VM_i| = 1$, $i=1,2,3,4$. Therefore $|VG| = 8$. □

Lemma 5.2 Suppose that G is a 3-connected cubic bipartite graph. Then:
(i)   if $|VG| \leq 18$, then G is hamiltonian;
(ii) if $|VG| \leq 14$, then for any $e \in EG$ there is a hamiltonian cycle containing e;
(iii) if $|VG| \leq 10$, then any set of three edges belongs to a hamiltonian cycle provided the set is not a cutset;
(iv) if $|VG| \leq 10$, then there is a hamiltonian cycle which avoids any given edge.

Proof: We prove (ii), (iii) and (iv) by inspecting the list of graphs in [B-C-C-S 76]. Part (i) can then be proved by using 3-cut reductions or 4-cycle reductions; see [mE 82]. □

Lemma 5.3 Let G be a 3-connected cubic graph, let $A \subseteq VG$ with $|A| = 5$ and let $e \in EG$. Then there exists a cycle C in G with $A \subseteq VC$ and $e \notin EC$.

Proof: By inspection, we see that the result holds for graphs with $|VG| \leq 8$; see [B-C-C-S 76] for a list of small cubic graphs. Now proceed by induction.

Case 1. G is cyclically 3-edge-connected. Since there exists a cyclic edge cutset S of size 3, form the two 3-cut reductions H and J with respect to S. The vertices $u \in VH \setminus VG, v \in VJ \setminus VG$. Without loss of generality assume $e \in EH$. Let $a = |A \cap VH|$.

If $1 \leq a \leq 4$, then there exists a cycle C_H in H which passes through $A \cap VH$ and u, and avoids e. Suppose, without loss of generality, that $uu_3 \notin EC_H$. By induction we may obtain a cycle C_J in J which passes through $A \cap VJ$ and v, and avoids vv_3. In G, $(C_H - \{u\}) \cup (C_J - \{v\})$ is a cycle through A which avoids e.

Small modifications of this argument give the result for $a = 0$ and $a = 5$.

Case 2. G is cyclically 4-edge-connected. If there exists an edge $f \in EG$ which is incident with no vertex of A, then form the f-reduction, G_f, of G. Since the required cycle exists in G_f, by induction, it exists in G.

If every edge of G is incident with a vertex of A, then $|VG| \leq 10$. Hence $|VG|=10$, G is bipartite and we can appeal to Lemma 5.2.   □

Lemma 5.4 Let G be a cyclically 4-edge-connected cubic graph with $|VG|\leq18$. Then G is hamiltonian or isomorphic to P or one of the Blanuša snarks.

Proof: See [Mc-R 86].   □

We can now prove the Nine Vertex Theorem which states that any 3-connected cubic graph has the nine vertex property of P. The proof given here is slightly different from that given in [H-M-P-T 82] and totally different from [K–L 82], where the notion of T-separators is used.

Theorem 5.5 Let G be a 3-connected cubic graph and let $A \subseteq VG$ with $|A|=9$. Then there is a cycle C in G such that $A \subseteq VC$.

Proof: The general method of proof is that of Lemma 5.3. First we note that all graphs G with $|VG| \leq 10$ have the required property. This result follows by testing the graphs in [B-C-C-S 76].

Case 1. G is not cyclically 4-edge-connected. Form 3-cut reductions H and J of G. Let $a = |A \cap VH|$ and assume, without loss of generality, that $a \geq 5$.

For $5 \leq a \leq 8$, there exists a cycle C_H through $A \cap VH$ and u, by the induction hypothesis. If C_H avoids uu_3 then by Lemma 5.3, there exists a cycle C_J through $A \cap VJ$ and v, which avoids vv_3. Combining C_H and C_J in G gives the required cycle through A.

If $a = 9$, then the cycle C_H through A may or may not use u. When $u \in VC_H$, assume $uu_3 \notin EC_H$. Then we employ Lemma 5.3 again to give a cycle C_J in J through v avoiding vv_3. So either C_H alone or a combination of C_H and C_J, gives a cycle in G through A.

Case 2. G is cyclically 4-edge-connected. If there exists an edge $f \in EG$ such that f is incident with no vertex of A, then form G_f, the f-reduction of G. By induction, there is a cycle of G_f through A and this gives the required cycle in G.

If every edge of G is incident with A, then $|VG| \leq 18$. If $|VG| = 18$, then G is bipartite and so is hamiltonian by Lemma 5.2. Otherwise G is P or hamiltonian by Lemma 5.4.   □

The theorem above is best possible in that graphs which are contractible to the Petersen graph do not have a cycle through an arbitrary set of 10 vertices. It was conjectured [H-M-P-T 82] that any 3-connected cubic graph which does not contain a cycle through some set of 10 vertices is contractible to the Petersen graph. This conjecture was settled in [E-H-L 84] and [K-L 82] where the 'Ten Vertex Theorem' was proved.

Theorem 5.6 Let G be a 3-connected cubic graph and let A be a subset of VG with $|A| = 10$. Then there is a cycle containing A unless G is contractible to the Petersen graph with A being mapped onto VP under the contraction.   □

Theorem 5.6 has been extended to 11 vertices [A-B-H-R 88] and 12 vertices [B-H 91]. However, it is clear because of the existence of a cubic hypohamiltonian graph on 18 vertices that we will not always have a cycle through a given set of vertices even if the graph is contractible to P in an appropriate way.

Actually we stop at 13 vertices. There is no cycle through the 13 black vertices of the graph in Figure 5.5. Although this graph is contractible to P, the black vertices are not mapped on to VP under any such contraction. It is conjectured in [B-H 91] that 13 vertices lie on a cycle, unless the graph is contractible to P or the graph of Figure 5.5, in the obvious ways.

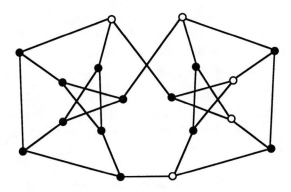

Figure 5.5

For planar 3-connected cubic graphs we can guarantee many more vertices lie on a cycle. In [H-M 86] it is shown that any 19 vertices lie on a cycle in such graphs. This is improved to 21 in [B-H-M 89]. However, there are many 3-connected cubic planar graphs which have 24 vertices that do not lie on a common cycle. Thus a theorem similar to Theorem 5.6 for planar graphs may be quite complicated because of the potential number of exceptional cases. It is likely, however, that any 23 vertices lie on a cycle in 3-connected cubic planar graphs.

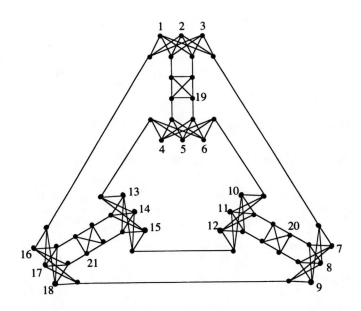

Figure 5.6

Define $f(k)$ $(k \geq 2)$ to be the largest positive integer such that any k-connected k-regular graph has a cycle through any specified set of at most $f(k)$ vertices. The Nine Vertex Theorem proves that $f(3) = 9$. In [dH 82] and [K-L 82] it was proved that $f(k) \geq k + 4$. This bound has been improved by Häggkvist and Mader [H-M 89] to $k + \lfloor \frac{1}{3}\sqrt{k} \rfloor$. Meredith [GM 73] gave examples of k–connected, k–regular non-hamiltonian graphs for each $k \geq 3$. These are obtained by replacing each vertex of the Petersen graph by a copy of $K_{k,k-1}$. In particular, since there is no cycle through all sets of size $k - 1$ in the bipartition of each replacement $K_{k,k-1}$, we have proved that $f(k) \leq 10k - 11$. McCuaig and Rosenfeld [M-R 84] improved this upper bound. By construction they prove that: (i) for each $k \geq 4$, $k \equiv 0 \pmod 4$, $f(k) \leq 6k - 4$; (ii) for each $k \geq 4, k \equiv 2 \pmod 4$ $f(k) \leq 8k - 5$. In Figure 5.6 we show a 4-connected, 4-regular graph which contains a set of 21 vertices that do not lie on a cycle.

Jackson [bJ 87] has produced infinite families of non-hamiltonian k–connected k–regular graphs for $k \equiv 0 \pmod 4$ which are smaller than those of Meredith.

Let $\delta(G)$ be the minimum degree of G. In [E-G-L 91] it has been shown that if G is k-connected, then G has a cycle of length at least $2\delta(G)$ through any specified set of k vertices.

Exercises

1. Prove Proposition 2.1.

2. Prove that $D(3,3)$ and $D(4,3)$ are hypohamiltonian while $D(2,4)$ is hamiltonian.

3. Show that $D(2,2,3)$ and $D(2,2,5)$ are hypohamiltonian.

4. Prove that $P(11,2)$ is hypohamiltonian; see Proposition 2.4.

5. Show that not all non-planar cubic hypohamiltonian graphs contain the Petersen graph.

6. Prove Lemma 2.5.

7. Show that $D(2,3) \cong P$.

8. (a) Show that $D(m,3)$ is hypohamiltonian for $m \geq 2$ and that $D(m,n)$ is hamiltonian for $m \geq 3, n > 3$.

(b)   Show that D(2,1,5) is isomorphic to J_5. Deduce that D(2,1,5) is
       hypohamiltonian. Further show that D(2,1,k) for $k \geq 5$ and odd, is
       isomorphic to J_k and hence deduce that D(2,1,k) is hypohamiltonian.

9.   For what k,m is P(k,m) hypohamiltonian? For what k,m is P(k,m)
      hamiltonian?

10.   Show that Coxeter's graph is hypohamiltonian.

11.   Let G be a hypohamiltonian graph on 17 vertices which contains a vertex
       of degree 6. Show that G has girth at most 4.

12.   Find three non-isomorphic hypohamiltonian graphs on 21 vertices.

13.   Show that F_8, F_{11} and F_{13} are flip-flops.

14.   Do there exist hypohamiltonian graphs of girth 3 on 19,20,22,25 vertices?

15.   What are the smallest hypohamiltonian graphs of girth 3 and of girth 4?

16.   Do there exist hypohamiltonian graphs containing K_4?

17.   Prove that, in a k-connected graph, there is a cycle through any k vertices.

18.   Prove that in a k-connected k-regular graph, any k + 1 vertices lie on some
       cycle.

19.   Suppose that G is a cubic graph containing a 4-cycle C with $\lambda_c(G) \geq 4$.
       Assume that G contains a 4-cycle reduction G_1 with respect to C with
       the property that for each $e \in EG_1$ there is a hamiltonian circuit containing
       e. Prove that G is hamiltonian.

20.   Suppose that G is cubic graph with $\lambda_c(G) \geq 4$. Assume that G contains a
       4-cycle $C = (u_1,u_2,u_3,u_4)$. Write $G_1 = (G - VC) \cup \{v_1v_2,v_3v_4\}$ i.e. G_1
       is a 4-cycle reduction with respect to C. Now suppose that G_1 contains a
       cyclic edge-cut S with $|S| \leq 3$. Prove that (i) v_1 and v_2 belong to the
       same component H of $G_1 - S$; (ii) v_3 and v_4 belong to $VG_1 \setminus VH$.

21.   Prove Lemma 5.2(i).

22.   Let $B \subseteq EG$, where G is a 3-connected cubic graph. If $|B| = 3$, is there a

cycle through all edges of B? What can be said if $|B| = 4$?

23. Under what conditions on a 3-connected cubic graph do any 12 vertices lie on a cycle?

[Hint: see [B-H 91].]

References

[A-B-H-R 88] R.E.L. Aldred, Bau Sheng, D.A. Holton and G.F. Royle, An eleven-vertex theorem for 3-connected cubic graphs, *J. Graph Th.*, 12, 1988, 561-570.

[B-H 91]   Bau Sheng and D.A. Holton, Cycles containing twelve vertices in 3-connected cubic graphs, *J. Graph Th.*, 15, 1991, 421-429.

[B-H-M 89]   Bau Sheng, D.A. Holton and B.D. McKay, Cycles in 3-connected cubic planar graphs III, submitted.

[jB 72]   J.A. Bondy, Variations on the hamiltonian theme, *Can. Math. Bull.*, 15, 1972, 57-62.

[B-C-C-S 76] F.C. Bussemaker, S. Cobeljic, D.M. Cvetkovic and J.J. Seidel, Computer investigation of cubic graphs, Technical University of Eindhoven, Maths Res. Rep. WSK-01, 1976.

[vC 73]   V. Chvàtal, Flip-flops in hypohamiltonian graphs, *Can. Math. Bull.*, 16, 1973, 33-41.

[C-E 83]   L. Clark and R. Entringer, Smallest maximally non-hamiltonian graphs, *Periodica Math. Hungarica*, 14, 1983, 57-68.

[C-S 77]   J.B. Collier and E.F. Schmeichel, New flip-flop constructions for hypohamiltonian graphs, *Disc. Math.*, 18, 1977, 265-271.

[C-S 78]   J.B. Collier and E.F. Schmeichel, Systematic searches for hypohamiltonian graphs, *Networks*, 8, 1978, 193-200.

[D-V 75]   J. Doyen and V. Van Diest, New families of hypohamiltonian graphs, *Disc. Math.*, 13, 1975, 225-236.

[E-G-L 91]   Y. Egawa, R. Glas and S.C. Locke, Cycles and paths through specified vertices in k-connected graphs, *J. Comb. Th.* B, 52, 1991, 20-29.

[mE 82]　　M.N. Ellingham, Constructing certain cubic graphs, Lecture Notes in Maths No. 952, Springer-Verlag, Berlin, 1982, 252-274.

[E-H-L 84]　M.N. Ellingham, D.A. Holton and C.H.C. Little, Cycles through 10 vertices in 3-connected cubic graphs, *Combinatorica*, 4, 1984, 265-273.

[sG 77]　　S. Gutt, Infinite families of hypohamiltonian graphs, *Acad. Roy. Belg. Bull. Cl. Sci.*, 63, 1977, 432-440.

[rH 87]　　R. Häggkvist, On the structure of non-hamiltonian graphs I, preprint.

[H-M 89]　R. Häggkvist and W. Mader, Circuits through prescribed vertices in k-connected k-regular graphs, submitted.

[wH 79]　　W. Hatzel, Ein Planaren hypohamiltonschan Graph mit 57 Knoten, *Math. Ann.*, 243, 1979, 213-216.

[H-D-V 67]　J.C. Herz, J.J. Duby and F. Vigué, Recherche systématique des graphes hypohamiltoniens, *in Théorie des Graphes*, (Proc. 1966 International Symp. Rom) Dunod, Paris, 1967, 153-160.

[H-G-R 64]　J.C. Herz, T. Gaudin and P. Rossi, Solution du problème No. 29, *Revue Francaise de Recherches Opérationelle*, 8, 1964, 214-218.

[H-M-P-T 82] D.A. Holton, B.D. McKay, M.D. Plummer and C. Thomassen, A nine point theorem for 3-connected graphs. *Combinatorica*, 2, 1982, 53-62.

[dH 82]　　D.A. Holton, Cycles through specified vertices in k-connected regular graphs, *Ars Combinatoria*, 13, 1982, 129-143.

[H-M 86]　D.A. Holton and B.D. McKay, Cycles in 3-connected cubic planar graphs II, *Ars Combinatoria*, 21A, 1986, 107-114.

[bJ 87]　　B.W. Jackson, Small r-regular r-connected nonhamiltonian graphs, *Ars Comb.*, 24, 1987, 77-83.

[K-L 82]　　A.K. Kelmans and M.V. Lomonosov, When m vertices in a k–connected graph cannot be walked round along a simple cycle, *Disc. Math*, 38, 1982, 317-322.

[wL 67]   W.F. Lindgren, An infinite class of hypohamiltonian graphs, *Amer. Math. Monthly*, 74, 1967, 1087-1089.

[M-R 84]   W.D. McCuaig and M. Rosenfeld, Cyclicability of r–regular r–connected graphs, *Bull. Austral. Math. Soc.*, 29, 1984, 1-11.

[Mc-R 86]   B.D. McKay and G.F. Royle, Constructing the cubic graphs on up to 20 vertices, *Ars Combinatoria*, 21A, 1986, 129-140.

[gM 73]   G.H.J. Meredith, Regular n-valent n-connected non-hamiltonian non-n-edge-colorable graphs, *J. Comb. Th.* B, 14, 1973, 55-60.

[rS 63]   R. Sousselier, Problème No. 29: Le Cercle des Irascibles, *Revue Française de Recherches Opérationelle*, 7, 1963, 405-406.

[cT 74a]   C. Thomassen, Hypohamiltonian and hypotraceable graphs, *Disc. Math.*, 9, 1974, 91-96.

[cT 74b]   C. Thomassen, On hypohamiltonian graphs, *Disc. Math.*, 10, 1974, 383-390.

[cT 76]   C. Thomassen, Planar and infinite hypohamiltonian and hypotraceable graphs, *Disc. Math.*, 14, 1976, 377-389.

[cT 78]   C. Thomassen, Hypohamiltonian graphs and digraphs, in Theory and Application of Graphs, Lecture Notes in Mathematics, No. 642, Springer, Berlin, 1978, 557-571.

[cT 81]   C. Thomassen, Planar cubic hypohamiltonian and hypotraceable graphs, *J. Comb. Th.* B, 30, 1981, 36-44.

8

Symmetry

0 . Prologue
I think that I shall never see
A poem showin' symmetry.
Anon.

The Petersen graph has exactly 120 automorphisms. All of its vertices are the same in the sense that any vertex can be mapped into any other by an automorphism (in fact by exactly 12 automorphisms). As we saw in Chapter 6 it has much more symmetry than this. All of its 120 paths of length 3 are the same. In fact there is exactly one automorphism which maps any one such path into any other. The Petersen graph is a Tutte graph since the constraint for the s–transitivity of a graph G given by $s \le \lfloor \frac{1}{2}(\gamma(G) + 2) \rfloor$ is satisfied as an equality when $G = P$. In this sense P is as symmetric as it can be.

The Petersen graph has the property that any two pairs of vertices which are the same distance apart are also the same in the sense above. We say P is **distance transitive**. More precisely if $u, v, x, y \in VP$ and $d(u,v) = d(x,y)$ then there is an automorphism of P which maps u to x and v to y. Very surprisingly there are only twelve finite connected cubic distance transitive graphs.

The automorphism group of P acts **primitively** on its vertices. Roughly speaking this means that the automorphism group acts transitively on the vertex set and there is no k-subset of vertices $(2 \le k < |VP|)$ which always stays together under the action of the automorphism group. We shall see this is a very unusual property for a vertex transitive graph having 2p (p a prime) vertices. There are only three other primitive distance transitive cubic graphs. These are K_4, the Coxeter graph and the Biggs-Smith graph on 102 vertices.

Although we have stressed the symmetry of the Petersen graph there is another way of measuring its symmetry which shows that it falls some way short of perfection. Every isomorphism between induced subgraphs of order at most three is not the restriction of an automorphism whereas, for example, the line graph of $K_{3,3}$ does have this property.

1. Distance transitive graphs

Let G be a connected graph. Then G is **distance transitive** (DT) if whenever u, v, x and y are vertices of G such that $d(u,v) = d(x,y)$, then there is an automorphism σ of G such that $\sigma(u) = x$ and $\sigma(v) = y$. There are just 12 finite cubic DT graphs one of which is of course the Petersen graph. In fact for each $r \geq 3$ there are only finitely many finite DT graphs of degree r. The first general proof of this [pC 82], depended on the classification of finite simple groups. A proof, independent of this classification, was given by [rW 85]. The proof of finiteness when $r = 3$, is not difficult; we sketch it below. It depends essentially on Tutte's result mentioned in the introduction and on being able to convert a bound on the order of the stabilizer of a vertex into a bound for the diameter. The proof for general r is modelled exactly on this proof.

Suppose that Γ is a permutation group acting on a set Ω. A **block** Δ is a subset $\Delta \subseteq \Omega$ such that for all $\sigma \in \Gamma, \sigma(\Delta) = \varnothing$ or Δ, see [hW 64]. Clearly $\Omega, \varnothing$ and $\{x\}$ ($x \in \Omega$) are blocks. All other blocks are said to be **non-trivial**. If the action of Γ is transitive on Ω and there are no non-trivial blocks, (Ω, Γ) is said to be a **primitive** permutation group. The automorphism group of P acts primitively on VP for example. In fact P is the only vertex-transitive graph with 2p vertices (pa prime) which has a primitive automorphism group and is not complete. If Γ acts transitively on Ω and non-trivial blocks exist, then (Ω, Γ) is said to be **imprimitive**.

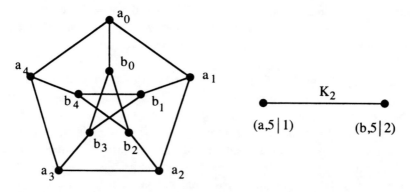

Figure 1.1

The graphs K_4, P, the Coxeter graph C_{28} and the sextet graph S(17) [B-H 83] (sometimes called the Biggs-Smith graph) have been singled out by Biggs [nB 73] as being of special interest. This is because they are the only DT cubic graphs whose automorphism group is primitive. We will call P, C_{28} and S(17), **the Biggs graphs.** The graph K_4 is not considered to be of sufficient interest to be included in this set. These graphs have similar properties. Indeed their structure can be described in a similar way.

This very nice description is due to Biggs. Let the symbol $(x,n \mid s)$, or simply $(n \mid s)$, denote the polygon with vertices $x_0, x_1,..., x_{n-1}$ and edges $x_i x_{i+s}$ for $i=0, 1,..., n - 1$ where the suffices are taken modulo n. Then the Petersen graph is constructed by taking five edges $a_i b_i$, $i = 0, 1,..., 4$ together with polygons $(a,5 \mid 1)$ and $(b,5 \mid 2)$. We show in Figure 1.1 the usual representation of P together with a symbolic representation. The symbol means that we take 5 copies of the graph K_2 (shown in the symbol) and form a cubic graph by joining the 'free ends of the K_2's' by the indicated polygons.

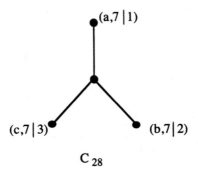

C_{28}

Figure 1.2

The Coxeter graph is shown symbolically in Figure 1.2. Here we take 7 copies of the graph $K_{1,3}$ (shown in the symbol) and join the 'free ends' by the indicated polygons.

The sextet graph $S(17)$, see Figure 1.3, is constructed in an analogous way.

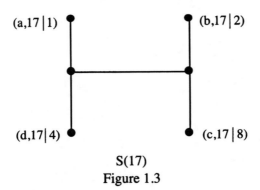

$S(17)$

Figure 1.3

Table 1 lists (see [B-S 71]) all cubic DT graphs along with some of their properties. Later, in Theorem 1.1 and the discussion following it, we indicate how these graphs are produced. The column $s(G)$ in the table indicates the strict s–transitivity of G.

graph	other descriptions	s(G)	|VG|	diam G	γ(G)
K_4	(3,3)-cage	2	4	1	3
$K_{3,3}$	(3,4)-cage	2	6	2	4
P	(3,5)-cage	3	10	2	3
Heawood	(3,6)-cage; Levi graph of PG(2,2)	4	14	3	6
Tutte 8-cage, T	Levi graph of Cremona-Richmond configuration [hC 50]	5	30	4	8
Cube		2	8	3	4
Pappus	See Figure 1.4	3	18	4	6
Desargues	P(10,3)	3	20	5	6
3-fold cover of T	see [aG 74]	5	90	8	10
Dodecahedron	Double cover of P	2	20	5	5
Coxeter C_{28}		3	28	4	7
Sextet graph S(17)	Biggs-Smith	4	102	7	9

The 12 cubic DT graphs

Table 1

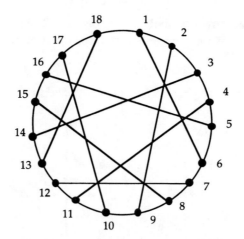

Pappus' graph

Figure 1.4

We need some preliminaries before outlining the proof of Theorem 1.1.

A graph F is **antipodal** if the relation R defined on VG by

> u R v if and only if d(u,v) = diam G or u = v

for all u,v ∈ VG, is an equivalence relation. For example, in Table 1, the
dodecahedron is antipodal - each R-equivalence class consisting of just two
elements. So in the dodecahedron each vertex has a vertex antipodal to it.
Regarding the dodecahedron as a Platonic solid this is geometrically obvious.
Another antipodal graph is the line graph of the Petersen graph (in Figure 1.5
vertices labelled with the same number belong to the same R-equivalence class).

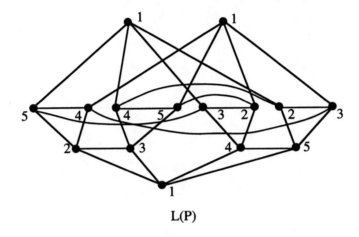

L(P)

Figure 1.5

Consider what this means for P; there exists a partition of EP into five
3–subsets such that within each 3-subset the three edges are mutually a distance 3
apart. The automorphism group of L(P) acts imprimitively on VL(P) and the
R–equivalence classes are the blocks of a system of imprimitivity.

Now let G be any DT graph of degree r > 2. If A(G) is imprimitive on VG,
then G is either bipartite or antipodal; see [dS 71] and Exercise 5. All the graphs
in Table 1 are bipartite with the exception of K_4, the Biggs' graphs which are
primitive, and the dodecahedron which is antipodal.

We can describe the structure of finite DT graphs in terms of their 'intersection
arrays'. For suppose G is a finite connected DT graph of degree r. Let
u,v∈VG. Define

$$s_{i,j}(u,v) = |\{w \in VG: d(u,w) = i, d(w,v) = j\}|.$$

Since G is distance transitive, the numbers $s_{i,j}(u,v)$, for i,j fixed, depend only on $d(u,v)$, i.e. if $d(x,y) = d(u,v)$ then $s_{i,j}(x,y) = s_{i,j}(u,v)$. Hence we may define the set of intersection numbers $s_{i,j,k}$ by

$$s_{i,j,k} = s_{i,j}(u,v), \tag{1}$$

where $d(u,v) = k$ $(1 \leq k \leq d = \text{diam } G)$. We write

$$c_j = s_{j-1,1,j}; \quad a_j = s_{j,1,j}; \quad b_j = s_{j+1,1,j} \quad (1 \leq j \leq d). \tag{2}$$

By definition

$$c_1 = 1, \quad c_i + a_i + b_i = r, \quad b_d = 0. \tag{3}$$

The numbers in (2) provide all the information needed to determine the numbers in (1); see Exercise 6. For P $a_1 = 0, a_2 = 2, b_2 = 0, b_1 = 2, c_1 = c_2 = 1, d = 2$; see Figure 1.6.

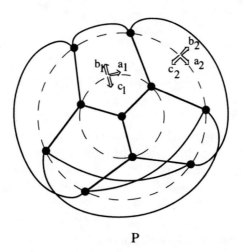

P

Figure 1.6

Now for $u \in VG$, let $N_i(u) = \{v: v \in VG, d(u,v) = i\}$ and let $k_i = |N_i(u)|$. Below we will regard u as the fixed vertex and 'measure' vertices by their distance from u. So for brevity, write $R_i = N_i(u)$ and regard R_i as the 'ring' of vertices distance i from u. So if $v \in R_i$, then c_i, a_i and b_i are the number of edges from v to the rings R_{i-1}, R_i and R_{i+1}, respectively; see Figure 1.7. Again c_i, a_i and b_i are the number of edges from v to vertices distance $i-1, i$ and $i+1$ from u. Thus, for example, if $a_1 = a_2 = ... = a_{s-1} = 0$ and $c_1 = c_2 = ... = c_s = 1$ then $\gamma(G) \geq 2s + 1$.

Clearly $k_i c_i = q(R_i, R_{i-1}) = q(R_{i-1}, R_i) = k_{i-1} b_{i-1}$; see Figure 1.7. So

$$k_i c_i = k_{i-1} b_{i-1}, \; i = 1, 2, ..., d. \tag{4}$$

Here we define $b_0 \equiv r, k_0 \equiv 1$ and $a_0 \equiv 0$.

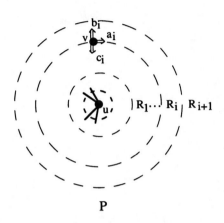

Figure 1.7

The sequence $\{b_i\}$ is nonincreasing, and the sequence $\{c_j\}$ is nondecreasing so, from (4), $\{k_i\}$ is unimodal; see Exercise 18.

The **intersection array** for the DT graph G is:

*	c_1	c_2	...	c_i	...	c_{d-1}	c_d
0	a_1	a_2	...	a_i	...	a_{d-1}	a_d
k	b_1	b_2	...	b_i	...	b_{d-1}	*

It is natural to study graphs in which the numbers c_j, a_j and b_j are independent of the vertices u and v given that $d(u,v) = j$. Such a graph is said to be **distance regular** (DR). So in a DR graph G, if $d(u,v) = k = d(x,y)$, for $u,v, x, y \in VG$, the numbers $s_{i,j}(u,v)$ and $s_{i,j}(x,y)$ (see (1)) are equal but now we do not assume that there is any automorphism which takes u to x and v to y. Indeed a DR graph may have no proper automorphisms at all. Hence a DT graph is a DR graph but not conversely. However there is only one cubic graph - the (3,12)-cage - which is DR and not DT [B-B-S 86]. In fact this graph is not even transitive.

The first intersection array one meets is quite often fairly indigestible, so it maybe helpful to give some examples. P, see Figure 1.6, has intersection array

*	1	1
0	0	2
3	2	*

The Pappus' graph, see Figures 1.4 and 1.8, has intersection array

*	1	1	2	3
0	0	0	0	0
3	2	2	1	*

So, from (4), $k_1 = 3, k_2 = 6, k_3 = 6, k_4 = 2$. This verifies that the Pappus graph has $k_0 + k_1 + k_2 + k_3 + k_4 = 18$ vertices. Since $a_1 = a_2 = 0$ and $c_1 = c_2 = 1$, the girth is at least 6 and as $c_3 = 2 > 1$, it is exactly 6. Since the array has 5 columns, the diameter is 4.

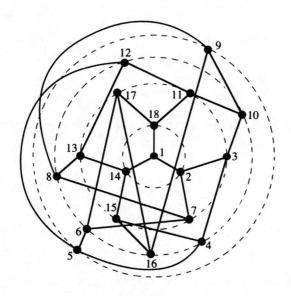

Figure 1.8

There is a considerable simplification if G is a cubic DR graph. From (3), $c_i + a_i + b_i = 3$ and, providing $i < d$, then $c_i \leq 2, a_i \leq 2$ and $b_i \leq 2$. It follows that **the columns** p_i ($i = 1, ..., d - 1$) **of the intersection array** are of type $(1,0,2), (1,1,1)$ or $(2,0,1)$. Suppose that there are α, β and γ columns of each type, respectively. Using the monotonicity conditions of $\{b_i\}$ and $\{c_j\}$ the intersection array is:

p_0	p_1	$p_2\cdots p_\alpha$	$p_{\alpha+1}\cdots p_{\alpha+\beta}$	$p_{\alpha+\beta+1}\cdots p_{d-1}$	p_d
*	1	1 ... 1	1 1 ... 1	2 2 ... 2	c_d
0	0	0 ... 0	1 1 ... 1	0 0 ... 0	$k - c_d$
3	2	2 ... 2	1 1 ... 1	1 1 ... 1	*
	$\leftarrow\ \alpha\ \rightarrow$		$\leftarrow\ \beta\ \rightarrow$	$\leftarrow\ \gamma\ \rightarrow$	

Therefore, given the numbers α,β,γ and c_d, the intersection array is determined and conversely. We will say the intersection array is of type $(\alpha,\beta,\gamma,c_d)$. In Table 2 we describe the intersection arrays for each of the DR graphs in Table 1 plus the (3,12)-cage. Obviously this does not describe the structure of the graphs completely or at least without some degree of further effort.

graph	α	β	γ	c_d
K_4	0	0	0	1
P	1	0	0	1
$K_{3,3}$	1	0	0	3
Heawood	2	0	0	3
(3,8)-cage	3	0	0	3
(3,12)-cage	5	0	0	3
cube	1	0	1	3
Pappus	2	0	1	3
Desargues	2	0	2	3
3-Fold Cover of T	4	0	3	3
dodecahedron	1	2	1	3
C_{28}	2	1	0	2
S(17)	3	3	0	3

Intersection array type of the cubic DR graphs

Table 2

Theorem 1.1 There are only a finite number of cubic DT graphs.

Proof: Suppose that G is a finite cubic DT graph. In particular, G is 1–transitive. Therefore (see Theorem 6.4.7), G is strictly s-transitive for some s such that $1 \le s \le 5$. Since, from Lemma 1.4.5, G is transitive, then

$$|A(G)| = |A_u(G)|.|VG| \tag{5}$$

where $A_u(G)$ is the stabilizer subgroup of the automorphism group A(G); u being some fixed vertex. From Lemma 6.4.6,

$$|A(G)| = |VG|.3.2^{s-1} \tag{6}$$

From (5) and (6)

$$|A_u(G)| = 3.2^{s-1} \tag{7}$$

The restriction of $A_u(G)$ to $N_i(u)$, $1 \le i \le d$ is again transitive, since G is DT. Write $A_0(G) = A_{\{u,v\}}(G)$, for $v \in N_i(u)$, i.e. $A_0(G)$ is the subgroup of $A(G)$ that fixes the set $\{u,v\}$ pointwise. Then, as above,

$$|A_u(G)| = |N_i(u)||A_0(G)|$$

$$= k_i|A_0(G)| \tag{8}$$

Hence, from (7) and (8), $k_i \mid 3.2^{s-1}$. Since $1 \le s \le 5, k_i \mid 48, i = 1, 2,..., d$.

Now suppose that G has an intersection array of type $(\alpha,\beta,\gamma,c_d)$. From (4), if $\alpha \ge 6$ then

$$k_6 = 2k_5 = 4k_4 = 8k_3 = 16k_2 = 32k_1$$

which is impossible since $k_6 \mid 48$. Hence $\alpha \le 5$.

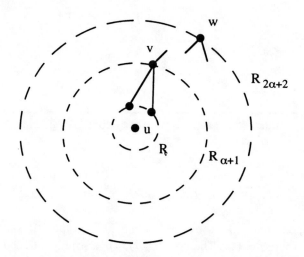

Figure 1.9

We now show that $d = \text{diam } G$ is bounded; in fact we show that $d \le 17$ although it would be an easy matter to improve this to $d \le 15$. Firstly assume that $\beta = 0$. If $\gamma = 0$, then, since $\alpha \le 5$, we have $d \le 6$. Thus suppose $\gamma \ge 1$. We show that $d \le 2\alpha + 1 \le 11$. Suppose otherwise, i.e. suppose that $d \ge 2\alpha + 2$. Choose $v \in N_{\alpha+1}(u) = R_{\alpha+1}$ and $w \in R_{2\alpha+2}$; see Figure 1.9.

Since $a_1 = a_2 = \ldots = a_\alpha = 0$ and $c_{\alpha+1} = 2$, u and v belong to a cycle of length $2\alpha + 2$. Since G is distance transitive and $d(v,w) = d(u,v)$, there must be a cycle of length $2\alpha + 2$ through v and w. But this is impossible since $b_{\alpha+1} = 1$. We deduce that $\beta > 0$. Assume that $d \geq 3\alpha + 3$. Choose $v \in R_{\alpha+1}$, $w \in R_{2\alpha+2}$ and $x \in R_{3\alpha+3}$; see Figure 1.10.

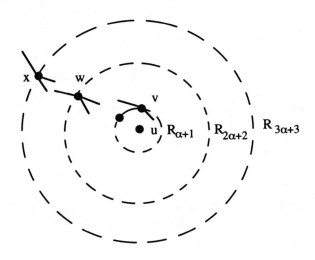

Figure 1.10

Because $\beta > 0$, then $a_{\alpha+1} = 1$ and consequently u and v lie on a cycle of length $2\alpha + 3$. Hence v and w must lie on a cycle of length $2\alpha + 3$. Since $b_{\alpha+1} = 1$, this implies $c_{2\alpha+2} = 2$. Therefore the column $p_{2\alpha+2}$ of the intersection array is $(2,0,1)$. In particular $a_{2\alpha+2} = 0$ and $b_{2\alpha+2} = 1$. Since $p_{2\alpha+2} = (2,0,1)$, $p_{3\alpha+3} = (2,0,1)$. Hence $a_{3\alpha+3} = 0$. It is now easy to check that x and w cannot together be in a cycle of length $2\alpha + 3$. This contradiction proves that $d \leq 3\alpha + 2 \leq 17$.

In all cases $d \leq 17$ and since there are $d + 1$ columns in any intersection array and these columns are of a restricted type, it follows that there is only a finite number of possible intersection arrays for a finite cubic DT graph.   □

In fact the proof we have given places severe constraints on the possible number of intersection arrays for DT graphs. Other constraints can also be imposed. For example, equation (4) tells us that the k_i's must be integers. Possibly the most useful and powerful constraint is due to Higman; see [nB 71]. Consider the tridiagonal matrix M

$$M = \begin{bmatrix} 0 & 1 & 0 & \cdots & \cdots & 0 & 0 \\ r & a_1 & c_2 & \cdots & \cdots & 0 & 0 \\ 0 & b_1 & a_2 & \cdots & \cdots & 0 & 0 \\ 0 & 0 & b_2 & \cdots & & \cdot & \cdot \\ \cdot & \cdot & \cdot & \cdot & \cdot & c_{d-1} & 0 \\ \cdot & \cdot & \cdot & \cdot & \cdot & a_{d-1} & c_d \\ 0 & 0 & 0 & & & b_{d-1} & a_d \end{bmatrix}$$

associated with an intersection array. M has $d + 1$ distinct real eigenvalues $r = \lambda_0 > \lambda_1 > \ldots > \lambda_d$ and so for each λ_i there is a unique left eigenvector u_i and a unique right eigenvector v_i both with first entry 1. Then the required constraint is that

$$\frac{<u_0, v_0>}{<u_i, v_i>}$$

is an integer, where $< , >$ denotes the usual inner product. Now using this constraint and others we have mentioned, it is easy to do a computer search and produce the classification of DT graphs given in Table 1.

There are only 15 DT graphs of degree 4. Smith, in [dS 73], [dS 74a] and [dS74b], used a computer to generate all feasible arrays with diameter less than 30 and then identified the DT graphs that correspond to each feasible array. Gardiner [aG 85] later gave an elementary proof of Smith's result 'by hand'.

Theorem 1.2 A DT graph of degree 4 is one of the following: K_5; $K_{2,2,2}$; $L(K_{3,3})$; $L(O(3))$; $L(H)$; $L(T)$; $(2.K_5)_3$; $K_{4,4}$; $4K_{4,4}$; Q_4, H^*; $O(4)$; $(2.O(4))_2$; $P_3(3)$; $P_6(3)$.   □

Here $K_{n,n,n}$ is the complete tripartite graph, $O(n)$ is the odd graph of degree n (see Chapter 9), H is the Heawood graph, H^* is the 'dual' of H, T is Tutte's 8-cage and Q_n is the n-dimensional cube. If G has diameter d, then r.G denotes an r-fold antipodal covering of G with diameter 2d, and $(r.G)_\gamma$ denotes an r-fold antipodal covering of G with diameter $2d + 1$ in which the parameter $c_{d+1} = \gamma$. For information on antipodal coverings see [aG 74]. $P_3(3)$ denotes the incidence graph of points and lines in the projective plane of order 3, and $P_6(3)$ denotes the incidence graph of the generalized hexagon associated with the group $G_2(q)$; see [rW 81].

The proofs of Theorem 1.1 and 1.2 both depend on a bound being found for $|A_u(G)|$, the order of the stabilizer of some vertex u; see equations (6) and (7).

This bound is then converted into a bound for the diameter. Cameron [pC 82] proved the following.

Theorem 1.3 There are only finitely many finite DT graphs of given degree $r > 2$. ☐

Cameron's argument follows exactly the same broad strategy as before. The required bound for the order of the vertex stabilizer follows from the conjecture of Sims on primitive permutation groups, which has been proved in [C-P-S-S 83]. This conjecture, asserts the existence of a function f such that if A is a primitive permutation group on a set X in which the stabilizer A_x of $x \in X$ has an orbit of size k (greater than one), then A_x has order at most f(k). So Theorem 1.3 is conditional on the general acceptance of the classification theorem of finite simple groups. In 1985, R. Weiss [rW 85] provided a proof of Theorem 1.3 which is independent of this theorem.

Ivanov has now listed all DT graphs of degree at most 11. Some of this work appears in [I-I-F 84] and [F-I-I 86]. Suppose that G is a DT graph and A(G) is its automorphism group, then we say (VG,A(G)) is a **distance transitive representation of the group** A(G). Ivanov [aI 86] has classified all distance transitive representations of S_n, the symmetric group on n symbols. Examples of such representations are the Johnson graphs, the Hoffman-Singleton graph and the odd graphs O(m). We meet the Johnson and odd graphs in the next chapter. O(3) is in fact, the Petersen graph.

Finally, Bannai and Ito, [B-I 87a], [B-I 87b], [B-I 88], [B-I 89], have begun to publish a sequence of papers in which it is eventually intended to show that there are only finitely many DR graphs of given degree r, or equivalently, that the diameters of DR graphs are bounded by a function of $r, r \geq 3$.

2 . Distance transitive line graphs

The following theorem proved in [nB 74a] shows that line graphs are seldom DT.

Theorem 2.1 Suppose that G is a connected graph with $\delta(G) \geq 3$. Then if the line graph L(G) of G is DT, G is an (r,g)-Moore graph for some r and g. ☐

In theory therefore all we need to do is to refer to our list of known Moore graphs and then to check whether their line graphs are DT. We summarise the known results in Table 3.

(r,g)-Moore graph	name	L(G)/name	intersection array
$r \geq 3, g = 3$	K_{r+1}	triangle graph	$\begin{array}{ccc} * & 1 & 4 \\ 0 & r-2 & 2r-6 \\ 2r-2 & r-1 & * \end{array}$
$r \geq 3, g = 4$	$K_{r,r}$	lattice graph	$\begin{array}{ccc} * & 1 & 2 \\ 0 & r-2 & 2r-4 \\ 2r-2 & r-1 & * \end{array}$
$r = 3, g = 5$	P	$L(P)$	$\begin{array}{cccc} * & 1 & 1 & 4 \\ 0 & 1 & 2 & 0 \\ 4 & 2 & 1 & * \end{array}$
$r = 7, g = 5$	Hoffman-Singleton, H	$L(H)$	$\begin{array}{cccc} * & 1 & 1 & 4 \\ 0 & 5 & 6 & 8 \\ 12 & 6 & 5 & * \end{array}$
$r = 57, g = 5$	existence unknown		
$r \geq 3, g \in \{6,8,12\}$	only known to exist for $r - 1$ a prime power	if they exist, their line graphs satisfy:-	$\begin{array}{cccc} * & 1 & 1 & 2 \\ 0 & r-2\ldots & r-2 & 2r-4 \\ 2r-2 & r-1\ldots & r-1 & * \end{array}$ $\qquad \leftarrow \quad \tfrac{1}{2}g + 1 \quad \rightarrow$

Line graphs which are DT

Table 3

All DT line graphs are primitive except for $L(K_4)$ and $L(P)$.

3. Rank 3 graphs

If a $(57,5)$-Moore graph exists then it is not DT. This was proved by Aschbacher [mA 71]. He proved that there does not exist a rank 3 permutation group with subdegree 57. We now explain what this means.

Let (Ω, Γ) denote a transitive permutation group Γ which acts on a set Ω. The **rank of** (Ω, Γ) is the number of orbits of Γ_ω. Since Γ is transitive this number is independent of the choice of ω in Ω. Γ acts on $\Omega \times \Omega$ in a natural way, i.e. $\sigma(\omega_1, \omega_2) = (\sigma\omega_1, \sigma\omega_2)$, $\sigma \in \Gamma$, $(\omega_1, \omega_2) \in \Omega \times \Omega$.

Let $D_0, D_1, D_2, \ldots, D_{s-1}$ be the orbits of Γ acting on $\Omega \times \Omega$. Write, for $\omega \in \Omega$,

$$D_i(\omega) = \{\omega': \omega' \in \Omega, (\omega, \omega') \in D_i\} \quad (i = 0, 1, ..., s - 1)$$

It follows that

$$D_0(\omega), D_1(\omega), D_2(\omega), ..., D_{s-1}(\omega)$$

are the orbits of Γ_ω. Thus s is the rank of (Ω, Γ). We adjust the notation so that $D_0(\omega) = \{\omega\}$ i.e. $D_0 = \{(\omega, \omega): \omega \in \Omega\}$. Denote by D_i^T the set

$$D_i^T = \{(\omega', \omega): (\omega, \omega') \in D_i\}.$$

Obviously D_i^T is itself an orbit, for $0 \leq i \leq s - 1$. Suppose now that $D_i^T = D_i$ for some i, $1 \leq i \leq s - 1$. Write $D = D_i$. Then D is said to be a **symmetric orbit**. Given such an orbit, we can associate a graph $G = G(D)$ with it. Thus $VG = \Omega$ and $\omega\omega' \in EG$ if and only if $(\omega, \omega') \in D$. Because D is symmetric, this graph is well defined. Furthermore, because Γ is transitive, G is regular of degree $|D(\omega)|$, where $D(\omega) = D_i(\omega)$, $\omega \in VG$.

The next problem is to determine when there is such a symmetric orbit. If Γ is of even order, then certainly we can choose such an orbit because Γ contains an element σ of order 2. Hence there exists ω, $\omega' \in \Omega$, $\omega \neq \omega'$ such that $\sigma(\omega, \omega') = (\omega', \omega)$. Therefore the orbit D containing (ω, ω') is symmetric. Equally the converse is true. That is, if there is a symmetric orbit, then Γ has even order.

We now restrict attention to rank 3 groups (Ω, Γ) such that Γ is of even order. The orbits of Γ_ω ($\omega \in \Omega$) are $D_0(\omega), D_1(\omega)$ and $D_2(\omega)$ and each of the orbits D_i are symmetric ($i = 0, 1, 2$). We call either of the associated graphs $G(D_i)$, $i = 1$, 2, a **rank 3 graph** associated with (Ω, Γ). Clearly the graphs $G(D_1)$ and $G(D_2)$ are the complements of one another.

Example. Consider the group (X, S_n) where S_n is the symmetric group acting on the n-set X. Let Ω denote the set of $\binom{n}{2}$ 2-subsets of X. There is a natural induced action of S_n defined on Ω by $\sigma(\{x_1, x_2\}) = \{\sigma(x_1), \sigma(x_2)\}$, $x_1, x_2 \in X$, $\sigma \in S_n$. This produces a new permutation group Γ acting on the set Ω. Furthermore this action is transitive. The induced action on $\Omega \times \Omega$ has just three orbits; the diagonal orbit D_0,

$$D_1 = \{(\{x_1, x_2\}, \{x_3, x_4\}): |\{x_1, x_2, x_3, x_4\}| = 4, x_i \in X, i = 1, 2, 3, 4\}$$
$$\text{and } D_2 = \{(\{x_1, x_2\}, \{x_3, x_4\}): \{x_1, x_2\} \in X, \{x_3, x_4\} \in X, |\{x_1, x_2, x_3, x_4\}| = 3\}.$$

Obviously D_1 and D_2 are symmetric, so we have rank 3 graphs $G(D_1)$ and $G(D_2)$.

As an example suppose that $n = 5$ and $X = \{1,2,3,4,5\}$. Write ij for $\{i,j\}$. Then

$$\Omega = \{12,13,14,15,23,24,25,34,35,45\}$$

and $D_1 = \{(ij,kg) : |\{i,j,k,g\}| = 4, 1 \leq i,j,k,g \leq 5\}$ and $D_2 = \{(ij,kg) : |\{i,j\} \cap \{k,g\}| = 1, i \neq j, k \neq g, 1 \leq i, j, k, g \leq 5\}$.

Hence $G(D_1)$ is the Petersen graph and $G(D_2)$ is the complement of P; see Figure 3.1.

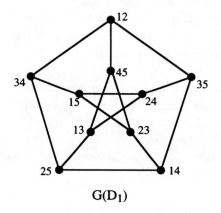

$$G(D_1)$$

Figure 3.1

We have established that, given a rank 3 group, there is an associated rank 3 graph on which the group acts as a group of automorphisms. However, it may not be the full automorphism group. In the case of the Petersen graph we know that P is strictly 3-transitive. Hence, using Lemma 6.4.6, $|A(P)| = 10.3.2^2 = 120 = |S_5|$. Therefore we know that the rank 3 group in this case is the full automorphism group $A(P)$.

In general, given that G is a DT graph of diameter d, its automorphism group $A(G)$ is a rank $d + 1$ group of even order. For clearly $A(G)$ is transitive and in its induced action on $VG \times VG$ it has $d + 1$ orbits - one for each distance $0,1,...,d$. Each such orbit D is symmetric because if $(x,y) \in D$ then $d(x,y)=d(y,x)$ and so, since G is DT, $(y,x) \in D$. On the other hand it does not follow that a rank $d + 1$ group of even order is associated with a DT graph of diameter d, since we are guaranteed only one symmetric orbit. This is why attention has focused on rank three groups of even order, since these are always associated with a rank 3 graph.

Suppose (Ω, Γ) is a rank 3 group and D_0, D_1 and D_2 are the orbits of Γ in its induced action on $\Omega \times \Omega$. Without loss of generality assume that $|D_1| \leq |D_2|$. Then $|D_1|$ is called the **subdegree** of (Ω, Γ). It is the degree of the associated rank 3 graph $G(D_1)$. Call $|\Omega|$, the **degree** of (Ω, Γ). Aschbacher proved that no rank 3 groups (Ω, Γ) exist with $|\Omega| = 57^2 + 1$ and of degree equal to $|VG(D_1)|$ and subdegree 57. Equivalently this means there exists no DT (57,5)-Moore graph. Higman [dH 64] had previously shown that if a rank 3 group had degree $r^2 + 1$ $(r \geq 3)$ and subdegree r, then $r \in \{2,3,7,57\}$. Equivalently, if there exists an (r,5)-Moore graph, then $r \in \{2,3,7,57\}$.

4. Strongly regular graphs

Distance regular graphs of diameter 2 are called **strongly regular.** They are usually described as follows. Suppose that G is a regular graph which is neither complete nor null. Assume that the number of vertices adjacent to u and v ($u \neq v$), $u,v \in VG$, depends only on whether or not u and v are adjacent, i.e. the number is independent of the particular pair of vertices. Then G is a **strongly regular graph with parameters** (n,a,c,d) where $n = |VG|$, a is the degree of G; if u and v are adjacent, c is the number of vertices adjacent to both u and v and if u and v are not adjacent then d is the number of vertices adjacent to both u and v. Hence the intersection array of a strongly regular graph is:-

$$
\begin{array}{ccc}
* & 1 & d \\
0 & c & a - d \\
a & a - c - 1 & *
\end{array}
$$

In our previous notation for DR graphs this means $a = r$, $a_1 = c$, $c_1 = 1$, $b_1 = a - c - 1$, $c_2 = d$ and $a_2 = a - d$. We make this change in our notation because there are now, unfortunately, at least three 'standard' notations for the parameters of strongly regular graphs. Rather than introduce another notation we stay with (n,a,c,d). In the new notation, using (4) of Section 1, for $u \in VG$.

$$|R_1| = |N_1(u)| = a \; ; \; |R_2| = |N_2(u)| = a (a - c - 1)/d. \tag{1}$$

The notation is illustrated in Figure 4.1.

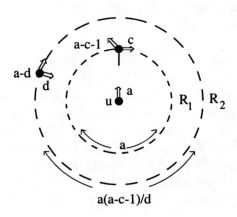

$$a(a-c-1)/d$$

Figure 4.1

It is straightforward to check, Exercise 10, that if G is a strongly regular graph with parameters (n,a,c,d), then its complement $\overline{G}$ is strongly regular with parameters $(\overline{n},\overline{a},\overline{c},\overline{d})$ where $\overline{n} = n, \overline{a} = n - a - 1, \overline{c} = n - 2a + d - 2, \overline{d} = n - 2a + c$.

For example, $L(K_5)$ is strongly regular with parameters $(10,6,3,4)$ and $P = \overline{L(K_5)}$ is strongly regular with parameters $(10,3,0,1)$. In Table 1, $K_{3,3}$ and P are the only strongly regular graphs since they are the only ones that have diameter 2. Other examples of strongly regular graphs are described in Table 4.

Graph	Name	Parameters
$L(K_m)(m>3)$	triangle graph	$(\frac{1}{2}m(m-1),2(m-2),m-2,4)$
$L(K_{m,m})(m>2)$	lattice graph	$(m^2,2(m-1),m-2,2)$
$kK_m(k,m>1)$		$(mk,m-1,m-2,0)$
$\overline{kK_m}(m>2)$	complete k-partite graph	$(mk,m(k-1),m(k-2),m(k-1))$
$P(q)$	Paley graph	$(q, \frac{1}{2}(q-1), \frac{1}{4}(q-5), \frac{1}{4}(q-1))$

Some strongly regular graphs

Table 4

In Table 4, $P(q)$ denotes the Paley graph on q vertices where q is a prime power and $q \equiv 1 \pmod 4$. Its vertex set is the Galois field $GF(q)$ with two vertices being joined if and only if their difference is a non-zero square.

Not all 4-tuples (n,a,c,d) can be the parameters for a strongly regular graph. There are some obvious constraints and we have already met one such in equation (1). This states that $\frac{1}{d}a(a - c - 1)$ is a non-negative integer. The fundamental integrality condition ([rB 63]) is:

Theorem 4.1 (Integrality condition). If there is a strongly regular graph with parameters (n,a,c,d) then

$$f = \frac{1}{2}\left\{(n-1) + \frac{(n-1)(d-c) - 2a}{\sqrt{(d-c)^2 + 4(a-d)}}\right\}$$

and

$$g = \frac{1}{2}\left\{(n-1) - \frac{(n-1)(d-c) - 2a}{\sqrt{(d-c)^2 + 4(a-d)}}\right\}$$

are non-negative integers.

Proof: Suppose that G is a strongly regular graph with parameters (n,a,c,d). Let M denote the adjacency matrix of G. Then, from Theorem 6.1.3 equation (1),

$$M^2 = aI + cM + d(J - I - M) \tag{1}$$

where J is the $n \times n$ matrix all of whose entries are 1. Trivially

$$MJ = JM = aJ. \tag{2}$$

So, I, J and M span a real algebra which from (2) is commutative. This algebra futhermore contains only symmetric matrices. Because of the symmetry and commutativity of this real algebra, there is an orthogonal matrix which simultaneously diagonalizes I, J and M; see [H-K 61]. Let J_0 be the $n \times 1$ matrix each of whose elements is 1. Then

$$MJ_0 = aJ_0 \text{ and } JJ_0 = nJ_0. \tag{3}$$

Hence, from (3), M has the eigenvalue a corresponding to the eigenvalue n of J. Moreover it is easy to check that these eigenvalues occur with multiplicity 1. Hence from (1), since I, M and J can be simultaneously diagonalized,

$$a^2 = a + ca + d(n - 1 - a)$$

i.e. $a(a - c - 1) = (n - a - 1)d.$ \hfill (4)

Any other eigenvalue of J is zero so if ρ is any other eigenvalue of M, ρ corrresponds to the eigenvalue 0 of J. Hence, from (1) and again since I, M and J can be simultaneously diagonalized,

$$\rho^2 = a + c\rho + d(-1 - \rho)$$

i.e. $\rho^2 = (a - d) + (c - d)\rho.$ \hfill (5)

Therefore M has exactly three eigenvalues a, λ and μ, where λ and μ are the roots of (5). So, writing $s = \sqrt{(c-d)^2 + 4(a-d)}$,

$$\lambda = \tfrac{1}{2}(c - d + s) \; ; \mu = \tfrac{1}{2}(c - d - s). \hfill (6)$$

By direct inspection of M, $m(a) = 1$. Write $m(\lambda) = f$ and $m(\mu) = g$, where $m(d)$ is the multiplicity of the eigenvalue d; see Section 6.1. Then, by definition,

$$m(a) + m(\lambda) + m(\mu) = n$$

i.e. $1 + f + g = n.$ \hfill (7)

Furthermore, since G has no loops,

$$0 = \text{trace } M$$

$$= am(a) + \lambda m(\lambda) + \mu m(\mu)$$

$$= a + f\lambda + g\mu. \hfill (8)$$

The values for f and g are now obtained directly from (6), (7), (8).   □

Theorem 4.2. A strongly regular graph has three eigenvalues.

Proof: This follows immediately from the proof of Theorem 4.1.   □

A partial converse of Theorem 4.1 is also true.

Theorem 4.3. A regular connected graph G is strongly regular if and only if it has just three eigenvalues.   □

We deduce from Theorem 4.1 two parameter types for which f and g are integers.

Type 1 $(n - 1)(d - c) = 2a.$

In this case $n = 1 + \frac{2a}{d-c} > 1 + a$. Hence $0 < d - c < 2$; Exercise 11. So $d - c = 1$ and $c = d - 1$, $a = 2d$ and $n = 4d + 1$. Any such strongly regular graph has parameters $(4d + 1, 2d, d - 1, d)$ for some d.

The graph $P(q)$ is of type 1.

In addition in this case, Van Lint and Seidel [V-S 66] proved that $4d + 1$ must be a sum of two squares. This is similar to the Bruck-Ryser condition for the existence of projective planes given in Section 5.3; see p.165.

Type 2 $(d - c)^2 + 4(a - d) = u^2$ for some integer u and u divides $(n - 1)(d - c) - 2a$ with the quotient being congruent to $(n - 1)$ mod 2.

For example $L(K_{m,m})$ is of type 2. Notice in particular that $L(K_{3,3}) \approx P(9)$. Hence $L(K_{3,3})$ is of type 1 **and** type 2.

Another constraint on the parameters of a strongly regular graph is given in [pC78].

Theorem 4.4 (Krein condition) Let G be a strongly regular graph. Suppose that both G and its complement are connected. Let G have eigenvalues a, λ and μ. Then

$$(\lambda + 1)(a + \lambda + 2\lambda\mu) \leq (a + \lambda)(\mu + 1)^2$$

$$(\mu + 1)(a + \mu + 2\lambda\mu) \leq (a + \mu)(\lambda + 1)^2. \qquad \square$$

There is a considerable literature on strongly regular graphs and we have only just scratched its surface. We recommend the interested reader to consult articles by Hubaut [xH 75] and Cameron [pC 78]. The Petersen graph invariably features in this literature. As an almost random example of such an appearance by P, Seidel [jS 68] determined all strongly regular graphs with smallest eigenvalue -2. One of these graphs is the Petersen graph. To give some of the flavour of Seidel's result it is necessary to say something more about the eigenvalues a, λ, μ of a strongly regular graph with parameters (n,a,c,d). From equation (5) and (6) we can deduce (Exercise 13) that

$$a > \lambda \geq 0 > -2 \geq \mu.$$

(The first inequality is standard; see [nB 74b], page 14.)

So Seidel was interested in when the last inequality was sharp. His list of those strongly regular graphs with $\mu = -2$ is given in Table 5.

Name	Parameters
$\overline{L(K_{m,m})}$	$(n^2, 2(n-1), n-2, 2)$
$L(K_{m,m})$	$(2n, 2(n-1), 2n-4, 2n-2)$
Shrikhande	$(16, 6, 2, 2)$
Triangle graph	$\left(\frac{1}{2}n(n-1), 2(n-2), n-2, 4\right)$
The 3 Chang graphs	$(28, 12, 6, 4)$
P	$(10, 3, 0, 1)$
Clebsch	$(16, 10, 6, 6)$
Schläfli	$(27, 16, 10, 8)$

Graphs with $\mu = -2$

Table 5

The Shrikhande graph [sS 59] is shown in Figure 4.2. The bounding vertices and edges are identified with those on the opposite side of the figure, as indicated by the arrows, so that the graph can be embedded in the torus.

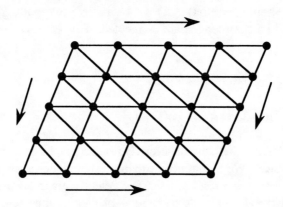

Figure 4.2

The three Chang graphs are discussed in [Cl 59], [Cl 60]. The complement of the Clebsch graph is illustrated in Figure 4.3 where the vertex i is joined to jk if and only if $i \in \{j,k\}$. We show the adjacencies of the vertex 3.

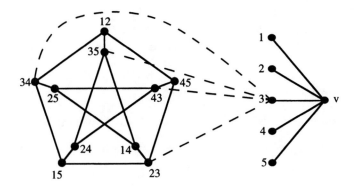

Figure 4.3

The vertices of the Schläfli graph are the 27 lines in a general cubic surface, with two vertices adjacent whenever their corresponding lines meet. The induced graph on the set of neighbours of some fixed vertex of the Schläfli graph is isomorphic to the Clebsch graph, which in turn has induced neighbourhoods isomorphic to $\overline{P}$; see Figure 4.3.

The Petersen, Clebsch and Schläfli graphs form a rank 3 tower. We explain what this means. Let $G(D^{(1)}{}_1)$ be a rank 3 graph associated with the group (Ω_1, Γ_1) where $D^{(1)}{}_0, D^{(1)}{}_1, D^{(1)}{}_2$ are the orbits of the induced action of Γ_1 on $\Omega_1 \times \Omega_1$. Now if $\omega \in \Omega_1$ write $D^{(1)}{}_1(x) = \Omega_2$. Consider (Ω_2, Γ_2) where Γ_2 denotes the restriction of Γ_1 to Ω_2 (Γ_2 acts transitively on Ω_2). Now we may repeat the argument if (Ω_2, Γ_2) is a rank 3 group to obtain another rank 3 graph $G(D^{(2)}{}_1)$ where $D^{(2)}{}_0, D^{(2)}{}_1, D^{(2)}{}_2$ are the orbits of the induced action of Γ_2 on Ω_2 and so on. The only ambiguity in this process is whether we choose $G(D^{(i)}{}_1)$ or $G(D^{(i)}{}_2)$. At the k-th step we have constructed a rank 3 tower of length k. In our case the Petersen, Clebsch and Schläfli graphs form a rank 3 tower of length 3 and, since we know the degrees of these graphs, there is no ambiguity in the construction of the tower.

We can put this another way. Suppose G_3 is isomorphic to the Schläfli graph which is rank 3. Write, for some $u \in VG_3$, $G_2 = \langle N_{G_3}(u) \rangle$. Choose $v \in VG_2$ and write $G_1 = \langle N_{G_3}(u) \cap N_{G_2}(v) \rangle$. Then G_2 and G_1 are isomorphic to the Clebsch and Petersen graphs, respectively. In a sense the Schläfli graph generalizes the Clebsch graph which in turn generalizes the Petersen graph.

The groups involved in the known towers are generally sporadic simple groups and some other 'exceptional' simple groups.

A final word, and it is hard to stop on this subject, Thomason [aT 89] has disproved a longstanding conjecture of Erdös using an infinite tower of rank 3 groups. Denote by $k_t(G)$ the number of complete subgraphs of order t in the graph G. Let

$$c_t(n) = \min\{k_t(G) + k_t(\overline{G}) : |VG| = n\}/\binom{n}{t}.$$

The Erdös conjecture, which is related to Ramsey's theorem, is that $\lim_{n \to \infty} c_t(n) = 2^{1-\binom{t}{2}}$. This latter number is the proportion of monochromatic K_t's in a random two edge-colouring of K_n. In fact Thomason has shown that $c_t < 2^{1-\binom{t}{2}}$ for $t \geq 4$ and indeed that $c_4 \leq \frac{1}{33}$.

5. Homogeneous graphs

The Petersen graph is a Tutte graph; see Section 6.4, p.206. This means that the Petersen graph is, in a certain sense, as symmetric as it could be. In other words if G is a strictly s-transitive cubic graph then $s \leq \lfloor \frac{1}{2}(\gamma(G) + 2) \rfloor$. Since $\gamma(P) = 5$ it follows that, as P is strictly 3-transitive, that this is the 'best' that we could hope for. However this is not the only way we can measure the symmetry of a graph.

We call a graph k-**homogeneous** if any isomorphism between subgraphs of size at most k is the restriction of an automorphism of the graph; a graph is said to be **homogeneous** if it is k-homogeneous for all $k \geq 1$. Sheehan [jS 74] introduced this idea and with Gardiner [aG 76] classified homogeneous graphs.

Theorem 5.1 G is a homogeneous graph if and only if G is one of
(i) tK_r for some $t \geq 1$, $r \geq 1$;
(ii) $\overline{tK_r}$, the complete t-partite graph with parts of size r for some $t \geq 2$, $r \geq 2$;
(iii) C_5;
(iv) $L(K_{3,3})$.   □

The proof of Theorem 5.1 is quite easy; Exercise 14. If G is homogeneous then, provided G is connected, G is a distance transitive graph of diameter at most 2. The classification then follows by an easy induction argument. Cameron [pC 80] showed that the list of finite 5-homogeneous graphs is the same as that for homogeneous graphs and Buczak [jB 80] gave a list of all 4-homogeneous graphs, using the classification of finite simple graphs. Buczak shows that the Schäfli graph and its complement are the only 4-homogeneous graphs which are not 5–homogeneous. Cameron and Macpherson [C-M 85] have now classified all 3–homogeneous graphs, again assuming the classification of finite simple graphs. In the next theorem, we give without explanation, their list of 3-homogeneous graphs.

Theorem 5.2 Assuming the classification of finite simple groups, G is 3–homogeneous if and only if G, up to complementation, is one of

(i) K_n (n ≥ 1);

(ii) $\overline{tK_r}$;

(iii) C_5;

(iv)   a graph on 100 vertices (described in [C-M 85]);

(v) $L(K_{m,m})$;

(vi)   the graph whose vertices are the maximal totally singular subspaces of the unitary space on PG(3,9), two vertices being adjacent if the corresponding subspaces intersect in a 1-space;

(vii)   the graph whose vertices are the points of V(2m,2) under a quadratic form of either type, with the zero vector adjacent to the set of non-zero singular vectors. V(2m,2) denotes an elementary abelian group of dimension 2m over GF(2). When m = 2 the quadratic form of type Ω^- yields the Clebsch graph;

(viii) the McLaughlin graph on 275 vertices.   □

Although we have not spelt out the details the reader will notice there are quite a few 3-homogeneous graphs. Since P is distance transitive it is certainly 2–homogeneous. Interestingly though, P is not 3-homogeneous. We see that <1,3,4> ≅ <2′,3,4> but there are no automorphisms σ whose restriction satisfies σ(<1,3,4>) = <2′,3,4>. In this sense then, the Petersen graph 'could be more symmetric'. However not even the Petersen graph can be perfect.

All these results for k-homogeneous graphs have now been superseded by work of Liebeck and Saxl [L-S 86] who have classified 2-homogeneous graphs (or equivalently rank 3 graphs). This classification again relies heavily on the classification of finite simple groups. We recommend the book by Yap [hY 86] for further reading.

Exercises

1.   (a) Show that A(P) acts primitively on VP.

     (b) Show that A(L(P)) acts imprimitively on V(L(P)).

     (c) Recall the labelling of P in Figure 1.1.3. Let Γ be the subgroup of A(P) generated by the automorphisms α = (1 2 3 4 5)(1′ 2′ 3′ 4′ 5′), β = (1 1′)(2 4′ 5 3′)(3 2′ 4 5′). Check that α maps the outer and inner cycles of P to themselves and that β exchanges them. Deduce that these cycles are blocks of imprimitivity of VP for β and hence that A(P) has an imprimitive transitive subgroup.

2.   Show that C_{28} is described by Figure 1.2.

3.   Draw the sextet graph starting from the representation in Figure 1.3.

4.   Show that P is the only vertex-transitive graph on 2p vertices, p a prime, which has a primitive automorphism group and is not complete.

5.   Let   G   be any distance transitive graph of degree   r $(r \geq 3)$ and diamG$\geq$3. Suppose that A(G) acts imprimitively on VG. Prove that G is either antipodal or bipartite.

     [Hint: Consider the action of A(G) on the sets R_i.]

6.   Prove that the elements of the intersection array of a distance regular graph determine the numbers $s_{i,j}(u,v)$.

7.   Check the intersection arrays of the graphs in Table 2.

8.   Give a condition on the elements of the intersection array of a distance regular graph which characterizes bipartite graphs.

9.   If G is a cubic DR graph, its intersection array columns are of type (1,0,2), (1,1,1) or (2,0,1). If G is a 4-regular DR graph what are its column types?

10. Let G be a strongly regular graph with parameters (n,a,c,d). Prove that $\overline{G}$ is strongly regular with parameters (n, n - a - 1, n - 2a + d - 2, n - 2a + c).

11. Prove that if G is a strongly regular graph with parameters (n,a,c,d) and G is of type 1, then d - c > 0.

12. Give some examples of strongly regular graphs with parameters (n,a,c,c) .

13. Prove that the eigenvalues λ and μ of a strongly regular graph satisfy $\lambda \geq 0 > -2 \geq \mu$. (This is not easy; see [jS 68].)

14. Prove Theorem 5.1.

15. Prove that $L(K_{3,3})$ is homogeneous.

16. Prove that $L(K_{m,m})$ is 3-homogeneous.

17. Prove that the Clebsch graph is 3-homogeneous. Is it 4-homogeneous? If not, why not?

18.   Let G be a distance-transitive graph and use the notation of Section 7. Prove that the sequence $\{c_j\}$ is nondecreasing and the sequence $\{b_j\}$ is nonincreasing.

[Hint: (i)   Choose $v \in R_{i+1}(u)$ (and here we emphasize the dependence on u) and $w \in R_1(u) \cap R_i(v)$.
(ii)   Prove that $R_{i-1}(w) \subseteq R_1(v) \subseteq R_i(u) \subseteq R_1(v)$.
(iii) Deduce that $\{c_j\}$ is nondecreasing.]

References

[mA 71]   M. Aschbacher, The nonexistence of rank three permutation groups of degree 3250 and subdegree 57, *J. Algebra*, 19, 1971, 538-540.

[B-I 87a]   E. Bannai and T. Ito, On distance-regular graphs with fixed valency, I, *Graphs and Combinatorics*, 3, 1987, 95-109.

[B-I 87b]   E. Bannai and T. Ito, On distance-regular graphs with fixed valency, III, *J. Algebra*, 107, 1987, 43-52.

[B-I 88]   E. Bannai and T. Ito, On distance-regular graphs with fixed valency, II, *Graphs and Combinatorics*, 4, 1988, 219-228.

[B-I 89]   E. Bannai and T. Ito, On distance-regular graphs with fixed valency, IV, *Europ. J. Combinatorics*, 10, 1989, 137-148.

[nB 71]   N. Biggs, Finite groups of automorphisms, *London Math. Soc. Lecture Note Series*, C.U.P., 1971.

[nB 73]   N.L. Biggs, Three remarkable graphs, *Can. J. Math.*, 25, 1973, 397-411.

[nB 74a]   N.L. Biggs, The symmetry of line graphs, *Utilitas Math.*, 5, 1974, 113-121.

[nB 74b]   N.L. Biggs, *Algebraic Graph Theory*, C.U.P., 1974.

[B-B-S 86] N.L. Biggs, A.G. Boshier and J. Shawe-Taylor, Cubic distance-regular graphs, *J. Lond. Math. Soc.*, 33, 1986, 385-394.

[B-H 83]   N.L. Biggs and M.J. Hoare, The sextet construction for cubic graphs, *Combinatorica*, 3, 1983, 153-165.

[B-S 71]   N.L. Biggs and D.H. Smith, On trivalent graphs, *Bull. Lond. Math. Soc.*, 3, 1971, 155-158.

[rB 63]   R.C. Bose, Strongly regular graphs, partial geometries, and partially balanced designs, *Pacific J. Math.*, 13, 1963, 389-419.

[jB 80]   J.M.J. Buczak, *Finite group theory*, D. Phil. thesis, Oxford University, 1980.

[pC 78]   P.J. Cameron, Strongly regular graphs, in *Selected Topics in Graph Theory*, (ed. by L.W. Beineke and R.J. Wilson), *Academic Press*, 1978.

[pC 80]   P.J. Cameron, 6-transitive graphs, *J. Comb. Th.* B, 28, 1980, 168–179.

[pC 82]   P.J. Cameron, There are only finitely many finite distance-transitive graphs of given valency greater than two, *Combinatorica*, 2, 1982, 9-13.

[C-M 85]   P.J. Cameron, and H.D. MacPherson, Rank three permutation groups with rank three subconstituents, *J. Comb. Th.* B, 39, 1985, 1-16.

[C-P-S-S 83] P.J. Cameron, C.E. Praeger, J. Saxl and G.M. Seitz, On the Sims conjecture and distance transitive graphs, *Bull. Lond. Math. Soc.*, 15, 1983, 499-506.

[Cl 59]   Chang Li-Chien, The uniqueness and non-uniqueness of the triangular association schemes, *Sci. Record Peking Math. (New Ser.)*, 3, 1959, 604-613.

[Cl 60]   Chang Li-Chien, Association schemes of partially balanced designs with parameters $v = 28$, $n_1 = 12$, $n_2 = 15$, and $p_{11}^2 = 4$, *Sci. Record Peking Math. (New Series)*, 4, 1960, 12-18.

[hC 50]   H.S.M. Coxeter, Self-dual configurations and regular graphs, *Bull Amer. Math. Soc.*, 56, 1950, 413-455.

[F-I-I 86]   I. Faradzhev, A.A. Ivanov and A.V. Ivanov, Distance-transitive graphs of valency 5, 6 and 7, *Europ. J. Combinatorics*, 7, 1986, 303-319.

[aG 74]   A. Gardiner, Antipodal covering graphs, *J. Comb. Th.* B, 16, 1974, 255-273.

[aG 76]   A. Gardiner, Homogeneous graphs, *J. Comb. Th.* B, 20, 1976, 94-102.

[aG 85]   A. Gardiner, An elementary classification of distance-transitive graphs of valency four, *Ars Combinatoria*, 19A, 1985, 129-141.

[dH 64]   D.G. Higman, Finite permutation groups of rank 3, *Math. Zeitschr.*, 86, 1964, 145-56.

[H-K 61]   K. Hoffman and R. Kunze, *Linear Algebra*, Prentice Hall, 1961.

[xH 75]   X.L. Hubaut, Strongly regular graphs, *Disc. Math.*, 13, 1975, 357-381.

[I-I-F 84] A.A. Ivanov, A.V. Ivanov and I.A. Faradzhev, Distance transitive graphs of valency 5, 6 and 7, *Zh. Vychisl. Mat. i. Mat. Fiz.*, 24, 1984, 1704-1718.

[aI 86]   A.A. Ivanov, Distance-transitive representations of the symmetric groups, *J. Comb. Th.* B, 41, 1986, 255-274.

[L-S 86]   M.W. Liebeck and J. Saxl, The finite primitive permutation groups of rank 3, *Bull. Lond. Math. Soc.*, 18, 1986, 165-172.

[jS 68]   J.J. Seidel, Strongly regular graphs with (-1,1,0) adjacency matrix having eigenvalue 3, *Lin. Alg. Appl.*, 1, 1968, 281-298.

[jS 74]   J. Sheehan, Smoothly embeddable subgraphs, *J. London Math. Soc.*, 9, 1974, 212-218.

[sS 59]   S.S. Shrikhande, The uniqueness of the L_2 association scheme, *Ann. Math. Statist.*, 30, 1959, 781-798.

[dS 71]   D.H. Smith, Primitive and imprimitive graphs, *Quart. J. Math. Oxford*, 22, 1971, 551-557.

[dS 73]   D.H. Smith, On tetravalent graphs, *J. London Math. Soc.*, 6, 1973, 659-662.

[dS 74a]   D.H. Smith, Distance transitive graphs of valency four, *J. London Math. Soc.*, 8, 1974, 377-384.

[dS 74b]   D.H. Smith, On bipartite tetravalent graphs, *Disc. Math.*, 10, 1974, 167-172.

[aT 89]   A. Thomason, A disproof of a conjecture of Erdös in Ramsey Theory, *J. Lond. Math. Soc.*, 39, 1989, 246-255.

[V-S 66]   J.H. Van Lint and J.J. Seidel, Equilateral point sets in elliptic geometry, *Proc. Nederl. Akad. Wetensch.* (A) 69 (= Indag. Math. 28), 1966, 335-348.

[rW 81]   R. Weiss, s-transitive graphs, in *"Algebraic Methods in Graph Theory"*,Vol I, II (Szeged 1978), (ed. L. Lovász and V.T. Sós), North Holland, Amesterdam - New York 1981, 827 - 847.

[rW 85]   R. Weiss, On distance transitive graphs, *Bull. Lond. Math. Soc.*, 17, 1985, 253-256.

[hW 64]   H. Wielandt, Finite Permutation Groups, *Academic Press*, New York, 1964.

[hY 86]   H.P. Yap, Some Topics in Graph Theory, *London Math Soc. Lecture Note Series*, 108, C.U.P., 1986.

9

The Petersen Graph in Diversity

0 . Prologue
We seek him here,
We seek him there,
Those Frenchies seek him everywhere.
Is he in heaven? Is he in hell,
That damned elusive Pimpernel?
(In *The Elusive Pimpernel* by the Baroness Orczy)

Like the Scarlet Pimpernel the Petersen graph turns up all over the place and often unexpectedly. This chapter is a by no means all-inclusive list of some of these venues.

There are exactly 19 connected cubic graphs on 10 vertices. The number of elements in the set, $C(n)$, of connected cubic graphs on n vertices grows rapidly with n; for example $|C(20)| = 510489$, $|C(30)| = 845480228069$. The Petersen graph is the only graph in $C(10)$ with 120 automorphisms; the only graph in $C(10)$ with girth 5; the only graph in $C(10)$ with diameter 2; the only bridgeless graph in $C(10)$ with chromatic index 4 and finally it is the only bridgeless non-hamiltonian graph in $C(10)$. These many ways in which P is unique within $C(10)$, are also reflected in the unique role that P plays within the theory of graphs. We now show some other sides to Petersen's character, and hope our discussions will not only support our central theme but also expose the reader to some other interesting areas of graph theory. This chapter makes no claim to being exhaustive. Its only claim is to enforce the well known caveat: graph theorists should always consider P and its generalizations before making conjectures.

The ubiquitous nature of the Petersen graph is further pursued in [C–W 85], [C–H–W 92].

1 . The dodecahedron and P
Let D be the graph of the dodecahedron (see Figure 1.1).

In D, each vertex i has a unique antipodal point i'. Write $(i')' \equiv i$. Then if $\sigma \in A(D)$,

$$(\sigma(i))' = \sigma(i'). \tag{1}$$

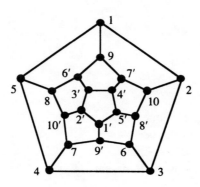

Figure 1.1

Now by identifying antipodal points in D (in the same way that projective 2–space is obtained by identifying antipodal points on the unit sphere in real 3–space) we obtain the Petersen graph.

The dodecahedron, considered as a 3-dimensional convex regular polyhedron, has 60 symmetries. Each vertex can be transformed into the position of vertex 1. Fixing the vertex 1, the 3 faces incident to this vertex may be cyclically rotated in 3 ways. On the other hand, when D is considered simply as the 1–skeleton of the dodecahedron there is the additional automorphism which fixes the vertex 1 and one of the faces incident with it, and interchanges the other two faces incident with 1. So D in this case has 120 automorphisms. From (1) each of these automorphisms is an automorphism of P and conversely. Incidentally D is strictly 2-transitive whereas P is strictly 3–transitive.

2. Locally Petersen graphs
If all the induced neighbourhoods of vertices of a graph G are isomorphic to a fixed graph H, then G is **locally** H. Formally, G is **locally** H if for all $v \in VG$, $\langle N(v) \rangle \cong H$.

For instance, any cycle is locally $\overline{K_2}$ (except C_3), the cube and P are locally $\overline{K_3}$ and K_n is locally K_{n-1}.

Write $T(n) \equiv \overline{L(K_n)}$ for $n \geq 3$, that is $T(n)$ is the complement of the **triangle graph**; see Section 8.4. Alternatively $T(n)$ may be described as follows. $T(n)$

has vertex set consisting of the $\binom{n}{2}$ 2–subsets chosen from $\{1,2,...,n\}$ with vertices X and Y (X≠Y) being joined if and only if X and Y are disjoint. In Figure 2.1 we illustrate the subgraph <N(12)> of $G_1 = \overline{T(7)}$. As usual we write ij rather than $\{i,j\}$.

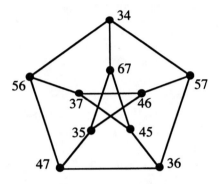

Figure 2.1

There exist exactly 3 non-isomorphic connected locally P graphs; see [jH 80]. These are G_1, G_2 and G_3, where G_2 and G_3 have 63 and 65 vertices respectively. The construction of G_2 and G_3 is group theoretic and is too complicated to give here. However, a flavour of their construction may be gained by considering that of G_1 from another viewpoint. Let G be the graph with vertices consisting of the 21 transpositions contained in the symmetric group S_7. Two distinct vertices are joined if and only if the corresponding transpositions commute. Since transpositions commute if and only if they are disjoint (i.e. they transpose non-intersecting pairs of elements), then it can be seen that G is isomorphic to G_1. The construction of G_2 and G_3 mimic that of G_1 although more group theory is required than we feel prepared to use here.

A graph G is **locally a polyhedron** if G is locally H where H is the 1–skeleton of some 3-dimensional convex regular polyhedron. The characterization of graphs that are locally a tetrahedron or locally an octahedron is easy and is left as Exercise 4. The graphs that are locally a cube or an icosahedron were described in [dB 83] and [B-B-B-C-A 85]. There are exactly two non-isomorphic graphs that are locally a cube. These are $\overline{L(K_{3,5})}$ and a graph H_{24} on 24 vertices. The latter is the 1-skeleton of a 4-dimensional regular polytype called the 24-cell. Thus let (e_1,e_2,e_3,e_4) be the standard ordered basis of $\mathbf{R}^4$. Then $VH_{24} = \{\pm e_i \pm e_j : i \neq j, 1 \leq i,j \leq 4\}$. Vertices $u,v \in VH_{24}$ are joined if and only if the inner product $<u,v> = 1$. This result has been generalized by Brouwer (unpublished) who proves that if G is a connected graph which is locally $\overline{L(K_{p,q})}$, where $p \geq q \geq 2$, with either $p > 3$ or $q > 2$ then either

(i) $G \cong \overline{L(K_{p+1,q+1})}$; or

(ii) $p = 4, q = 2$ and $G \cong H_{24}$; or

(iii) $p = q = 3$ and G is a Johnson graph, $J(6,3)$ consisting of the $\binom{6}{3}$
       3subsets of $\{1,2,...,6\}$ with vertices u and v $(u \neq v)$ being joined if
       and only if $|u \cap v| = 2$; see Section 9.8.

The exceptional cases for p and q are easy to describe - the case (p,1) for
p>3, allowing infinitely many non-isomorphic solutions.

There are precisely three locally icosahedral graphs; one on 120 vertices and
graphs derived from this graph on 40 and 60 vertices.

The problem of describing all graphs that are locally a dodecahedron seems to be
quite difficult. In [pV 85], the first example of such a graph was described. The
structure of this graph is given group theoretically. Unlike the locally polyhedra
graphs considered earlier, in this case there are examples of infinite graphs which
are locally a dodecahedron; see [hC54], [jT 74].

Graphs that are locally n-cycles have been characterized by Vince [aV 81]. Hall
[jH85] lists all locally H graphs, where H has at most eleven vertices.

We saw in Section 8.4 that the Clebsch graph is locally $\overline{P}$. It might be an
interesting problem to characterize locally $\overline{P}$ graphs.

3.   Spectral properties of P

Suppose that G is a connected graph on n vertices and $M \equiv M(G)$ is its
adjacency matrix. The **spectrum,** spec(G), of G is the spectrum (i.e. the set of
eigenvalues λ together with their multiplicities $m(\lambda)$) of M. We write
$spec(G) = \{\lambda_1^{s_1}, \lambda_2^{s_2}, ..., \lambda_n^{s_n}\}$ to mean that λ_i is an eigenvalue of M,
$m(\lambda_i) = s_i$, $s_i \geq 0$, $i = 1, 2,..., n$, and $\lambda_1 \geq \lambda_2 \geq ... \geq \lambda_n$. When G is a regular
graph, $\lambda_1 = \deg u$ $(u \in VG)$; see the proof of Theorem 6.1.3.

It is easy to verify directly that $spec(P) = \{3, 1^5, -2^4\}$. In particular therefore,
spec(P) is integral. There are only 12 other connected cubic graphs which have
integral spectra; see [B-C 75], [aS 78]. In order to describe the structure of these
graphs we require another definition.

Suppose that H and K are any two graphs. Then $H \otimes K$ is the graph with
$V(H \otimes K) = VH \times VK$ and (h,k) joined to (h',k') if and only if $hh' \in EH$ and
$kk' \in EK$. This is called the **conjunction** of H and K. It is easy to prove that
if G has an integral spectra, then so does $G \otimes K_2$. The graphs in Figure 3.1

have integral spectra. The other connected cubic graphs with integral spectra are
$G_i \otimes K_2$ (i = 1,2,3,4,5), $K_{3,3}$ and the (3,8)-cage.

In Figure 3.1, $S(K_4)$ denotes a subdivision of K_4. It is interesting to note that
$G_1 \otimes K_2 = P(4,1)$, $G_2 \otimes K_2 = P(3,1) \otimes K_2 = P(6,1)$ and $G_3 \otimes K_2 = P \otimes K_2 = P(10,3)$. We might suspect that this process of taking products to
produce connected graphs with integral spectra might continue indefinitely. This is
not the case however (see Exercise 5).

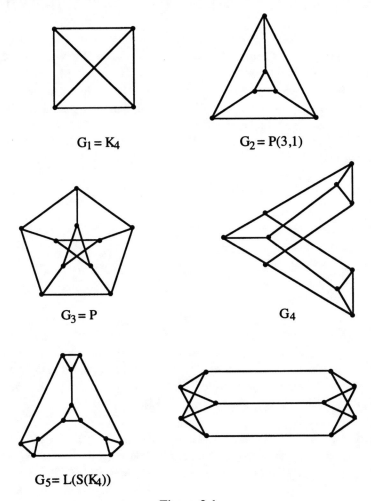

Figure 3.1

It is easy enough to calculate spec(P) directly. However, by calculating the
spectrum indirectly we can illustrate many of the more elementary techniques that
were used in the investigation of the spectral properties of graphs. Once again, as

in Section 8.4, the key to the arguments is finding a matrix which commutes with the adjacency matrix M of P. Then, providing we know enough about the eigenspace of this new matrix, we can evaluate spec(M). (See [nB 74] for a nice development of the theory of this section.)

Proposition 3.1 Suppose that G is a regular graph on n vertices and spec(G) = $\{\lambda_1{}^{s1}, \lambda_2{}^{s2}, ..., \lambda_n{}^{sn}\}$. Then

$$\text{spec}(\overline{G}) = \{(n-1-\lambda_1)^{s1}, -(1+\lambda_n)^{sn}, -(1+\lambda_{n-1})^{sn-1}, ..., -(1+\lambda_2)^{s2}\}.$$

Proof: Write $M \equiv M(G)$ and $\overline{M} \equiv M(\overline{G})$. Let J be the n × n matrix all of whose elements are 1. Then, since G is regular, $MJ = JM$ and $\overline{M}J = J\overline{M}$. Let I be the n × n unit matrix. Then $M + \overline{M} = J - I$. So $M\overline{M} = (J - I - \overline{M})\overline{M} = J\overline{M} - \overline{M} - \overline{M}\,\overline{M} = \overline{M}J - \overline{M}I - \overline{M}\,\overline{M} = \overline{M}(J - I - \overline{M}) = \overline{M}M$. Therefore, since M and $\overline{M}$ commute, it follows that they may be simultaneously diagonalized. Therefore if X is an eigenvector associated with λ_i,

$$MX = \lambda_i X \tag{1}$$

and

$$\overline{M}X = \mu_i X, \tag{2}$$

for some eigenvalue μ_i of $\overline{M}$.

But from (1) and (2)

$$(J - I)X = (M + \overline{M})X = (\lambda_i + \mu_i)X.$$

Hence $\lambda_i + \mu_i \in \text{spec}(J - I) = \{n - 1, -1^{n-1}\}$. It is a standard result that $\lambda_1 = r$ where r is the degree of regularity of G. Hence $\lambda_1 + \mu_1 = n - 1$ where μ_1 is the largest eigenvalue of $\overline{M}$. The result now follows.   □

Proposition 3.2 Suppose that G is a regular graph of degree r, with n vertices and m (m ≥ n) edges and that spec(G) = $\{\lambda_1{}^{s1}, \lambda_2{}^{s2}, ..., \lambda_n{}^{sn}\}$. Then spec(L(G)) = $\{(r - 2 + \lambda_1)^{s1}, (r - 2 + \lambda_2)^{s2}, ..., (r - 2 + \lambda_n)^{sn}, -2^{m-n}\}$.

Proof: The proof in [hS 67] is fairly long and we leave it as Exercise 6.   □

Proposition 3.3 Spec(P) = $\{3, 1^5, -2^4\}$.

Proof: It is easy to check that $\text{spec}(K_n) = \{n - 1, - 1^{n-1}\}$. So $\text{spec}(K_5) = \{4,-1^4\}$. From Proposition 3.2, $\text{spec}(L(K_5)) = \{6,1^4,-2^5\}$ and from Proposition 3.1, $\text{spec}(P) = \text{spec}(\overline{L(K_5)}) = \{3,1^5,-2^4\}$. □

A beautiful application of Proposition 3.3 is given in Exercise 38.

We can describe some further properties of P now that we know its spectrum. Suppose $c_k(G)$ is the number of closed trails of length k, in a graph G on n vertices. Then it is easy to see that $c_k(G)$ is equal to the trace, $\text{tr}M^k$, of M^k, since the (i,i)-th element of M^k, is the number of closed trails of length k starting and terminating at v_i. Hence, see Exercise 7,

$$c_k(G) = \text{tr}M^k = \sum_{i=1}^{n} s_i\lambda_i^k, \tag{3}$$

where $\text{spec}(G) = \{\lambda_1^{s_1},\lambda_2^{s_2},...,\lambda_n^{s_n}\}$.

Therefore from (3) and Proposition 3.3,

$$c_1(P) = 0, c_2(P) = 30, c_3(P) = 0,$$

indicating, respectively, that P has no loops, 15 edges and no triangles. Admittedly this is not startlingly new or valuable information. More interesting perhaps is that P has 2000 spanning trees; P is said to have **complexity** 2000. This fact can be derived from the next result. But firstly we define the characteristic polynomial $\varphi(G,\lambda)$ of a regular graph G of degree r by

$$\varphi(G,\lambda) = \prod_{i=1}^{n} (\lambda-\lambda_i)^{s_i}.$$

Proposition 3.4 Let G be a regular graph of degree r on n vertices. Then the number of spanning trees in G is $\frac{1}{n}\prod_{i=2}^{n} (\lambda_1-\lambda_i)^{s_i}$, where $\lambda_1 = r$. □

The proof of this can be found in [gK 47].

From Propositions 3.3 and 3.4, P has $\frac{1}{10}(3 - 1)^5 (3 + 2)^4 = 2000$ spanning trees.

There is a vast literature on graph spectra; a nice survey has been made in Cvetković et al. [C-D-G-T 88].

4. Orientations

In 1939, Robbins [hR 39] proved that a graph G has a strongly connected orientation if and only if G is connected and bridgeless. By an orientation we mean any directed graph G obtained from G by specifying, for each edge, an order on its ends. This ordering is indicated by an arrow; see Figure 4.1. A **strongly connected orientation** means that for any two vertices u and v there is a directed uv-path.

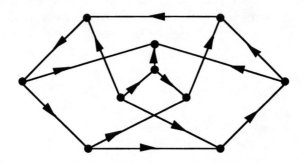

Figure 4.1

If G is regarded as a model of a two-way traffic system, then Robbins' Theorem gives necessary and sufficient conditions for the introduction of a one–way system that still allows access from any point x of the system to any other point y. However, these conditions in no way compare the smallest distance $d(x,y)$ in the two-way system $\vec{d}$, to the smallest distance $d(x,y)$ in the one-way system. In fact, as we all know from experience, there may be quite a discrepancy. For example, if G is a cycle of length n, then there are always vertices x and y such that $\vec{d}(x,y) = 1$ and $\vec{d}(y,x) = n - 1$ for any strongly connected orientation. Chvátal and Thomassen [C-T 78] obtained a comparison between the lengths of cycles and directed cycles.

Theorem 4.1 Every graph G can be orientated so that if an edge uv belongs to a cycle of length k in G, then uv or vu belongs to a directed cycle of length at most h(k) in the oriented graph G, where

$$h(k) = (k-2)2^{\lfloor \frac{1}{2} (k-1) \rfloor} + 2 \quad (k \geq 3).$$

Proof: The proof, which is not easy, can be found in [C-T 78].   □

In any bridgeless graph G of diameter d, every edge belongs to a cycle of length at most $2d + 1$; Exercise 8. Suppose that G is a bridgeless graph of diameter d and $\vec{G}$ is some orientation of G. Suppose $x,y \in VG$ and let $x = u_0, u_1, u_2, \ldots, u_n = y$ be an xy-path of length $n \leq d$ in G. Then in $\vec{G}$, since there exists a directed cycle of length at most $h(2d + 1)$ between u_i and u_{i+1} (or u_{i+1} and u_i), $d(u_i, u_{i+1}) \leq h(2d+1) - 1$. Therefore

$$\vec{d}(x,y) \leq n(h(2d+1)-1) \leq d((2d-1)2^d+1). \tag{1}$$

Hence the directed diameter of G is at most this number.

Define the function f as follows: every bridgeless graph G of diameter d has an orientation $\vec{G}$ with the directed diameter of G at most $f(d)$. Then, from (1),

$$f(d) \leq d((2d-1)2^d+1).$$

This bound is improved upon in [C-T 78], where it is shown that

$$\tfrac{1}{2}d^2 + d \leq f(d) \leq 2d^2 + 2d.$$

Exact values for $f(d)$ seem very difficult to establish. As a first step, we can show that $f(2) = 6$ with the Petersen graph providing the lower bound, i.e. every orientation $\vec{P}$ of P has diameter at least 6. This can be proved fairly easily.

Lemma 4.2 Every strongly connected orientation of the Petersen graph contains a directed cycle of length 5.

Proof: Let $\vec{P}$ be a strongly connected orientation of P. Since $\vec{P}$ is strongly connected it contains a directed cycle. Choose a shortest directed cycle C and assume its length $\lambda(C)$ is at least 6. In P there are no cycles of length 7 and any cycle of length greater than 7 has a diagonal. Hence we may assume that $\lambda(C) = 6$. Since all 6-cycles in P are similar, we may assume that $C=(1,2,3,4,5,6)$; see Figure 4.2.

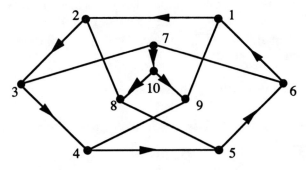

Figure 4.2

Because $\vec{P}$ is strongly oriented, we may assume that either $7 \to 10, 10 \to 8$ and $10 \to 9$ or that $7 \to 10, 10 \to 8, 9 \to 10$. Remembering that $\vec{P}$ is strongly oriented and C is a smallest directed cycle, it is now easy to argue to a contradiction.   □

Theorem 4.3 Every orientation of P has diameter at least 6.

Proof: Let $\vec{P}$ be an orientation of P which has diameter at most 5. Then, since it has a diameter, $\vec{P}$ is a strongly connected orientation of P. Therefore, by Lemma 4.2, we may assume that C = (1,2,3,4,5) is a directed cycle of $\vec{P}$; see Figure 4.3.

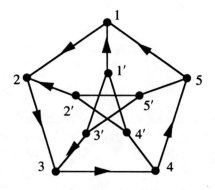

Figure 4.3

Write $E_0 = \{1\ 1', 2\ 2', 3\ 3', 4\ 4', 5\ 5'\}$. Then there are essentially two ways in which the edges of E_0 can be oriented.

Firstly suppose that three consecutive edges, $1\ 1', 2\ 2'$ and $3\ 3'$ say, are each directed towards C, as in Figure 4.3. Then $\vec{d}(1,3') \geq 6$. A similar contradiction is obtained if these three edges are all directed away from C. Finally we may suppose that just two consecutive edges are directed in the same sense; for example $3' \to 3$ and $4' \to 4$ forcing $5 \to 5', 2 \to 2'$ and $1' \to 1$. However $\vec{d}(3,1') \leq 5$ forces $5' \to 3' \to 1'$ and $\vec{d}(3,4') \leq 5$ forces $5' \to 2' \to 4'$. But then $\vec{d}(1,3') \geq 6$ which is a contradiction.   □

Theorem 4.3 proves that $f(2) \geq 6$. Next we prove that $f(2) \leq 6$.

Theorem 4.4 Every bridgeless graph of diameter 2 has an orientation of diameter at most 6.

Proof: Let G be a bridgeless graph of diameter 2 and let $\vec{G}$ be an orientation of G. From (1) (and the argument preceding it), if every edge of G is in a triangle

$$\vec{d}(x,y) \leq 3 \times 2 = 6$$

for all $x,y \in V\vec{G}$.

We may therefore assume that there exists some edge $uv \in E(\vec{G})$ which belongs to no triangle. Write $X = N(u) \setminus \{v\}$ and $Y = N(v) \setminus \{u\}$; see Figure 4.4. Let X_1 and Y_1 be, respectively, the elements of X and Y joined, respectively, to no elements of Y and X. Write $X_2 = X \setminus X_1$, $Y_2 = Y \setminus Y_1$ and $Z = V\vec{G} \setminus (X \cup Y \cup \{u,v\})$.

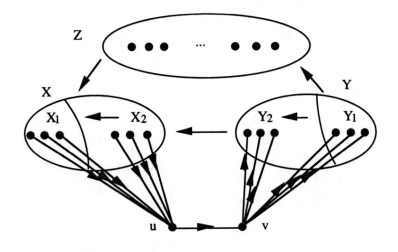

Figure 4.4

It is easy to check that the orientation described most conveniently by

$$u \to v \to Y \to Z \to X \to u \quad \text{and} \quad Y_1 \to Y_2 \to X_2 \to X_1$$

has diameter at most 6. (Here $A \to B$ means every edge joining a vertex $a \in A$ to a vertex $b \in B$ is directed from a to b.) □

5. Hamiltonian paths and the Petersen graph

Lovász [lL 70] conjectured that every connected vertex-transitive graph has a hamiltonian path. There are in fact only four known connected graphs which are vertex-transitive and do not have hamiltonian cycles. Thomassen [cT78b p.163] conjectures that there are only a finite number of such graphs. The four known graphs are P, the Coxeter graph C_{28}, see Figure 5.1, and the graphs obtained from these by 'inflating' each vertex to a triangle; see Figure 5.2. It does seem very strange that, in the period since 1978, no other exceptional graphs have been discovered. Such an arbitrary construction as 'inflation' surely cannot be definitive.

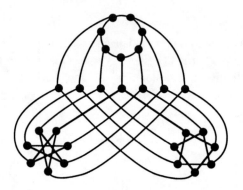

Figure 5.1

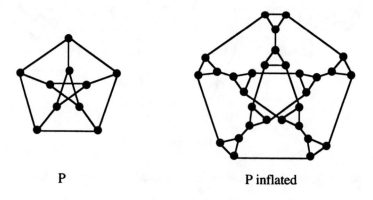

P   P inflated

Figure 5.2

Lovász's conjecture has been verified for many classes of vertex-transitive graphs. For example, if G is a connected vertex-transitive graph with a prime number of vertices, then G is hamiltonian; Theorem 5.1 below, see [jaT 67]. Obviously, since we still only know of four connected non-hamiltonian vertex–transitive graphs, all the constructive results that verify the Lovász conjecture produce the stronger conclusion that the graph is hamiltonian.

Theorem 5.1 Suppose that G is a connected vertex-transitive graph with a prime number p ($p \geq 3$) of vertices. Then G is hamiltonian.

Proof: Let $VG = \{v_1, v_2, ..., v_p\}$, $p \geq 3$ a prime. Since G is vertex-transitive, Lemma 1.4.5 gives

$$|A(G)| = |A_1(G)|p, \tag{1}$$

where $A(G)$ and $A_1(G)$ are, respectively, the automorphism group and the stabilizer of the vertex v_1. From (1), $p \mid |A(G)|$ and hence, from elementary group theory, there exists $\sigma \in A(G)$ such that σ is of order p. As G is connected and $p \geq 3$, $E(G) \neq \emptyset$. Suppose $v_1 v_2 \in EG$. Since p is prime, there is no loss of generality in assuming that σ (or some power of σ) satisfies $\sigma(v_1) = v_2$. Hence we may suppose that $\sigma = (v_1 v_2 v_3 ... v_p)$. Thus $(v_1, v_2, ..., v_p)$ is a hamiltonian cycle. $\square$

In a series of papers by Alspach, Marušič and Parsons, Theorem 5.1 has been extended and generalized; see [bA 79], [dM 81], [M-P 82] and [M-P 83].

Theorem 5.2 Let $p \geq 3$ be a prime. Suppose that G is a connected vertex-transitive graph of order n. Then

(i) G has a hamiltonian path if $n \in \{p, p^2, p^3, 2p, 3p, 4p, 5p\}$;

(ii) G has a hamiltonian cycle if $n \in \{p, p^2, p^3, 2p, 3p\}$ except when $p = 5$ and $G \cong P$;

(iii) G has a hamiltonian decomposition if (a) $n = p$ or (b) $n = 2p$ and $p \equiv 3$ (mod 4). $\square$

By a **hamiltonian decomposition**, we mean that if G is regular of degree r, then EG is the union of $\lfloor \frac{1}{2}r \rfloor$ disjoint hamiltonian cycles, together with a 1–factor if r is odd.

Lipman [mL 85] takes a different approach. He considers graphs with a certain type of automorphism group rather than a certain order. So he proves that (i) if a connected graph has a transitive nilpotent group acting on it, then the graph has a hamiltonian path; (ii) a connected vertex-transitive graph with a prime power number of vertices has a hamiltonian path. So, in particular, every known connected vertex-transitive graph with a prime power number of vertices at least equal to three, is hamiltonian. (See [mH 59] for a definition of transitive nilpotent group.) Another way of attacking the Lovász conjecture is suggested by Theorem 5.3 (see [lB 79]).

Theorem 5.3 Every connected vertex-transitive graph of order $n \geq 4$ has a cycle of length at least $\sqrt{3n}$.

Proof: Let G be a connected vertex-transitive graph with $n \geq 4$ vertices. The result is trivially true if G is regular of degree 2. Therefore suppose that G is regular of degree $r \geq 3$. A result of Watkins [mW 70], states that such a vertex-transitive graph is 3-connected. From this it is easy to deduce that any two longest cycles have at least 3 vertices in common; see Exercise 9. Let t be the length of a longest cycle and let H be the set of t-subsets of VG whose elements are vertices of a longest cycle. By transitivity we may assume that each vertex of G is in some fixed number, s say, of elements of H. Therefore

$$t|H| = ns. \tag{1}$$

Now consider a fixed element $X \in H$. Each element of X is in exactly s elements of H, whereas each element of H contains at least 3 elements of X. Therefore

$$3|H| \leq ts. \tag{2}$$

So, from (1) and (2) $t^2 \geq 3n$.   □

Yet another direction in which some progress has been made with the Lovász conjecture involves Cayley graphs. Suppose that A is a group. Choose $S \subseteq A$ so that (i) $1 \notin S$; (ii) if $x \in S$ then $x^{-1} \in S$. Define the **Cayley graph** $G \equiv G(A,S)$ **associated with** (A,S) by:

$$VG = A; \quad EG = \{\{g,h\}: g^{-1}h \in S\}.$$

Conditions (i) and (ii) guarantee that G is a graph. G is connected if and only if S generates A. A Cayley graph is vertex-transitive and, in view of the Lovász conjecture, attention has focused on the hamiltonicity of Cayley graphs. Moreover, it seems very likely that the great majority of transitive graphs are Cayley graphs. At the very least this is true for graphs with at most 26 vertices; see [M-R 90]. In fact P, C_{28} and their inflations, are not Cayley graphs; see Exercise 10. Because of their 'group' construction it is somewhat easier to get to grips with the structure of Cayley graphs. Lovász has in fact conjectured that all connected Cayley graphs are hamiltonian. Some evidence to support this conjecture is provided by the following typical result; see [C-Q 83] and [dM 83].

Theorem 5.4
(i)   If G is a connected Cayley graph of order pq, where p and q are distinct primes, then G is hamiltonian.
(ii)   Every connected Cayley graph $G(A,S)$, where A is an abelian group of order at least 3, has a hamiltonian cycle.
(iii) Every connected Cayley graph associated with a semi-direct product of a cyclic group of prime order by an abelian group of odd order, has a hamiltonian cycle.   □

Witte and Gallian provide a useful survey in [W-G 84] of hamiltonian cycles in Cayley graphs; see also [A-Z 89a].

In yet another direction in [L-G-H 75], it is proved that every connected vertex-transitive graph on an even number of vertices, has a 1-factor. Petersen's theorem states that every non-empty regular graph of even degree has a 2-factor; see Theorem 4.3.3. We know that if G is a non-empty connected vertex-transitive graph, then it is regular of degree r, say. If r is even, then G has a 2-factor. If r is odd, then $|VG|$ is even and G has a 1-factor L. Hence $G - EL$ is regular of degree $r - 1$ and so has a 2-factor which is not necessarily connected. In all cases G has a 2-factor.

A very nice generalization of Lovász's conjecture is due to Bondy; see Exercise 11. Remember that a hamiltonian cycle is a 2-factor with just one component. Bondy [jB 78], conjectures that 'every non-empty connected vertex-transitive graph has a 2-factor with at most 2 components'.

Marušič [dM 88], asks two interesting questions. Firstly does there exist a connected vertex-transitive graph with $2p$ vertices (p a prime), other than P, that is not 1-factorizable?

An **n-circulant** is a graph with $2n$ vertices and an automorphism having two orbits of size n . So secondly, does there exist a regular, connected n-circulant, other than P, which is not 1-factorizable?

The role of the Petersen graph in the theory of hamiltonian graphs, as we saw in Theorem 1.5.8, is exceptional in both senses of the word. We provide now a small sample, from amongst the many, of this phenomenon.

Suppose that G is a graph. A cycle C of G is **dominating**, if $VG \setminus VC$ is an independent set. Let $\sigma_2(G)$ be the minimum size of $N(u) \cap N(v)$, taken over all pairs of independent vertices u, v ($u \neq v$) of G.

Theorem 5.5 Suppose that G is 2-connected and contains some dominating cycle. Then G contains a dominating cycle of length at least min{VG,$2\sigma_2(G)$}, unless G = P.

Proof: See [B-V 91].   □

Theorem 5.6 Let G be a 3-connected graph on n vertices such that $\sigma_2(G) \geq \frac{1}{2}(n+1)$. Then G is hamiltonian.

Proof: See [bJ 91].   □

The Petersen graph is an example of a non-hamiltonian 3-connected graph for which $\sigma_2(G) = \frac{1}{2}n$. G. Chen (see [bJ 91]) conjectures that P is the only such example. Jackson, in the same paper, also conjectures that if G is a 2–connected graph on n vertices such that $\sigma_2(G) \geq \frac{1}{2}n$, then either G is hamiltonian, or G has a spanning subgraph of a specified and very limited type, or G is the Petersen graph.

Now suppose that G is a finite k-connected graph (k≥2) on n vertices and with independence number $\alpha(G) = \alpha$. In 1972, Chvátal and Erdös [C-E 72], proved that if $\alpha \leq k$, then G is hamiltonian; see also [J-O 90]. Bigalke and Jung have proved the next result.

Theorem 5.7 If G is 1-tough and $\alpha \leq k+1, k \geq 3$, then either G is hamiltonian or G = P.

Proof: See [B-J 79].   □

Finally, G is **pancyclic** if G has cycles of each length t, $3 \leq t \leq$ IVGI. Benhocine and Fouquet have proved the following result.

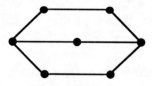

Figure 5.3

Theorem 5.8 If $\alpha \leq k+1$, then the line graph of G is pancyclic unless G is either a cycle of length 4, 5, 6 or 7, or a specified graph with seven vertices and eight edges, see Figure5.3, or G = P.

Proof: See [B-F 87].   □

6. Kneser's conjecture

Kneser [mK 55] conjectured that 'if the $\binom{2n+k}{n}$ n-subsets ($n \geq 1$, $k \geq 1$) of a (2n+k)-set are partitioned into $k + 1$ classes, then one of these classes contains two disjoint n-subsets'.

It is easy to split the n-subsets into $k + 2$ classes so that the assertion of this conjecture is not true. Let $S = \{1,2,...,2n+k\}$. Suppose M_i is the set of all n−subsets with smallest element i. Write $M = M_{k+2} \cup ... \cup M_{k+n+1}$. Then $(M_1, M_2,...,M_{k+1}, M)$ is a partition of the n-subsets of S into $k + 2$ classes such that any two n-subsets in the same class intersect.

The conjecture may be rephrased as follows. Let $S = \{1,2,...,2n+k\}$. Define the graph $KG \equiv KG(n,k)$ by (i) VKG is the set of n-subsets of S and (ii) two vertices are joined if and only if they are disjoint.

Kneser's conjecture in this formulation translates to:

$$\chi(KG(n,k)) = k + 2.$$

The graphs $KG(n,k)$ are often called **Kneser graphs.** Notice that $KG(2,1) \cong P$.

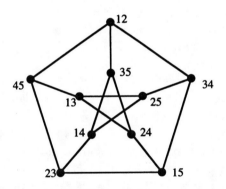

Figure 6.1

Certainly $\chi(P) \geq 3$. In Figure 6.1 the set $\{i,j\}$ is denoted by ij. We have $M_1 = \{\{1,2\},\{1,3\},\{1,4\},\{1,5\}\}$, $M_2 = \{\{2,3\},\{2,4\},\{2,5\}\}$ and $M = \{\{3,4\},\{3,5\},\{4,5\}\}$, with $S = \{1,2,3,4,5\}$, $n = 2$ and $k = 1$. Taking these sets as the colouring classes we see that $\chi(P) \leq 3 = k + 2$. In general the sharpness of Kneser's conjecture shows that $\chi(KG(n,k)) \leq k + 2$. In [lL 78], Lovász proved that equality holds. We outline a proof due to Bárány [iB 78]. We first state two classical geometric results, the first due to Borsuk [kB 33] and the second to Gale [dG 56]. We note that $\|x\|$ is the usual metric in $\mathbf{R}^n$.

Lemma 6.1 Let $S_k = \{x \in \mathbb{R}^{k+1}: \|x\| = 1\}$ ($k \geq 1$) be the k-dimensional sphere. If S_k is the union of $k + 1$ subsets which are open in S_k, then one of these sets contains antipodal points.   □

Lemma 6.2 Suppose that $n \geq 1, k \geq 1$. There exists a $(2n + k)$-subset S of S_k, such that any open hemisphere contains an n-subset of S.   □

Theorem 6.3 $\chi(KG(n,k)) = k + 2$ for $n \geq 1, k \geq 1$.

Proof: We have yet to prove that $\chi(KG(n,k)) \geq k + 2$. Suppose VKG is the set of n-subsets of a $(2n + k)$-set S. Arbitrarily colour the vertices with colours $1,2,...,k + 1$. Now 'identify' S with the set S in the statement of Lemma 6.2.

Let $C_j = \{x \in S_k :$ the open hemisphere with centre x contains some n-subset of S which is coloured j$\}$ for $j = 1,2,....,k+1$.

Obviously C_j is open and $S_k = C_1 \cup C_2 \cup ... \cup C_{k+1}$. By Lemma 6.1 we may choose C_t ($1 \leq t \leq k+1$), so that C_t contains antipodal points, h_1 and h_2 say. Let H_1 and H_2 be, respectively, the open hemispheres with centres h_1 and h_2. Since h_1 and h_2 are antipodal, $H_1 \cap H_2 = \varnothing$. But, by the definition of C_t, H_1 and H_2 contain n-subsets, u_1 and u_2 say, of S, such that u_1 and u_2 are both coloured t and furthermore $u_1 \cap u_2 = \varnothing$, since $H_1 \cap H_2 = \varnothing$. This means that, under the identification, $u_1u_2 \in EKG$. Therefore the colouring is not a 'good' colouring, thus proving that $\chi(KG(n,k)) \geq k + 2$.   □

Theorem 6.3 has been extended by Frankl [pF 85]. Suppose that S is an n-set and suppose that k and t ($n>k>t>0$) are integers. The **generalized Kneser graph** $K = K(n,k,t)$ is defined by (i) VK is the set of k-subsets of S and (ii) vertices u and v are joined if and only if $|u \cap v| < t$. Be careful here. The parameters of the generalized Kneser graph do not correspond directly to the parameters of the Kneser graph itself. The exact correspondence is clarified by the next assumption. We may assume that $n + t > 2k$, otherwise K is edgeless; see Exercise 12.

Let $t(n,k)$ be the Turán number i.e. $t(n,k)$ is the minimum number of edges which are possible in a graph G of order n and $\alpha(G) < k$, where $\alpha(G)$ is the independence number of G. By Turán's theorem [bB 78, p.294],

$$t(n,k) = (k-1) \binom{s}{2} + rs,$$

where $s = \lfloor n / (k-1) \rfloor$ and $r = n - (k-1)s$.

The extremal graph is

$$T(n,k) \equiv rK_{s+1} \cup (k\text{-}1\text{-}r)K_s.$$

Let $S = VT(n,k)$. Then S is an n-set and we identify it with the set S in the definition of $K(n,k,2)$. We now colour the vertices of K. Let $e \in ET(n,k)$. Colour $v \in VK$ with the colour 'e' if and only if v is a k-subset of S which contains both of the end-vertices of e. All of the vertices coloured 'e' are independent, since if u and v are coloured e, $u \neq v$, $|u \cap v| \geq 2$; both u and v containing the two end-vertices of e. Also because $\alpha(T(n,k)) < k$, every k–subset of $S = VT(n,k)$ contains an edge e joining a pair of its elements. Hence every k-subset of S, i.e. vertex of $K(n,k,2)$, is coloured. Hence it follows that we have a 'good' colouring of K. So $\chi(K(n,k,2)) \leq t(n,k)$. Frankl has also proved that if n is large enough, equality holds.

A generalization of Kneser's conjecture, due to Erdös {pE 76], has been proved by Alon, Frankl and Lovász in [A-F-L 86]; see also [kS 90]. Also, in [A-F-L 86], the result of Frankl has been generalized.

Finally, some interesting remarks on cycles in Kneser graphs are to be found in [yK 91].

7. Odd Graphs

Let S be a finite set of cardinality $2k - 1$, $k \geq 2$. The **odd graph** $O(k)$ is defined by (i) $VO(k)$ is the set of (k-1)-subsets of S and (ii) two vertices are joined if and only if they are disjoint.

Notice that $O(3) = KG(2,1) = P$ and, more generally, $KG(n,1) = O(n + 1)$ $(n \geq 1)$. When Petersen [jP 98] introduced $O(3)$ in 1898 he did not define it in terms of 2-subsets. This definition was given by Kowalewski [aK 17] in 1917, where both $O(3)$ and $O(4)$ were studied.

From Theorem 6.3, $\chi(O(k)) = 3$ $(k \geq 2)$. There is therefore no problem with determining the number of colours needed to 'good' colour the vertices of $O(k)$. The interesting problem arises when we try to colour the edges. Since for any fixed (k-1)-subset of S, i.e. a vertex of $O(k)$, there are k (k-1)-subsets of S disjoint from it, $O(k)$ is regular of degree k. Hence by Vizing's Theorem (Theorem 3.1.1)

$$k \leq \chi'(O(k)) \leq k + 1. \tag{1}$$

$O(k)$ has $\binom{2k-1}{k-1}$ vertices and this number is odd if k is a power of 2 and therefore in this case, $\chi'(O(k)) = k + 1$; Exercise 13. We know that $\chi'(O(3)) = 4$ and by [nB 79], $\chi'(O(i)) = i$, $i \in \{5,6,7\}$. Funnily enough, the value of $\chi'(O(6))$ was of interest to the footballers of Croam [M-L 73].

'In the little English hamlet of Croam the consuming passion of the inhabitants is Association Football. In fact the members of the village football team have become so ruthless in their will to win that no other team will play against them.

'Thus the eleven footballers of Croam (who are incidentally the only able-bodied men in the village) are forced to arrange their own matches between two teams of five, with the eleventh man as referee. Further, such is the bitterness and recrimination which follows even these matches, that it has proved necessary to rule that only one match can be played with the same teams and the same referee. This rule was originally regarded with some misgivings, as it was felt that it might seriously limit the number of matches which could be played. However, a villager who has a head for figures worked out that there are 1386 different ways of splitting the eleven men into two teams of five and a referee. This number is thought to be adequate but not generous, for the footballers of Croam are dedicated men.

'But there is a second rule which these men, united by their love of football, but embittered by isolation, have been forced to make in order to keep the peace. No five men will play together as a team more than once on any given day of the week. Therein lies the problem. Can all the possible matches be played under this restriction? Can all the matches be played if Sunday games are allowed?' See Exercise 14.

Determining the chromatic index of $O(k)$ is difficult in general and so is determining whether or not $O(k)$ is hamiltonian. We know that $O(3)$ is non-hamiltonian and after much computation ([M-L 72], [mM 76]) it has been shown that $O(i)$ is hamiltonian for $i \in \{4,5,6,7,8\}$; indeed $O(4), O(5)$ and $O(6)$ have hamiltonian decompositions.

Let $O(k)$ be the odd graph associated with the $(2k - 1)$-set S. Any permutation of S induces an incidence preserving permutation of the vertices of $O(k)$. Thus the symmetric group S_{2k-1} on $(2k-1)$ symbols, is a subgroup of the automorphism group $A(O(k))$ of $O(k)$, in the same way that S_5 is a subgroup of $A(P)$. However we can prove more.

Theorem 7.1 $A(O(k)) \cong S_{2k-1}$.

Proof: Let $s \in S$ and let $V(s) = \{v \in VO(k): s \in v\}$. Then $V(s)$ is independent and $|V(s)| = \binom{2k-2}{k-2}$. Now, see [E-K-R 61], $\alpha(O(k)) = \binom{2k-2}{k-2}$ and any independent set with this number of vertices is equal to $V(s)$ for some $s \in S$. Let $\theta \in A(O(k))$. Then θ induces a permutation of the sets $V(s)$, $s \in S$. Define $\bar{\theta}: S \to S$ by

$$\bar{\theta}(s) = t \text{ where } \theta(V(s)) = V(t).$$

Since θ is a permutation, $\bar{\theta}$ is a permutation of S. Now verify that the map $\theta \to \bar{\theta}$ ($\theta \in A(O(k))$) is an isomorphism of $A(O(k))$ onto S_{2k-1}; Exercise 15. □

Example. We illustrate this proof for the case $k = 3$; see Figure 6.1. We have $V(1) = \{\{1,2\},\{1,3\},\{1,4\},\{1,5\}\}$, ..., $V(5) = \{\{1,5\},\{2,5\},\{3,5\},\{4,5\}\}$. If $\theta \in A(P)$ then θ permutes the set $\{V(1),...,V(5)\}$. For example suppose the induced action of θ is the cyclic permutation $(V(1)V(2)...V(5))$, then $\bar{\theta} = (12345) \in S_5$.   □

If $u,v,x,y \in VO(k)$ and $d(u,v) = d(x,y)$, then $|u \cap v| = |x \cap y|$. Hence there exists a permutation σ of S, whose induced action $\bar{\sigma}$ satisfies $\bar{\sigma}(u) = x$ and $\bar{\sigma}(v) = y$ ($\bar{\sigma} \in A(O(k))$). In other words, $O(k)$ is a DT graph; see Section 8.1. Its diameter is $k - 1$ and it has intersection array; see Exercise 16 and 17:

(i)   **k odd**

						k		
*	1	1	2	2	...	$\frac{1}{2}(k-1)$	$\frac{1}{2}(k-1)$	
0	0	0	0	0	...	0	$\frac{1}{2}(k+1)$	
k	k-1	k-1	k-2	k-2	...	$\frac{1}{2}(k+1)$	*	

(ii)   **k even**

						k			
*	1	1	2	2	...	$\frac{1}{2}k - 1$	$\frac{1}{2}k - 1$	$\frac{1}{2}k$
0	0	0	0	0	...	0	0	$\frac{1}{2}k$
k	k-1	k-1	k-2	k-2	...	$\frac{1}{2}k+1$	$\frac{1}{2}k+1$	*

The intersection arrays determine the eigenvalues of $O(k)$ and the eigenspaces; see [nB 74]. Indeed the eigenvalues of $O(k)$ are $\lambda_i = (-1)^i(k - i)$ with $m(\lambda_i)$ given by

$$m(\lambda_i) = \binom{2k-1}{i} - \binom{2k-1}{i-1} \quad (0 \le i \le k-1).$$

In particular O(k) always has integral spectrum; see Section 3.

The odd graphs have many other interesting properties; see [nB 79], [cP 81] and [maM 91]. For example: (i) O(k) is strictly 3-transitive and (ii) any DR graph with intersection array equal to the intersection array of O(k) is isomorphic to O(k) [aM 82].

We conclude this section with a conjecture due to Meredith and Lloyd [M-L 72].
Conjecture 7.2 The odd graph O(k), k ≥ 4, has a hamiltonian decomposition.

8 .   The Johnson graphs and even graphs
The Johnson graph J(n,k), $1 \le k \le \frac{1}{2}n$ is defined by (i) VJ(n,k) consists of the k-subsets of an n-set S and (ii) vertices u and v are joined if and only if $|u \cap v| = k - 1$.

We observe that $J(5,2) = \overline{P}$. J(n,k) seems to share many of the properties of the Kneser graphs. S_n acts distance transitively on J(n,k); see [aI 86]. Its intersection array is

*	1	4	9	...	k^2
0	n-2	2(n-4)	3(n-6)	...	k(n-2k)
k(n-k)	(k-1)(n-k-1)	(k-2)(n-k-2)	(k-3)(n-k-3)	...	*

The diameter of J(n,k) is k and, like the odd graphs; see [pT 86], its parameters determine the graph uniquely, except when (n,k) = (2,8), in which case there are exactly four non-isomorphic graphs; see [lC 59]. The vertex set of O(k) coincides with the vertex set of J(2k-1,k-1).

Another graph in the same mould is the **even graph** E(m). The graph E(m) is defined by (i) VE(m) consists of the partitions of a 2m-set S into two equal parts (each of size m) and (ii) vertices $\{A_1,A_2\}$ and $\{B_1,B_2\}$ are joined whenever either $|A_i \cap B_i| = m - 1$ or $|A_i \cap B_{3-i}| = m - 1$ holds for i = 1, 2. Again S_{2m} acts distance transitively on E(m). The even graph E(m) is isomorphic to the antipodal quotient of J(2m,m). Its intersection array is given in [aI 86].

The Kneser, odd and Johnson graphs are unified by a generalization due to Chen and Lih [C-L 87]. They define a **uniform subset graph,** G(n,k,t), to have the k–subsets of an n-set as vertices and two vertices are joined if and only if the corresponding k-subsets intersect in exactly t elements. So the Kneser graph KG(n,k) is G(2n+k,n,0), the odd graph O(k) is G(2k-1,k-1,0), and the Johnson graph J(n,k) is G(n,k,k-1). Of course, G(5,2,0) = G(5,3,1) = P.

Chen and Lih make the following conjecture.

Conjecture 8.1 The graph G(n,k,t) is hamiltonian for any admissable (n,k,t) except (5,2,0) and (5,3,1).

Heinrich and Wallis [H-W 78], proved that G(n,k,0) is hamiltonian if (a) n$\geq$k+[ks/(s - 1)], where s = $2^{1/k}$; (b) k = 1, n $\geq$ 3; (c) k = 2, n $\geq$ 6; and (d) k = 3, n $\geq$ 7. Chen and Lih settle their conjecture for the cases (n,k,k-1), (n,k,k-2) and (n,k,k-3) as well as for suitably large n, when k is given and t is zero or one. This is not strong enough though, to help us with the odd graph conjecture.

9.   Embeddings

We have seen in Chapter 6 how the Petersen and Heawood graphs are both Moore (and Tutte) graphs. Another connection is that the Heawood graph is **the** toroidal dual of K_7 and the Petersen graph is **the** projective plane dual of K_6. We justify the definite articles below. Incidentally since the icosahedron is the planar dual of the dodecahedron and we have observed in Section 1 that the dodecahedron may be 'projected' onto P, there is also a close connection between the icosahedron, the dodecahedron, the Heawood graph and P.

Suppose that G is a finite graph. Regard G as a 1-dimensional simplicial complex. Then an **embedding of G into a surface** S is a continuous map f: G $\rightarrow$ S such that G and f(G) are homeomorphic via f. In fact it was Petersen [jP 2, p.420] who proved that every graph G can be embedded in some surface S .

Embeddings f_1,f_2: G $\rightarrow$ S are **equivalent** if there exists an automorphism σ of G and a homeomorphism h: S $\rightarrow$ S such that $hf_1 = f_2\sigma$. When G has only one equivalence class of embeddings into S, then G is **uniquely embeddable** in S. Whitney [hW 32] proved that every planar 3-connected graph has a unique embedding in the sphere. This is a consequence of his theorem that every 3–connected planar graph has a unique dual, since the dual of a planar connected graph corresponds to one of its embeddings in a sphere. Extending these results, Negami [sN 83], [sN 84], [sN 85], [sN 88],see also [sK 91], has shown that 'every 6-connected toroidal graph is uniquely embeddable and every 5-connected projective planar graph which contains a subdivision of K_6 as a subgraph, is uniquely embeddable. Therefore we are justified in describing the Heawood graph, see Figures 9.1 and 9.2, as **the** toroidal dual of K_7 and the Petersen graph, see Figures 9.3 and 9.4, as **the** projective plane dual of K_6.

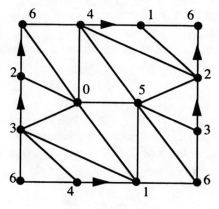

Toroidal K_7

Figure 9.1

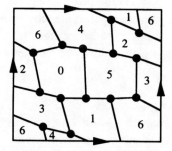

The Heawood graph as the toroidal dual of K_7

Figure 9.2

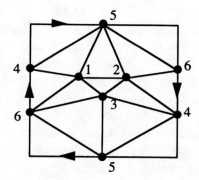

An embedding of K_6 in a projective plane

Figure 9.3

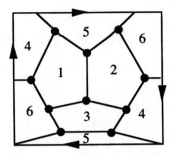

P as the projective plane dual of K_6

Figure 9.4

Of course the motivation behind Heawood's construction is that it shows that at least 7 colours are necessary to face colour all toroidal maps. The genus $g(G)$ [aW 73], of a graph G, is the minimum genus among all orientable surfaces in which G can be embedded. The non-orientable genus $\tilde{g}(G)$ is similarly defined. Remarkably, Babai [lB 91], see also [cT 91], has shown that there are only finitely many finite connected vertex-transitive graphs of given orientable genus g, greater than 2.

A graph G is **toroidal** if $g(G) = 1$ and **projective planar** if $\tilde{g}(G) = 1$. If S_k is an orientable surface of genus k, then $\chi(S_k)$ is the maximum number of colours required to properly colour the faces of any map on S_k. $\chi(\tilde{S}_k)$ is defined similarly, where $\tilde{S}_k$ is the non-orientable surface of genus k. Heawood [pH 90] proved that

$$\chi(S_k) \le \lfloor \tfrac{1}{2}(7 + \sqrt{1+48k}\,) \rfloor \quad (k \ge 0). \tag{1}$$

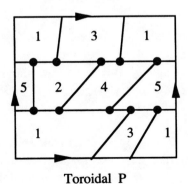

Toroidal P

Figure 9.5

Ringel and Youngs *et al* later established equality in (1) for all $k > 0$. From (1), $\chi(S_1) \le 7$. The toroidal Heawood graph then shows $\chi(S_1) = 7$.

Ringel [gR 59] proved that $\chi(\tilde{S}_k) = \lfloor \frac{1}{2}(7 + \sqrt{1+24k}) \rfloor$ for $k = 1$ and $k \ge 3$; Franklin [pF 34] had previously proved that $\chi(\tilde{S}_2) = 6$. Thus $\chi(\tilde{S}_1) = 6$. The projective planar embedding of P shows that $\chi(\tilde{S}_1) \ge 6$.

We have illustrated an embedding of P in the projective plane. Consider the embedding of P in the torus shown in Figure 9.5.

The dual of this toroidal embedding is a multigraph which is 'almost' isomorphic to K_5; see Figure 9.6.

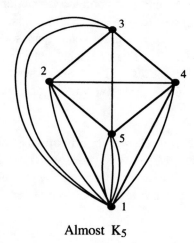

Almost K_5

Figure 9.6

Since $\chi(P) = 3$ and $\gamma(P) = 5$, where $\gamma(P)$ is the girth of P, it follows by considering subdivisions G of P, that there exist toroidal graphs G with $\gamma(G) \ge m$ (for any $m \ge 6$) and $\chi(G) = 3$. On the other hand, it is easy to prove [aW 73] that if G is toroidal and $\gamma(G) \ge 6$, then $\chi(G) \le 3$. Therefore in this sense the subdivisions G of P with $\gamma(G) \ge 6$, are extremal.

Although only peripherally related to our theme, we should mention here an important result in embedding graphs on surfaces. Wagner's theorem, Theorem 1.6.6, states that K_5 and $K_{3,3}$ are the only minor- minimal non-planar graphs. A graph H is a **minor** of G, if H can be obtained from a subgraph of G by contracting edges. For example $K_{3,3}$ and K_5 are both minors of P. A graph G is a **minor-minimal** non-planar graph if it contains no proper minor which is non-planar. So P is not minor-minimal, whereas both K_5 and $K_{3,3}$ are. Now Glover, Huneke and Wang [G-H-W 79] and Archdeacon [dA 80], [dA 81] demonstrated that there were exactly 35 minor-minimal non-projective planar graphs. For higher genus the length of the list grows rapidly. Indeed Erdös in the 1930's conjectured that the list was always finite for non-orientable surfaces and this was proved by Archdeacon and Huneke [A–H81]. Finally it was proved by Robertson and Seymour to be true for all surfaces; see [R-S 85].

Let G be a graph. A **delta-wye exchange of** G is a graph obtained from G by deleting the edges of a triangle (v_1,v_2,v_3) in G and adding a new vertex u and edges uv_i, i = 1, 2, 3. A **wye-delta exchange** of G is a graph obtained from G by performing the reverse operation.

Any (finite) graph can obviously be embedded in 3-space. But when can a graph be embedded so that no two disjoint circuits are 'linked'? (The precise meaning of 'linked' will not be pursued here, see [R-S-T 92]). Robertson, Seymour and Thomas have proved that a graph has a linkless embedding if and only if it contains as a minor none of the seven graphs on 15 vertices which can be obtained by delta-wye and wye-delta exchanges from K_6; one of these graphs is the Petersen graph; see Exercise 30.

10. The cycle double cover conjecture
The cycle double cover conjecture, CDCC, has attracted considerable attention over the last few years. (Excellent references are the articles by Jaeger [fJ 85] and Bondy [jB 90].) Any counterexample to the CDCC must be a snark; recall that it was conjectured in Chapter 3 that P is the only minor-minimal snark. Seymour mentions this conjecture in [pS 79]; the same conjecture was made by Szekeres [gS 73].

The cycle double cover conjecture (CDCC) Let G be a bridgeless finite graph. Then there exists a sequence of cycles in G such that each edge of G is in exactly 2 of the cycles.

In this conjecture we should especially notice the word 'sequence'. The conjecture states that there exists a sequence of cycles $C_1, C_2, ..., C_m$ in G such that for each $e \in EG$, $e \in EC_i$ for exactly 2 distinct values of i, $i \in \{1, 2, ..., m\}$. It may well be that if these two values are i_1 and i_2 then C_{i_1} is the same cycle as C_{i_2}. Such a sequence is called a **cycle double cover,** CDC, of G. We saw in Chapter 1 Exercise 37, that every Eulerian graph admits a cycle decomposition. The CDCC is a natural way of extending this idea to graphs with vertices of odd degree.

It is easy to see that if a 2-connected planar graph is embedded in the plane, then trivially there exists such a cycle double cover. We simply take the set of face boundaries. We can generalize this idea.

If a graph G is embedded in a surface S, the complement of G, relative to S, is a collection of open sets called **regions**. Some, but not necessarily all, of these regions may be topological disks (2-**cells**). If all of these regions are 2-cells, we say that it is a 2-cell embedding. Figure 10.1(a) illustrates an embedding of K_4 in the torus which is not 2-cell while Figure 10.1(b) shows a 2-cell embedding.

A (2-cell) embedding of a graph G on a surface S is said to be **strong** if each face boundary is a cycle. Figure 10.1(a) shows an embedding of K_4 in the torus which is not strong; the edge e appears in only one face. The embedding shown in 10.1(b) is strong.

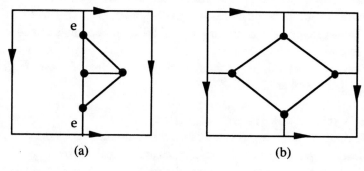

(a)   (b)

Figure 10.1

Strong embedding conjecture (SEC) Every 2-connected graph has a strong embedding in some surface.

Obviously if the SEC is true then so is the CDCC. We simply take the family of cycles which are face boundaries. In the case of cubic graphs every double cover is the family of face boundaries of a strong embedding on some surface. Accordingly for cubic graphs, the CDCC and SEC are equivalent.

Jaeger, see [fJ 85], shows that a minimum counterexample to the CDCC is a snark; see Exercise 18. Goddyn [lG 85] has shown that a minimum counterexample to the CDCC has girth at least 7. Tarsi [mT 86], see also Goddyn [lG 89], has proved the conjecture for graphs with a hamiltonian path. The CDCC is true for 4–edge-connected graphs by a theorem of Jaeger [fJ 79]. Alspach and Zhang [A–Z89b], have proved the conjecture for 2-connected cubic graphs containing no subdivision of P. Catlin [pC 89] proved that under certain edge-connectivity restrictions, a graph either has a cycle double cover of a certain specified type or is subcontractible to P. This result is related to Tutte's Conjecture 2.6.3.

Finally, suppose G is a cubic graph and $e \in EG$. Denote by G(e) the unique cubic graph which has G - e as a subdivision. A snark G which has the property that for each $e \in EG, \chi'(G(e)) = 4$, is called **strong**. Celmins [uC 87] has shown that a minimum counterexample to the CDCC must be a strong snark. In summary, we are looking for a strong snark of girth at least 7. No snarks of girth at least 7 are known.

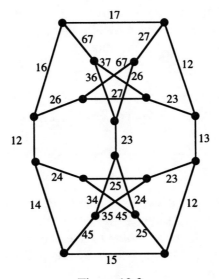

Figure 10.2

While it is true that a 3-edge-colourable 3-regular graph has a covering with even cycles where every edge is used twice (just take the three 2-factors induced by the 3–edge-colouring), the converse is not true. In Figure 10.2 the even cycle covering is indicated by labelling the edges with 2 digits, each digit representing a cycle. The example is from [hF 84].

The Petersen graph, however, has no such covering with even cycles. That is, one cannot add to the CDCC the condition that the cycles are even. Interestingly however, the flower snarks do have a cycle double cover using only even cycles; see Exercise 19.

We now give two conjectures; see [hF 84]. Recall that a cycle is dominating when its vertex set meets every edge of the graph.

Conjecture 10.1 (H. Fleischner) Every snark has a dominating cycle.

Conjecture 10.2 (C. Thomassen) Every 4-connected line graph is hamiltonian.

If G is a snark then L(G) is 4-connected. A hamiltonian cycle of L(G) corresponds to a dominating cycle in G; see Exercise 20. Therefore Conjecture 10.1 is a special case of Conjecture 10.2. In [sZ 91], Zhan has proved that every 7–connected line graph is hamiltonian-connected.

Cycle double covers have been studied from many different viewpoints. For example, Bondy [jB 90] considers path analogues of the CDCC; see also [hL 90]. Yet another direction is to consider the smallest number of cycles which provide a CDC for certain types of graphs. In this area, Bondy [jB 90] conjectures as below.

Conjecture 10.3 Let G be a 2-connected cubic graph on n vertices, $n > 6$. Then G admits a CDC consisting of at most $\frac{1}{2}n$ cycles.

This conjecture is sharp. The Petersen graph cannot be covered with fewer than five cycles.

We conclude by hinting at the connection between cycle decompositions and flows; see Section 5.1. Recall that a cycle decomposition **C** of G is a set of cycles such that each edge of G belongs to at least one cycle of **C** The length of **C** is the sum of the lengths of the cycles in **C** and is denoted by $\lambda(C)$. Raspaud proved the next theorem.

Theorem 10.4 Let G be a graph different from K_4, with a nowhere-zero 4–flow. Then G has a cycle decomposition **C** with $\lambda(C) \leq |EG| + |VG| - 3$.

Proof: See [aR 91] and also [cZ 90].   □

It is easy to see that the complete bipartite graph $G = K_{3,3m}$ has a nowhere-zero 3-flow and that every CDC of G has length at least $|EG| + |VG| - 3$. This shows that the theorem is sharp, since if a graph has a nowhere-zero 3-flow then, by Lemma 5.1.5, it has a nowhere-zero 4-flow.

Then again, the problem of determining $\lambda(C)$ for certain classes of graphs, is intimately related to the Chinese Postman problem with subdivisions of the Petersen graph playing a key role; see [cZ 90].

11. Geodetics

A connected graph G is **geodetic,** if given any pair of distinct vertices u and v, there exists a unique uv-path of length d(u,v). For example, trees, odd cycles, complete graphs and Moore graphs of girth 5, are geodetic. In particular, the Petersen graph is geodetic. Ore [oO 62] suggested that such graphs be characterized. However it seems to be a difficult problem to do this in any meaningful way. It is easy to prove that G is geodetic if and only if every block of G is geodetic; see Exercise 22. (A **block** is any maximal 2-connected induced subgraph.) Accordingly we will assume that a geodetic graph is 2-connected. A nice characterization of planar geodetic graphs is given by Watkins [mW 64] and Stemple and Watkins [S-W 68]. A geodetic graph is planar if and only if it is a subdivision of K_n for $3 \le n \le 4$.

It seems that the most interesting geodetics have diameter 2. For example the Moore graphs of girth 5 are geodetics of diameter 2. In fact, Parthasarathy and Srinivasan [P-S 85] conjectured that a diameter 2 geodetic graph of minimum degree 3, must contain P as an induced subgraph. Subsequently, using a nice construction based on a certain affine plane, Scapellato [rS 89], produced a counterexample to this conjecture. At any rate, this class of geodetics seems the hardest to classify and most constructions of geodetics [bZ 75], [jaP 80], [C–P82], [jaP 84], [rS 86] are based on geodetics of diameter 2. The Moore graphs are the only regular geodetic graphs of diameter 2. But firstly we digress.

The Friendship Theorem asserts that any set of people, in which any two persons have a unique friend in common, contains a politician - that is a person who is a friend to everyone.

In order to put this in graph theoretic terms we need to define the graphs F_n^*. For n odd, let $F_n^* = K_1 + \frac{1}{2}(n - 1)K_2$ - see Figure 11.1. Clearly, the vertex of degree n - 1 is the politician.

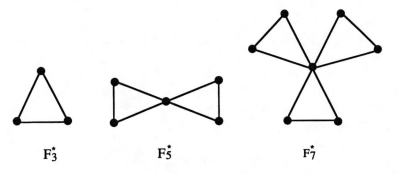

$$F_3^* \qquad\qquad F_5^* \qquad\qquad\qquad F_7^*$$

Figure 11.1

Theorem 11.1 (Friendship Theorem). If G is a graph on $n \geq 1$ vertices such that each distinct pair of vertices has a unique common neighbour, then G is isomorphic to the friendly graph F_n^*.

Proof: Suppose that G satisfies the conditions of the statement of the theorem. Firstly assume that G is regular of degree r. Then the conditions of the theorem imply that G is strongly regular with parameters $(n,r,1,1)$. Hence by Theorem 8.4.1 and using the notation of that theorem,

$$f - g = - r/\sqrt{r-1} \quad (r \geq 2)$$

is an integer. Hence $r = 0$ or 2 and G is K_1 or K_3. In either case G is a friendly graph.

If G is not regular, a routine counting shows that $G \cong F_n^*$ for some n; see Exercise 23.   □

Lemma 11.2 Suppose that G is a geodetic graph with $n \geq 3$ vertices. If G is regular and has diameter 2, then $\gamma(G) = 5$.

Proof: Suppose that G is a geodetic graph with $n \geq 3$ vertices, which is regular of degree r and has diameter 2. Suppose that G contains a triangle i.e. $\gamma(G) = 3$. Then every edge must lie on a triangle; see Exercise 24. Since G is a geodetic of diameter 2, this means that each distinct pair of vertices has a unique common neighbour. So G is a friendly graph. But F_3^* is the only regular friendly graph with $n \geq 3$ vertices. However F_3^* has diameter 1 which is a contradiction.

Now suppose that G does not contain a triangle. Since G is a geodetic graph, it cannot contain an induced 4-cycle. Therefore $\gamma(G) \geq 5$. But since the diameter of G is 2 and G is regular of degree r (≥ 2 since the diameter of G is 2), G contains a 5-cycle. Hence $\gamma(G) = 5$.   □

Theorem 11.3 Suppose that G is a geodetic graph which is regular and has diameter 2. Then G is a Moore graph of girth 5.

Proof: Suppose that G satisfies the conditions of the theorem. Then from Lemma 11.2, $\gamma(G) = 5$. Since G is regular with $\gamma(G) = 5$ and diameter equal to 2, G is a Moore graph.   □

If we no longer insist that two people who know each other must have a unique friend, then we have an appropriate model for geodetic graphs which are K_4-free (i.e. do not contain a subgraph isomorphic to K_4) with diameter 2; notice that, by definition, a geodetic graph cannot contain an induced subgraph isomorphic to K_4-e. Thus an **acquaintance set,** is a set of people in which any two people who do not know each other have a unique friend in common and any two distinct people who do know each other have at most one common friend.

Let $\Im(n)$ be the set of graphs G on n vertices such that (i) G does not contain a 4-cycle and (ii) subject to (i), G is edge-maximal.

Obviously if $G \in \Im(n)$ then diam $G \leq 3$. Write $\Im(n,d) \equiv \{G \in \Im(n):$ diamG=d$\}$, $d \in \{1,2,3\}$.

Lemma 11.4 G is a K_4-free geodetic of diameter 2 on n vertices if and only if $G \in \Im(n,2)$.

Proof: If G is a K_4-free geodetic of diameter 2, then it is easy to show that any two distinct vertices that are non-adjacent are contained in a 5-cycle. This guarantees edge-maximality. Everything else follows immediately from the definitions. ▢

Sometimes the equivalence established in Lemma 11.4 has been glossed over in the literature. As a result there are overlaps. The graphs in $\Im(n,2)$ have a nice characterization.

Let Q be a finite projective plane with a polarity π; see Section 5.3. In this context, a **polarity** is an involutory function which transforms points to lines and lines to points, and preserves incidence; see [jH 79]. Then $G \equiv G(Q,\pi)$ is the graph defined by (i) VG is the set of points of Q; (ii) p is joined to q if and only if $p \in \pi(q)$, p,q $\in$ VG, p ≠ q. G is called a **projective graph**; see [rB 46]).

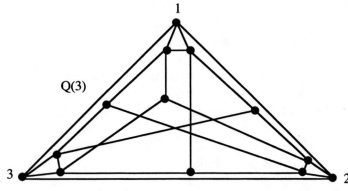

Figure 11.2

Let $PG(2,q)$ be the finite projective plane over the Galois field $GF(q)$, where q is a prime power. A polarity π is given by $\pi(a,b,c) = [a,b,c]$, where $[a,b,c]$ denotes the set of points satisfying $ax + by + cz = 0$. Let $Q(q) = G(PG(2,q),\pi)$. Then, from the definition, $Q(q) \in \Im(q^2 + q + 1)$. We illustrate $Q(3)$ in Figure 11.2. Observe that $Q(3) \setminus \{1,2,3\} \cong P$.

The next theorem could be attributed to a committee of authors ([hS 72], [jS 74] and [B-E-F 79]).

Theorem 11.5 $\Im(n,2)$ is the union of the following classes of graphs:
(i) $F_n^*(n \geq 4)$,
(ii) $K_1 + (sK_2 \cup K_1)$ $(n = 2(s + 1), s \geq 1)$,
(iii)   girth five Moore graphs,
(iv)   projective graphs.

Proof: The proof is not difficult. We sketch only the outlines of the argument. Suppose that $G \in \Im(n,2)$. If G is regular then, from Theorem 11.3, G is a girth five Moore graph. If G contains a vertex of degree $n - 1$, then G is a friendly graph (or an 'almost' friendly graph of type (ii) in the statement of the theorem).

So suppose G is not regular and G, using routine counting, does not contain a vertex of degree $n - 1$. Let $\Delta(G) = \Delta \leq n - 2$. The next stage is to prove that
(a)   if an edge uv lies in a triangle, then $\deg u = \deg v = \Delta$,
(b)   if an edge uv does not lie in a triangle, then $\{\deg u, \deg v\} = \{\Delta-1, \Delta\}$.

So each vertex of G has degree $\Delta - 1$ or Δ. We form a projective plane Q from G as follows: the points of Q are the vertices of G, and the lines L_v, $v \in VG$ are defined by

$$L_v = \begin{cases} N(v) & \text{if } \deg v = \Delta \\ N(v) \cup \{v\} & \text{if } \deg v = \Delta - 1 \end{cases}$$

where $N(v)$ is the neighbour set of v. This is a projective plane because of (a) and (b). The mapping π defined by $\pi(v) = L_v$, $v \in VG$ is a polarity of Q and $G = G(Q,\pi)$. Thus G is a projective graph. $\square$

The problem of classifying the elements of $\Im(n,3)$ seems to be much harder. It will be touched upon in the next section.

12. Geodetics and an extremal problem

The Petersen graph belongs to $\Im(10)$. It has 10 vertices and does not contain a C_4. Subject to these conditions it is edge-maximal. However it does not have the maximum number of edges consistent with these constraints. The graph G in Figure 12.1 also belongs to $\Im(10)$ and yet has 16 edges. We know, by checking the list of graphs in the statement of Theorem 11.5, that $G \in \Im(10,3)$.

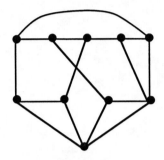

Figure 12.1

We shall see that, even though P is not an extremal graph in this sense, the line graph L(P) of P is extremal; see Figure 12.2.

Write $t(n) = \max\{|EG| : G \in \Im(n)\}$. From [E-R-S 66] and [wB 66], $Q(q) \in \Im(q^2 + q + 1)$ and

$$t(q^2 + q + 1) \geq \tfrac{1}{2}q(q + 1)^2, \tag{1}$$

where q is a prime power.

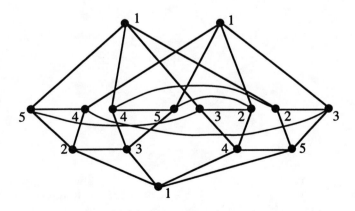

Figure 12.2

A classical argument [iR 58] (see the proof of Theorem 15.2) shows that for all $n \geq 2$,

$$t(n) \leq \frac{1}{4}n(1 + \sqrt{4n-3}).$$ (2)

Füredi [zF 83] proved that for $q \geq 2$ and even,

$$t(q^2 + q + 1) \leq \frac{1}{2}q(q + 1)^2.$$ (3)

Therefore when $q \geq 2$ is a power of 2,

$$t(q^2 + q + 1) = \frac{1}{2}q(q + 1)^2.$$ (4)

The order of $t(n)$ is therefore well understood. From (1) and (2), since $t(n)$ is increasing, we have $t(n) \sim \frac{1}{2}n\sqrt{n}$ (that is, the ratio of $t(n)$ and $\frac{1}{2}n\sqrt{n}$ tends to 1 as n tends to infinity). Let

$$\mathfrak{S}_{max}(n,d) \equiv \{G \in \mathfrak{S}(n,d): |EG| = t(n)\} \qquad (d \in \{2,3\})$$

$$\mathfrak{S}_{max}(n) \quad \equiv \{G \in \mathfrak{S}(n) : |EG| = t(n)\}$$

$$\equiv \mathfrak{S}_{max}(n,2) \cup \mathfrak{S}_{max}(n,3) \qquad (n \geq 4)$$

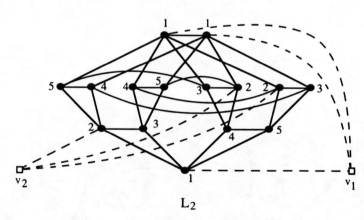

$$L_2$$

Figure 12.3

Füredi proved that if q is a prime power, then $Q(q)$ is the only element of $\mathfrak{S}_{max}(q^2 + q + 1)$. In [C-F-S 87], an attempt is made to examine $\mathfrak{S}_{max}(n)$ for small values of n. Write $L_0 = L(P)$. We define graphs L_i ($i = 1,2,3,4,5$) as follows:

(i) $VL_i \equiv VL_{i-1} \cup \{v_i\}$;

(ii) $EL_i \equiv EL_{i-1} \cup \{iv_i : i$ is any vertex labelled i in Figure 12.2$\}$.

We show L_2 in Figure 12.3.

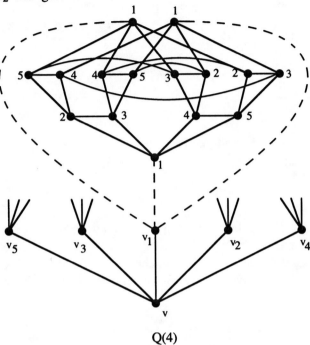

Q(4)

Figure 12.4

In [C-F-S 87] it is shown that L_i is the only element of $\mathfrak{S}(15+i)$, $i = 0, 1, 2, 3$.
Finally, define the graph L with vertex set $VL = VL_5 \cup \{v\}$ and edge set
$EL = EL_5 \cup \{vv_i : i = 1,2,3,4,5\}$. Then $L \in \mathfrak{S}_{max}(21)$. But $Q(4)$ is the
only element of $\mathfrak{S}_{max}(21)$. Hence $L \cong Q(4)$. Therefore we have a nice way of
describing the projective graph $Q(4)$; see Figure 12.4. Compare this with the
description of $Q(3)$ in Figure 11.2; $Q(4) \setminus \{v_1,v_2,v_3,v_4,v_5,v\} \cong L(P)$;
$Q(3) \setminus \{1,2,3\} \cong P$. In this context see also [rS 89].

13. Generalized Petersen and permutation graphs

The generalized Petersen graph was defined in Exercise 38 of Chapter 1. The
Petersen graph, $P(5,2)$ [C-P 72], is the only generalized Petersen graph which
has chromatic index equal to 4. In 1983 Alspach [bA 83] completed the
determination of the parameters k,m for which $P(k,m)$ is hamiltonian. For
minor technical reasons we will assume that $m < \lfloor \frac{1}{2}k \rfloor$; the definition simply
requires $m < \frac{1}{2}k$. References to the many authors who contributed to the solution
of this problem can be found in Alspach's paper.

Theorem 13.1 The generalized Petersen graph $P(k,m)$ is non-hamiltonian if and only if $m = 2$ and $k \equiv 5 \pmod 6$. $\square$

In fact, Schwenk [aS 89] has gone one better and counted the actual number of hamiltonian cycles. The **Fibonacci numbers** $f_n, n \geq 0$, are defined by $f_0 = f_1 = 1, f_{n+1} = f_n + f_{n-1}, n \geq 1$.

Theorem 13.2 The number of hamiltonian cycles in $P(k,2), k \geq 3$ is

(i) $2f_{\frac{1}{2}k+2} - 2f_{\frac{1}{2}k-2} - 2$ for $k \equiv 0, 2 \pmod 6$;

(ii) k for $k \equiv 1 \pmod 6$;

(iii) 3 for $k \equiv 3 \pmod 6$;

(iv) $k + 2f_{\frac{1}{2}k+2} - 2f_{\frac{1}{2}k-2} - 2$ for $k \equiv 4 \pmod 6$;

(v) 0 for $k \equiv 5 \pmod 6$. $\square$

A graph G is **n-extendable** if it contains a set of n independent edges and every set of n independent edges can be extended to (i.e. is a subset of) a perfect matching. Schrag and Cammack [S-C 89] conjecture that $P(k,m)$ is 2-extendable if and only if $k \neq 5m$. They settle the conjecture for $m \leq 7$ and $k \geq 3m+5$.

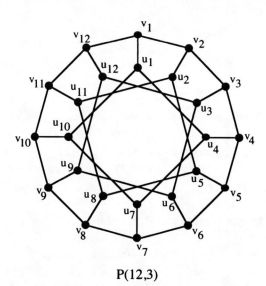

P(12,3)

Figure 13.1

Let G be a graph with n vertices. Suppose $VG = \{1, 2, ..., n\}$ $(n \geq 4)$. Choose $\sigma \in S_n$, the symmetric group acting on VG . The σ-**permutation graph of G**, $P_\sigma(G)$, consists of disjoint copies of G, G_1 and G_2 , along with the n edges obtained by joining i in G_1 with $\sigma(i)$ in G_2 ($i = 1,2,...,n$). If the initial graph G is an n-cycle and i is joined to $i + 1$ and $i - 1$ in G (modulo n) then $P_\sigma(G)$ is called a **cycle permutation graph** and we denote it by $C(n,\sigma)$. The connection between $P(k,m)$ and $C(n,\sigma)$ is given in the next theorem. In Figure 13.1 we illustrate $P(12,3)$. This generalized Petersen graph is also isomorphic to $C(12,\sigma)$ for some σ; see Exercise 25. We write (s,t) to denote the greatest common divisor of integers s and t.

Theorem 13.3 A generalized Petersen graph $P(k,m)$ is a cycle permutation graph if and only if one of the following is true:

(i) $(k,m) = 1$;

(ii) $k \equiv 0 \pmod 4$, $m \equiv \pm 1 \pmod 4$ and $(\frac{1}{4}k, \frac{1}{4}(m \pm 1)) = 1$;

(iii) $k \equiv 0 \pmod 5$, $m \equiv \pm 2 \pmod 5$, $(\frac{1}{5}k, \frac{1}{5}(m \pm 2)) = 1$ and $(\frac{1}{5}k, \frac{1}{5}(2m \pm 1)) = 1$.

Proof: (See [S-R 84].) Remember that a cubic graph with $2n$ vertices is a cycle permutation graph if and only if it has two vertex disjoint n-cycles as induced subgraphs. Now the argument proceeds by case by case analysis.   □

The toughness and binding numbers of certain permutation graphs are to be found in [G-P-S 91] and [P-R-S 90].

14. Generalized Petersen edge-colourings and groups

Vizing's theorem, Theorem 3.1.1, states that if G is a graph, then

$$\Delta(G) \le \chi'(G) \le \Delta(G) + 1.$$

Graphs G for which $\chi'(G) = \Delta(G)$ are said to be of **class 1** [F-W 77]. Otherwise they are of **class 2.** Cubic class 2 graphs are of considerable interest. For example snarks are examples of such graphs; see Chapter 3.

Castagna and Prins [C-P 72] proved that every generalized Petersen graph other than P itself was a class 1 graph. Of course, if a generalized Petersen graph is hamiltonian, then it is a class 1 graph.

A graph G is said to be **uniquely $\chi'(G)$-edge-colourable** if the edge partition induced by the $\chi'(G)$ colour classes is unique up to permutations of the colours. Now if G is a cubic class 1 graph and G is uniquely 3-edge-colourable, then G has exactly three cycles. This is because if the colour classes are R, B and G, the 3 hamiltonian cycles are induced by $R \cup B$, $R \cup G$ and $B \cup G$; any other hamiltonian cycle would induce a different colouring.

On the other hand, Thomason [aT 82] proved that even though $P(6t+3,2)$ ($t \geq 0$) has exactly three hamiltonian cycles (see also Theorem 13.2), $P(6t+3,2)$ ($t \geq 2$) is not uniquely 3-edge-colourable.

Finally, Frucht, Graver and Watkins [F-G-W 71] analysed the automorphism groups of $P(k,m)$. As we know already, the automorphism group of $P(5,2)$ is isomorphic to S_5. $P(10,2)$ is the graph of the dodecahedron so its automorphism group is isomorphic to the direct product of the alternating group A_5 and S_2. $P(10,3)$ has its automorphism group isomorphic with $S_5 \times S_2$. In [F-G-W 71] it was shown that in general the automorphism group of $P(k,m)$ is easy to describe.

15. Pandora's Box
The material in this section can be found in [rJ 85].

The distance $d(e,e^*)$ between distinct edges e, e^* of P is defined by

$$d(e,e^*) = \begin{cases} 0 & \text{if they are incident} \\ 1 & \text{if they are not incident but there exists an} \\ & \text{edge which is simultaneously incident to both} \\ & e \text{ and } e^* \\ 2 & \text{otherwise.} \end{cases}$$

For $F \subseteq EP$ write $d(F) = i$ if, for all $e, e^* \in F$, $e \neq e^*$, $d(e,e^*) = i$ ($i = 0,1,2$). We say that F is an i-**maximal set** if $d(F) = i$ and F is maximal subject to this property. So a 0-maximal set F consists of three edges incident to some vertex. A 0-maximal set is called a **claw**; there are 10 claws in P. The other i-maximal sets can be described as follows:-

(0)   there are exactly 6 1-maximal sets F with $|F| = 5$; these are the 6 1–factors of P.

(1)   there are exactly 20 1-maximal sets F with $|F| = 3$; such a set F consists of 3 non-consecutive edges of a 6-cycle C with the property that there exists a unique vertex v of P at distance 2 from each of the vertices of C. The **twin** F^* **of this** F is the other set of 3 non-consecutive edges in C.

(2)   there are exactly 5 2-maximal sets F with $|F| = 3$; such a set F corresponds to a set of 3 vertices of the line graph of P which are mutually distance 3 apart; see Figure 12.2.

To verify these claims and the claims below, it might be helpful to refer to the drawings of P in Figure 1.1.7.

We now construct PG(3,2) from the edges and i-maximal sets of P:

(3)   the set **P** of points is EP,

(4)   the set **L** of lines is defined by **L** = L(0) ∪ L(1) ∪ L(2)

where (i) L(0) is the set of claws; (ii) L(1) is the set of 20 1-maximal sets described in (1); (iii) L(2) is the set of 5 2-maximal sets described in (2).

If m ∈ L(i) we say that m is a **type i line.**

(5)   the set **PL** of planes consists of the 15 sets F ⊆ E(P) such that |F| = 7 and EP \ F are the edges of an 8-cycle. Such a plane is shown in Fig. 15.1 where the dotted edges are the edges of an 8-cycle.

At least we have 15 points, 35 lines and 15 planes so one can see why the next theorem has a chance of being true.

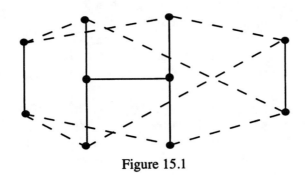

Figure 15.1

Theorem 15.1 (**P**,**L**,∈) ≅ PG(3,2).

Proof: The following claims are more than sufficient to establish the required conditions on '∈'. Its verification is by inspection of the Figures 12.2 and 15.1.

Claims
(i)   each pair of points is in exactly one line and this line is in every plane containing this pair of points;
(ii) each point belongs to 7 lines; more precisely each point belongs to 2, 4 and 1 lines of types 0, 1 and 2, respectively;
(iii) each point belongs to 7 planes;
(iv) each plane contains 7 lines; more precisely each plane contains 2, 4 and 1 lines of types 0, 1 and 2, respectively;

(v)   each pair of lines in a plane intersect in exactly one point;

(vi) there are 15.7.4 triples (p,m,F), p ∈ **P**, m ∈ **L**, F ∈ **PL** such that m∈F, p ∈ F, p ∉ m;

(vii) there are 35.12 pairs (p,m), p ∈ **P**, m ∈ **L**, p ∉ m; each such pair belongs to exactly one plane.   □

We now show how the Hoffman-Singleton graph can be constructed from **P**; see Section 6.1.

Firstly let us say that the lines m, m' ∈ **L** are **vertex disjoint** if the edges of m have no vertices in common with those of m'.

We define the graph G as follows. VG = **P** ∪ **L**. The adjacencies of G are given by:

(i)   each point is adjacent to each line to which it belongs;

(ii) each line is adjacent to each point that it contains;

(iii) each type 0 line is adjacent to the two type 2 lines disjoint from it and to the two type 1 lines which are vertex disjoint from it;

(iv) each type 1 line m is adjacent to the type 0 line g vertex-h disjoint from it and in addition m is adjacent to the three type 1 lines h disjoint from both m and g and not equal to the twin m* of m;

(v)   each type 2 line is adjacent to the four type 0 lines disjoint from it;

(vi) there are no other adjacencies.

These adjacencies are illustrated in Figure 15.2 where (say) a directed arc weighted 2 from L(0) to L(1) indicates that each type 0 line is joined to two type 1 lines; the numbers in each circle indicate the number in the associated set.

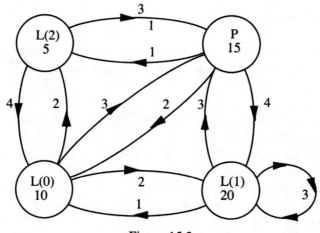

Figure 15.2

Theorem 15.2 G is isomorphic to the Hoffman-Singleton graph.

Proof: By inspection adjacency is symmetric, so G is a graph. Now the Hoffman-Singleton graph is the unique graph on 50 vertices which is regular of degree 7 and has girth 5. Certainly G has 50 vertices and is regular of degree 7. Now by inspection (albeit considerable) G has diameter 2. This is sufficient to show that the girth of G is 5. To show this we use the classical argument referred to in Section 12. Assuming G has diameter 2, then there are $\binom{50}{2} - \frac{50.7}{2}$ non-adjacent pairs of vertices of G, which is exactly the same number $\left[50.\binom{7}{2}\right]$ of paths of length 2 in G. Since G is diameter 2 there exists therefore, exactly one path of length 2 between each pair of non-adjacent vertices and hence G contains no cycles of length 3 or 4. Clearly G has girth 5. $\square$

Exercises

1. Show how P is derived from D.

2. Show that D is strictly 2-transitive.

3. (a) Determine all graphs which are locally K_n.
    (b) Find a graph which is locally $K_1 \cup K_2$. Find all such graphs.

4. Characterize those graphs which are locally a tetrahedron. Characterize those graphs which are locally an octahedron.

5. Suppose that G is a connected graph. When is $G \otimes K_2$ disconnected? When is $(G \otimes K_2) \otimes K_2$ disconnected?

6. Suppose that G is a regular graph of degree r on n vertices and that $spec(G) = \{\lambda_1{}^{s1}, \lambda_2{}^{s2}, ..., \lambda_n{}^{sn}\}$. Prove that $spec(L(G)) = \{(r-2+\lambda_1)^{s1}, (r-2+\lambda_2)^{s2}, ..., (r - 2 + \lambda_n)^{sn}, -2^{m-n}\}$.

    [Hint: (i) Let $C = [c_{ij}]$ be the vertex-edge incidence matrix i.e. $c_{ij} = 1$ if and only if 'i' is incidence with edge 'j' and $c_{ij} = 0$ otherwise. Let M be the usual adjacency matrix of G. Prove that $M = CC^T - rI$.

    (ii) Let N be the usual adjacency matrix of L(G). Prove that $N = C^T C - 2I_q$, where $q = |EG|$.

    (iii) Now use the result that if X is any real $n \times m$ matrix with $n \leq m$, then $spec(X^T X) = spec\ XX^T \cup \{0^{m-n}\}$ and each eigenvalue is non-negative [hS 67].]

7. Prove that the number of closed trails of length k in a graph G is equal to the trace of $M(G)^k$.

8. Prove that in any 2-edge-connected graph G of diameter d, every edge belongs to a cycle of length at most $2d + 1$.

9. Prove that in a 3-connected graph any two longest cycles have at least three common vertices.

10. (a) Prove that P is not a Cayley graph.

    [Hint: Assume that $P = (A,S)$.
    (i)   Verify that, since the diameter of P is 2, then $A = S \cup S^2$.
    (ii) There are two groups of order 10, Z_{10}, and D_5 where D_5 is the dihedral group.
    (iii) Enumerate the possible sets S for both groups.]

    (b) Prove that C_{28} is not a Cayley graph.
    (c) Prove that the inflations of P and C_{28} are not Cayley graphs.

11. Prove that P, C_{28} and their inflations all have 2-factors with at most 2 components.

12. Prove that $K(n,k,t)$ is edgeless if $n + t \le 2k$.

13. Suppose that $k \ge 2$ is an integer. Prove that if k is a power of 2, $\binom{2k-1}{k-1}$ is odd.

14. Give an official ruling on the Croam villagers' problems.

15. Verify that the map $\theta \to \bar{\theta}$ described in Theorem 7.1 is an isomorphism from $A(O(k))$ onto S_{2k-1}.

16. Prove that $O(k)$ is distance transitive.

17. Determine the intersection arrays for the odd, even and Johnson graphs.

18. Prove that a minimum counterexample to the cycle double cover conjecture is a snark.

[Hint: Take G to be a minimum counterexample.

    (i)   Assume that G has an edge-cut of size 2; contract one of these edges to obtain a smaller graph G'; prove that any cycle double cover for G' gives a cycle double cover for G.

    (ii)   Use the result of Fleischner [hF 76] which states that if $v \in VG$ and deg $v \geq 4$, then there exist edges e and f incident to v with the following property: by deleting e and f and adding a new edge joining the ends of e and f distinct from v, one obtains a **bridgeless** graph G'. Prove that a cycle double cover for G' yields a cycle double cover for G.]

19.   Prove that the flower snarks have a cycle double cover using only even cycles.

20.   Prove that if G is a graph such that L(G) is hamiltonian, then G itself has a dominating cycle.

21.   Two graphs are **homeomorphic** if they are isomorphic or can be made so by the subdivision (repeated if necessary) of edges of either graph into two by the insertion of vertices. Construct a geodetic graph which is homeomorphic to the Petersen graph (see [jaP 77]).

22.   Prove that a graph G is geodetic if and only if every block of G is geodetic.

23.   Prove that if G is a graph on $n \geq 1$ vertices such that each distinct pair of vertices has a unique common neighbour and G is **not** a regular graph, then $G \cong F_n^*$.

[Hint: Suppose $u,v \in VG$ and $uv \notin EG$. Prove that deg u = deg v.]

24.   Suppose that G is a regular geodetic graph with $n \geq 3$ vertices and girth equal to 3. Prove that every edge of G is in a triangle.

25.   Prove that $P(12,3) \cong C(12,\sigma)$ for some $\sigma \in S_{12}$.

26.   Prove that $P(n,2)$ is hamiltonian if and only if $n \not\equiv 5 \pmod 6$.

27.   Prove that $P(n,3)$ is hamiltonian for $n \neq 5$.

28.   In the definition of $P(k,m)$, $m \neq \frac{1}{2}k$. Now suppose we allow $m = \frac{1}{2}k$. Determine when $P(2m,m)$ is hamiltonian.

29.   Determine the crossing number (see Section 2.6) of P. What can you say
       about the crossing number of $C(n,\alpha)$ when α is a transposition; see
       [rR84].

30.   Determine the seven graphs on 15 edges which can be obtained by delta-
       wye and wye-delta exchanges from K_6.

31.   Let G be a bridgeless graph. The **cyclicity** of G, [cL 87], is the
       minimum number of cycles whose union is EG. The elements of a set of
       cycles are **consistently orientable,** if G can be oriented so that each is a
       directed cycle. The **dicyclicity of** G is the minimum number of
       consistently orientable cycles whose union is EG.

       Prove that the cyclicity of P is 3 and the dicyclicity of P is 4. (It seems
       to be a difficult but interesting problem to characterize those graphs whose
       cyclicity and dicyclicity are equal.)

32.   Suppose that G is a graph and $S \subseteq V(G)$. Let $h(G - S)$ denote the
       number of components of G - S. G is t-tough if $|S| \geq t.h(G - S)$ for all
       $S \subseteq V(G)$ satisfying $h(G - S) \geq 2$.

       Prove that if G is hamiltonian, then G is 1-tough. Prove that P is
       $\frac{4}{3}$–tough. Prove that even if G is $\frac{3}{2}$ –tough it need not be hamiltonian.

       [Hint: Consider an inflation of P.]

33.   Prove that P(9,2) is uniquely 3-edge-colourable.

       Is P(9,2) the only nonplanar cubic uniquely 3-edge-colourable graph?
       Fiorini and Wilson [F-W 78] conjecture that this is the case.

34.   If a graph G is connected and S is a cycle of G, then S is a **separating
       cycle,** if the deletion of S from G disconnects G; see [cT 78a] and
       [T–T81]. Prove that $K_{r,r+1}, K_1 + C_n$ and P, have separating cycles. Do
       they have separating paths?

35.   Let G and H be connected graphs. The **Ramsey number** $r(G,H)$ is the
       smallest integer n such that in any 2-edge-colouring - say red and blue - of
       the edges of K_n, there exists either a red subgraph isomorphic to G or a
       blue subgraph isomorphic to H. Use P to show that $r(C_4,K_5-e) \geq 11$.

36. Let X and Y be 5×5 circulant matrices with first rows respectively $(0,1,0,0,1)$ and $(0,0,1,1,0)$. Verify that the adjacency matrix, see Section 6.1, for P can be expressed as

$$\begin{bmatrix} X & I \\ I & Y \end{bmatrix}$$

37. Let J be the 5×5 matrix, all of whose elements are 1. Let J^1 and J_1 be respectively the 5×1 and 1×5 matrices all of whose elements are 1. Then, replacing 1 by 0, O, O^1 and O_1 are similarly defined. Let G be the graph with adjacency matrix

$$\begin{bmatrix} X & X & I & J\text{-}I & J^1 \\ X & X & J\text{-}I & I & J^1 \\ I & J\text{-}I & Y & Y\text{+}I & O^1 \\ J\text{-}I & I & Y\text{+}I & Y & O^1 \\ J_1 & J_1 & O_1 & O_1 & O \end{bmatrix}$$

Use G to show that $r(K_5\text{-}e, K_5\text{-}e) \geq 22$; see [E-R 88], [C-E-H-M-S 89].

38. Suppose that G and H are graphs. Write $H \mid G$ if G is the edge-disjoint union of copies of H.

(a) Prove that $P \nmid K_{10}$ [L-S 87].

[Hint: By contradiction. Suppose that A_i is the adjacency matrix of a Petersen graph (i=1,2,3)
   (i)   verify that $A_1 + A_2 + A_3 = J\text{-}I$; see the proof of Theorem 6.1.3.
   (ii) verify that the vector I_1 of all ones is an eigenvector for each A_i and that there is a five-dimensional space of eigenvectors, with eigenvalue 1, orthogonal to I_1; see Proposition 3.3.
   (iii) deduce that since the orthogonal component of I_1 has dimension 9 there exists a common eigenvector w with eigenvalue 1 for A_1 and A_2; its coordinates sum to 0.
   (iv) now let w act on both sides of the equation in (i).]

(b) Let G be the (7,5)-cage. Does $G \mid K_{50}$? (Biggs [nB 92] conjectures that the proof above generalizes to show that $G \nmid K_{50}$.)

[Comments. (i) Häggkvist [rH 89, p135] conjectures that if H is any cubic graph with (3r+1) vertices (r≥5) then $H \mid K_{3r+1}$; and (ii) $K_{11} \mid K_{111}$. This latter is equivalent to proving the non-existence of the projective plane of order 10; see Section 5.3].

39. Let f(d) be the maximum order of a generalized Petersen graph of diameterd. Prove
    (i) f(2) = 10;
    (ii) f(3) = 14 (consider P(7,2)).
    [Comment: Beenker and Van Lint [B-V 88] have proved that $f(d) = 4d^2 - 12d + 10$ (d≥4).]

40. Suppose that G is a graph and $C = (v_1, v_2, ..., v_m)$, $m \ge 3$, is a cycle of G. Then the edge $v_i v_j$ is a **diagonal** of C if $i \ne j-1, j+1$ (mod m). Let σ(C) be the number of diagonals of C and σ(G) the maximum of σ(C) taken over all cycles C of G. Prove that if $G \in \{P, K_{n,m}\}$ m > n, then $\sigma(G) \le \delta(\delta-2)$, where $\delta = \delta(G)$.

    [Comments: (i) Tian and Zang [T-Z 91] have proved that if G is a 2–connected graph with $|VG| \ge 2\delta + 1, \delta \ge 3$, then $\sigma(G) \ge \delta(\delta-2) + 1$, unless $G \in \{P, K_{n,m}\}$ m > n. Deduce that this result is sharp. (ii) This result confirms a conjecture of Gupta, Kahn and Robertson [G-K-R 80]; see also [pA85].]

41. Let G be a graph. Let B be any bipartite subgraph of G with the maximum possible number of edges and let $b(G) = |EB| / |EG|$. Prove that if G is either P or the dodecahedron, then $b(G) = \frac{4}{5}$.

    [Comments: (i) Hopkins and Staton [H-S 82] proved that if G is cubic and triangle-free, then $b(G) \ge \frac{4}{5}$. Bondy and Locke [B-L 86] showed that P and the dodecahedron are the only graphs for which equality holds. (ii) Bondy and Locke also showed that the only cubic graphs which contain a set of 5-cycles covering every edge of G exactly m times, for some fixed m, are those same two graphs.]

42. A graph is **perfect** [vC 87] if, for each of its induced subgraphs F, the chromatic number χ(F) of F equals the independence number $\alpha(\overline{F})$ of $\overline{F}$. Lovász [lL 72] proved that every minimal imperfect graph H has precisely $\alpha(H)\alpha(\overline{H}) + 1$ vertices.
    (i) Find all the minimal imperfect subgraphs of P.
    (ii) Prove that P is not perfect.

References

[A-F-L 86]   N. Alon, P. Frankl and L. Lovász, The chromatic number of Kneser hypergraphs, *Trans. Amer. Math. Soc.*, 298, 1986, 359–370.

[bA 79]   B.R. Alspach, Hamiltonian cycles in vertex-transitive graphs of order 2p, Proc. Tenth S.E. Conf. Boca Raton, *Congressus Numer.* XXIII, 1979, 131-139.

[bA 83]   B.R. Alspach, The classification of Hamiltonian generalized Petersen graphs, *J. Comb. Th.* B, 34, 1983, 293-312.

[A-Z 89a]   B.R. Alspach and C. Zhang, Hamiltonian cycles in cubic Cayley graphs on dihedral groups, *Ars Combinatoria*, 28, 1989, 101-108.

[A-Z 89b]   B.R. Alspach and C. Zhang, Cycle coverings of graphs, submitted.

[dA 80]   D. Archdeacon, A Kuratowski theorem for the projective plane, Ph.D. Thesis, Ohio State Univ., 1980.

[dA 81]   D. Archdeacon, A Kuratowski theorem for the projective plane, *J. Graph Th.* 5, 1981, 243-246.

[A-H 81]   D. Archdeacon and P. Huneke, On irreducible graphs for non-orientable surfaces, preprint, 1981.

[pA 85]   P. Ash, The maximum number of diagonals of a cycle in a block, *Disc. Math.*, 55, 1985, 305-309.

[lB 79]   L. Babai, Long cycles in vertex-transitive graphs, *J. Graph. Th.*, 3, 1979, 301-304.

[lB 91]   L. Babai, Vertex-transitive graphs and vertex-transitive maps, *J. Graph Th.*, 15, 1991, 587-627.

[rB 46]   R. Baer, Polarities in finite projective planes, *Bull. Amer. Math. Soc.*, 52, 1946, 77-93.

[iB 78]   I. Bárány, A short proof of Kneser's conjecture, *J. Comb.*

Th. A, 25, 1978, 325-326.

[B-V 88]   G. Beenker and J Van Lint, Optimal generalized Petersen graphs, *Phillips J. Res*., 43, 1988, 129-136.

[B-F 87]   A. Benhocine and J-L. Fouquet, The Chvátal-Erdös condition and pancyclic line graphs, *Disc.Math*., 66, 1987, 21-26.

[B-J 79]   A. Bigalke and H. Jung, Über Hamiltonsche Kreise und Unabhängige Ecken in Graphen, *Monatsch. Math*., 88, 1979, 195-210.

[nB 74]   N.L. Biggs, *Algebraic Graph Theory*, Cambridge Tracts in Mathematics, No.67, C.U.P., 1974.

[nB 79]   N.L. Biggs, Some odd graph theory, Second Int. Conf. on Combinatorial Maths, Annals New York Acad. Sci., 319, New York Acad. Sci., New York, 1979, 71-81.

[nB 92]   N.L. Biggs, Private communication.

[B-B-B-C-A 875] A. Blokhuis, A.E. Brouwer, D. Buset, A.M. Cohen and M. Arjeh, The locally icosahedral graphs, *Finite Geometries* (Winnipeg, Man. 1984), Lecture Notes in Pure and Applied Math., 103, Dekker, New York, 1985, 19-22.

[bB 78]   B. Bollobás, *Extremal Graph Theory*, Academic Press, New York, 1978.

[jB 78]   J.A. Bondy, Hamilton cycles in graphs and digraphs, Proc. of the Ninth South Eastern Conference on Combinatorics, Graph Theory and Computing. (Florida Atlantic Univ., Boca Raton, Fla., 1978), *Congress Numer*, XXI, Utilitas Math., Winnipeg, Man., 1978, 3-28.

[jB 90]   J.A. Bondy, Small cycle double covers of graphs, in *Cycles and Rays*, (ed. G. Hahn, G. Sabidussi and R. Woodrow), Kluwer Academic Publ., Boston, 1990, 21-40.

[B-E-F 79] J.A. Bondy, P. Erdös and S. Fajtlowicz, Graphs of diameter 2 with no 4-cycles, Internal Report, University of Waterloo, 1979.

[B-L 86]   J.A. Bondy and S. Locke, Largest bipartite subgraphs in triangle-free graphs with maximum degree three, *J. Graph Th.*, 10, 1986, 477-504.

[kB 33]   K. Borsuk, Drei Sätze über die n-dimensionale euklidische Sphäre, *Fund. Math*, 20, 1933, 177-190.

[B-V 91]   H. Broersma and H. Veldman, Long dominating cycles and paths in graphs with large neighbourhood unions, *J. Graph Th.*, 15, 1991, 29-38.

[wB 66]   W.G. Brown, On graphs that do not contain a Thomsen graph, *Can. Math. Bull*, 9, 1966, 281-285.

[dB 83]   D. Buset, Graphs which are locally a cube, *Disc. Math.*, 46, 1983, 221-226.

[B-C 75]   F.C. Bussemaker and D.M. Cvetkovič, There are exactly 13 connected, cubic integral graphs, Memorandum 1975-15, Eindhoven University of Technology, 1975.

[C-P 72]   F. Castagna and G. Prins, Every generalized Petersen graph has a Tait Colouring, *Pacific J. of Math*, 40, 1972, 53-58.

[pC 89]   P. Catlin, Double cycle covers and the Petersen graph, *J. Graph Th.*, 13, 1989, 465-483.

[uC 87]   U. Celmins, On conjectures relating to snarks, Ph.D. Thesis, Waterloo, 1987.

[lC 59]   L. Chang, The uniqueness and non-uniqueness of the triangular association schemes, *Sci. Record. Math.* New Ser. 3, 1959, 604-613.

[C-W 85]   G. Chartrand and R.J. Wilson, The Petersen graph, Proc. 1st Colorado Conference on Graph Theory, (ed. F. Harary and J. Maybie), Wiley, New York, 1985, 69-100.

[C-H-W 92]   G. Chartrand, H. Heira and R.J. Wilson, The Ubiquitous Petersen Graph, preprint.

[C-L 87]   B-L. Chen and K-W. Lih, Hamiltonian uniform subset graphs,
                     J. Combin. Th. B, 42, 1987, 257-263.

[C-Q 83]   C.C. Chen and N. Quimpo, Hamiltonian Cayley graphs of
                     order pq, *Combinatorial Mathematics X*, (Ed. L.R.A. Casse)
                     Lecture Notes in Mathematics No. 1036, Springer-Verlag,
                     Berlin, 1983, 1-5.

[vC 87]   V. Chvátal, Perfect graphs, *Surveys in Combinatorics*, (C.
                     Whitehead ed.), L.M.S. Lecture Notes Series No.123,
                     C.U.P., 1987, 43-52.

[C-E 72]   V. Chvátal and P. Erdös, A note on Hamiltonian circuits,
                     Disc. Math., 2, 1972, 111-113.

[C-T 78]   V. Chvátal and C. Thomassen, Distances in orientations of
                     graphs, *J. Comb. Th.* B, 24, 1978, 61-75.

[C-E-H-M-S 89]   C.R.J. Clapham, G. Exoo, H. Harborth, I. Mengersen and J.
                     Sheehan, The Ramsey numbers of K_5 - e, *J. Graph Th.*, 13,
                     1989, 7-15.

[C-F-S 87]   C.R.J. Clapham, A. Flockhart and J. Sheehan, Graphs without
                     4–cycles, *J. Graph Th.*, 13, 1989, 29-47.

[C-P 82]   R.J. Cook and D.G. Pryce, A class of geodetic blocks, *J.
                     Graph Th.*, 6, 1982, 157-168.

[hC 54]   H. Coxeter, Regular honeycombs in hyperbolic space, *Proc.
                     Int. Congress Mathematicians*, Amsterdam, 3, 1954, 155-169.

[C-D-G-T 88]   D.M. Cvetkovič, M. Doob, I. Gutman and A. Torgašev,
                     Recent Results in the Theory of Graph Spectra, Annals of
                     Discrete Math., 36, North-Holland, Amsterdam, 1988.

[pE 76]   P. Erdös, Problems and results in combinatorial analysis, in
                     Colloq. Int. Th. Combin. Rome 1973, Acad. Naz. Lincei,
                     Rome, 1976, 3-17.

[E-K-R 61]   P. Erdös, C. Ko and R. Rado, Intersection theorems for
                     systems of finite sets, *Quart. J. Math.*, 12, 1961, 313-320.

[E-R-S 66]   P. Erdös, A. Rényi and V.T. Sós, On a problem of graph theory, *Studia Sci. Math. Hungar.*, 1, 1966, 215-235.

[E-R 88]   G. Exoo and D. Reynolds, Ramsey numbers based on C_5-decompositions, *Disc. Math.*, 71, 1988, 119-127.

[F-W 77]   S. Fiorini and R.J. Wilson, *Edge-colourings of graphs*, Research Notes in Mathematics No. 16, Pitman, London, 1977.

[F-W 78]   S. Fiorini and R.J. Wilson, Edge colorings of graphs, *Selected Topics in Graph Theory* (ed. L. Beineke and R.J. Wilson), Academic Press, New York, 1978, 103-126.

[hF 76]   H. Fleischner, Eine gemeinsame Basis für die Theorie der Eulerschen Graphen und den Satz von Petersen, *Monatsh. Math.*, 81, 1976, 267-278.

[hF 84]   H. Fleischner, Cycle decompositions, 2-coverings, removable cycles and the four-color disease, *Progress in Graph Theory*, Academic Press, Toronto, 1984, 233-246.

[pF 85]   P. Frankl, On the chromatic number of the general Kneser-graph, *J. Graph Th.*, 9, 1985, 217-220.

[pF 34]   P. Franklin, A six colour problem, *J. Math., Phys.*, Massachussets Inst. Tech., 13, 1934, 363-369.

[F-G-W 71]   R. Frucht, J. Graver and M.E. Watkins, The groups of the generalised Petersen graphs, *Proc. Camb. Phil. Soc.*, 70, 1971, 211-218.

[zF 83]   Z. Füredi, Graphs without quadrilaterals, *J. Comb. Th.* B, 34, 1983, 187-190.

[dG 56]   D. Gale, Neighbouring vertices on a convex polyhedron in *Linear Inequalities and Related Systems* (ed. H.W. Khun and A.W. Tucker), Princeton Univ. Press, Princeton, N.J., 1956.

[G-H-W 79]   H. Glover, P. Huneke and C. Wang, 103 graphs that are irreducible for the projective plane, *J. Comb. Th.* B, 27,

1979, 332-370.

[lG 85]   L. Goddyn, A girth requirement for the double cycle cover conjecture, in "Cycles in Graphs", North-Holland Math. Stud., 115, North Holland, Amsterdam-New York, 1985, 13-26.

[lG 89]   L. Goddyn, Cycle double covers of graphs with hamilton paths, *J. Combin. Th.* B, 46, 1989, 253-254.

[G-P-S 91] D. Guichard, B. Piazza and S. Stueckle, On the vulnerability of permutation graphs of complete and complete bipartite graphs, *Ars Combinatoria*, 31, 1991, 149-157.

[G-K-R 80] R. Gupta, J. Kahn and N. Robertson, On the maximum number of diagonals of a circuit in a graph, *Disc. Math.*, 32, 1980, 37–43.

[rH 89]   R. Häggkvist, Decompositions of complete bipartite graphs, *Surveys in Combinatorics*, (ed. J. Siemens), London Math. Soc., Lecture Note Series 141, C.U.P., 1989, 115-146.

[jH 80]   J.J. Hall, Locally Petersen graphs, *J. Graph. Th.*, 4, 1980, 173-187.

[jH 85]   J.J. Hall, Graphs with constant link and small degree or order, *J. Graph. Th.*, 9, 1985, 419-444.

[mH 59]   M. Hall Jr., *The Theory of Groups*, Macmillan, New York, 1959.

[pH 90]   P.J. Heawood, Map-colour theorem, *Quart. J. Math*, 24, 1890, 332-338.

[H-W 78]   K. Heinrich and W.D. Wallis, Hamiltonian cycles in certain graphs, *J. Austral. Math. Soc.* A, 26, 1978, 89-98.

[jH 79]   J.W.P. Hirschfield, Projective Geometries over Finite Fields, Clarendon Press, Oxford, 1979.

[H-S 82]   G. Hopkins and W. Staton, Extremal bipartite subgraphs of cubic triangle-free graphs, *J. Graph Th.*, 6, 1982, 115-121.

[aI 86]   A.A. Ivanov, Distance-transitive representations of the symmetric groups, *J. Comb. Th.* B, 41, 1986, 255-274.

[bJ 91]   B. Jackson, Neighborhood unions and hamiltonian cycles, *J. Graph Th.*, 15, 1991, 443-451.

[J-O 90]   B. Jackson and O. Ordaz, Chvátal-Erdös conditions for paths and cycles in graphs and digraphs, a survey, *Disc. Math.*, 84, 1990, 241-254.

[fJ 79]   F. Jaeger, Flows and generalized coloring theorems in graphs, *J. Comb. Th.* B, 26, 1979, 205-216.

[fJ 85]   F. Jaeger, A survey of the cycle double cover conjecture, in *Cycles in Graphs*, (ed. B.R. Alspach and C.D. Godsil), Annals of Discrete Math. 27, North-Holland, Amsterdam, 1985, 1-12.

[rJ 85]   R. Jeunissen, The Petersen graph as a box of Pandora, *Nieuw Arch. Wisk.*, 3, 1985, 219-233.

[gK 47]   G. Kirchoff, Über die Auflösung der Gleichurgen, auf welche man bei der Untersuchung der linearen Verteilung galvanischer Sträne geführt wird, *Ann. Phys. Chem.*, 72, 1847, 497-508.

[sK 91]   S. Kitakubo, Bounding the number of embeddings of 5-connected projective-planar graphs, *J. Graph Th.*, 15, 1991, 199-205.

[mK 55]   M. Kneser, Aufgabe 300, *J. Deutsch. Math. Verein.*, 58, 1955.

[yK 91]   Y. Kohayakawa, A note on induced cycles in Kneser graphs, *Combinatorica*, 11, 1991, 245-251.

[aK 17]   A. Kowalewski, W.R. Hamilton's Dodekaederaufgube als Buntordnungsproblem Sitzungsber. *Akad. Wiss. Wien* (Abt. IIa) 126, 67-90, 963-1007.

[hL 90]   H. Li, Perfect path double covers in every simple graph, *J. Graph Th.*, 14, 1990, 645-650.

[mL 85]   M. Lipman, Hamiltonian cycles and paths in vertex-transitive graphs with abelian and nilpotent groups, *Disc. Math.*, 54,

1985, 15-21.

[cL 87]   C.H.C. Little, Problem Section, The Symposium on Cycles
               and Rays, Montreal, 1987.

[L-G-H 75]   C.H.C. Little, D.D. Grant and D.A. Holton, On defect-d
               matchings in graphs, *Disc. Math.*, 13, 1975, 41-54.

[L-S 87]   O. Lossers and A. Schwenk, Solution of advanced problems,
               6434, *American Math. Monthly*, 94, 1987, 885-886.

[lL 70]   L. Lovász, Unsolved problem II in "Combinatoral Structures
               and their Applications" (Proceedings of the Calgary
               International Conference on Combinatorial Structures and their
               Applications, 1969), (ed. R. Guy, H. Hanani, N. Sauer and J.
               Schonheim), Gordon and Breach, New York, 1970.

[lL 72]   L. Lovász, A characterization of perfect graphs, *J. Comb. Th.
               B*, 13, 1972, 95-98.

[lL 78]   L. Lovász, Kneser's conjecture, chromatic number and
               homotopy, *J. Comb. Th. A*, 25, 1978, 319-324.

[dM 81]   D. Marušič, On vertex symmetric digraphs, *Disc. Math.*, 36,
               1981, 69-81.

[dM 83]   D. Marušič, Hamiltonian circuits in Cayley graphs, *Disc.
               Math.*, 46, 1983, 49-54.

[dM 88]   D. Marušič, Research problems 92, 93, *Disc. Math.*, 70,
               1988, 215.

[M-P 82]   D. Marušič and T.D. Parsons, Hamiltonian paths in vertex-
               symmetric graphs of order 5p, *Disc. Math.*, 42, 1982, 227-
               242.

[M-P 83]   D. Marušič and T.D. Parsons, Hamiltonian paths in vertex-
               symmetric graphs of order 4p, *Disc. Math.*, 43, 1983, 91-96.

[mM 76]   M. Mather, The rugby footballers of Croam, *J. Comb. Th. B*,
               20, 1976, 62-63.

[M-R 90]   B.D. McKay and G. Royle, Construction of all simple graphs
             with at most 26 vertices and transitive automorphism group,
             Ars Combinatoria, 30, 1990, 161-176.

[M-L 72]   G.H.J. Meredith and E.K. Lloyd, The hamiltonian graphs O_4
             to O_7, *Combinatorics*, (ed. D.J.A. Welsh and D.R.
             Woodall), IMA, Southend-on-Sea, 1972, 229-236.

[M-L 73]   G.H.J. Meredith and E.K. Lloyd, The footballers of Croam,
             J. Comb. Th. B, 15, 1973, 161-166.

[maM 91]   M. Mollard, Cycle-regular graphs, *Disc. Math.*, 89, 1991,
             29–41.

[aM 82]   A. Moon, Characterization of the odd graphs O_k by
             parameters, *Disc. Math.*, 42, 1982, 91-97.

[sN 83]   S. Negami, Uniqueness and faithfulness of embedding of
             toroidal graphs, *Disc. Math.*, 44, 1983, 161-180.

[sN 84]   S. Negami, Uniquely and faithfully embeddable projective-
             planar triangulations, *J. Comb. Th.* B, 36, 1984, 189-193.

[sN 85]   S. Negami, Unique and faithful embeddings of projective-
             planar graphs, *J. Graph Th.*, 9, 1985, 235-243.

[sN 88]   S. Negami, Re-embeddings of projective-planar graphs, *J.
             Comb. Th.* B, 44, 1988, 276-299.

[oO 62]   O. Ore, The Theory of Graphs, *American Math. Soc. College*,
             38, Amer. Math. Soc., Providence, R.I., 1962.

[P-S 84]   K.R. Parthasarathy and N. Srinivasan, Geodetic blocks of
             diameter three, *Combinatorica*, 4, 1984, 197-206.

[P-S 85]   K.R. Parthasarathy and N. Srinivasan, On geodetic blocks of
             diameter 2: some structural properties, *Ars Combinatoria*,
             20, 1985, 49-60.

[jP 98]   J. Petersen, Sur le théorèm de Tait, *Intermèd. Math.* 5,
             1898, 225-227.

[jP 2]   J. Petersen, Les 36 officiers, *Annuaire des Mathématiciens*, 1902, 413-427.

[P-R-S 90]   B. Piazza, R. Ringeisen and S. Stueckle, On the vulnerability of cycle permutation graphs, *Ars Combinatoria*, 29, 1990, 289–296.

[jaP 77]   J. Plesnik, Two constructions of geodetic graphs, *Math. Slovaca* , 27, 1977, 65-71.

[jaP 80]   J. Plesnik, A construction of geodetic blocks, *Acta Fac. Rerum Natur. Univ. Comen. Math.*, 36, 1980, 47-60.

[jaP 84]   J. Plesnik, A construction of geodetic graphs based on pulling subgraphs homeomorphic to complete graphs, *J. Comb. Th. B*, 36, 1984, 284-297.

[cP 81]   C.E. Praeger, Primitive permutation groups and a characterization of the odd graphs, *J. Comb. Th.* B, 31, 1981, 117-142.

[aR 91]   A. Raspaud, Cycle covers of graphs with a nowhere-zero 4–flow, *J. Graph Th.*, 15, 1991, 649-654.

[iR 58]   I. Reiman, Über ein Problem von K. Zarankiewicz, *Acta Math. Acad. Sci. Hungar.*, 9, 1958, 269-278.

[rR 84]   R.D. Ringeisen, On cycle permutation graphs, *Disc. Math.*, 51, 1984, 265-275.

[gR 59]   G. Ringel, *Farbungsprobleme auf Flächen und Graphen*, Deutsches Verlag, Berlin, 1959.

[hR 39]   H. Robbins, A theorem on graphs with an application to a problem of traffic control, *Amer. Math. Monthly*, 46, 1939, 281-283.

[R-S 85]   N. Robertson and P.D. Seymour, Graph minors - a survey, *Surveys in Combinatorics*, Lond. Math. Soc. Lecture Note Series No. 103, C.U.P., 1985, 153-171.

[R-S-T 92]   N. Robertson, P.D. Seymour and R. Thomas, Linkless embeddings of graphs in 3-space, preprint.

[hS 67]   H. Sachs, Über Teiler, Faktoren und charakteristische Polynome von Graphen, II, *Wiss. Z. Tech. Hochsch. Ilmenau*, 13, 1967, 405-412.

[kS 90]   K. Sarkaria, A generalized Kneser conjecture, *J. Comb. Th.* B, 49, 1990, 236-240.

[rS 86]   R. Scapellato, Geodetic graphs of diameter two and some related structures, *J. Comb. Th.* B, 41, 1986, 218-229.

[rS 89]   R. Scapellato, Geodetic graphs of diameter two without Petersen subgraphs, *Ars Combinatoria*, 27, 1989, 201-202.

[S-C 89]   G. Schrag and L. Cammack, On the 2-extendability of the generalized Petersen graphs, *Disc. Math.*, 78, 1989, 169-177.

[aS 78]   A.L. Schwenk, Exactly thirteen cubic connected graphs have integral spectra, *Theory and Applications of Graphs*, (ed. Y. Alavi and D.R. Lick) Lecture Notes in Mathematics No. 642, Springer-Verlag, Berlin, 1978, 516-533.

[aS 89]   A. Schwenk, Enumeration of hamiltonian cycles in certain generalized Petersen graphs, *J. Comb. Th.* B, 47, 1989, 53-59.

[pS 79]   P. Seymour, Sums of circuits, *Graph Theory and Related Topics*, (ed. J. Bondy and U.S.R. Murty) Academic Press, New York, 1979, 431-355.

[hS 72]   H.L. Skala, A variation of the Friendship Theorem, *Siam J. Appl. Math.*, 23, 1972, 214-220.

[jS 74]   J.G. Stemple, Geodetic graphs of diameter two, *J. Comb. Th.* B, 17, 1974, 266-280.

[S-W 68]   J.G. Stemple and M.E. Watkins, On planar geodetic graphs. *J. Comb. Th.*, 4, 1968, 101-117.

[S-R 84]   S. Stueckle and R.D. Ringeisen, Generalized Petersen graphs
                  which are cycle permutation graphs, *J. Comb. Th.* B, 37,
                  1984, 142-150.

[gS 73]   G. Szekeres, Polyhedral decomposition of cubic graphs, *Bull.
                  Austral. Math. Soc.*, 8, 1973, 367-387.

[mT 86]   M. Tarsi, Semi-duality and the cycle double cover conjecture,
                  J. Comb. Th. B, 41, 1986, 332-340.

[pT 86]   P. Terwilliger, The Johnson graph $J(d,r)$ is unique if
                  $(d,r) \neq (2,8)$, *Disc. Math.*, 58, 1986, 175-189.

[aT 82]   A. Thomason, Cubic graphs with 3 hamiltonian cycles are not
                  always uniquely colourable, *J. Graph. Th.*, 6, 1982, 219-
                  221.

[cT 78a]   C. Thomassen, On separating cycles in graphs, *Disc. Math.*,
                  22, 1978, 57-73.

[cT 78b]   C. Thomassen, *Selected Topics in Graph Theory*,
                  (ed.L.W.Beineke and R.J. Wilson), Academic Press, 1978.

[cT 91]   C. Thomassen, Tilings of the torus and the Klein bottle and
                  vertex-transitive graphs on a fixed surface, *Trans. Amer.
                  Math. Soc.*, (to appear).

[T-T 81]   C. Thomassen and B. Toft, Non-separating induced cycles in
                  graphs, *J. Comb. Th.* B, 31, 1981, 199-224.

[T-Z 91]   F. Tian and W. Zang, The maximum number of diagonals of a
                  cycle in a block and its extremal graphs, *Disc. Math.*, 89,
                  1991, 51-63.

[jT 74]   J. Tits, Buildings of spherical type and finite BN-parts,
                  Lecture Notes in Math. No. 386, Springer, Berlin, 1974.

[jaT 67]   J. Turner, Point-symmetric graphs with a prime number of
                  points, *J. Comb. Th.*, 3, 1967, 136-145.

[pV 85]   P. Vanden Cruyce, A finite graph which is locally a dodecahedron, *Disc. Math.*, 54, 1985, 343-346.

[aV 81]   A. Vince, Locally homogeneous graphs from groups, *J. Graph. Th.*, 5, 1981, 417-422.

[mW 64]   M.E. Watkins, A characterization of the planar geodetic graphs and some geodetic properties of certain nonplanar graphs, Dissertation, Yale University, 1964.

[mW 70]   M.E. Watkins, Connectivity of transitive graphs, *J. Comb. Th.*, 8, 1970, 23-29.

[aW 73]   A.T. White, *Graphs, Groups and Surfaces,* North-Holland, London, 1973.

[W-G 84]   D. Witte and J. Gallian, A survey: Hamiltonian cycles in Cayley graphs, *Disc.Math.*, 51, 1984, 293-304.

[hW 32]   H. Whitney, Congruent graphs and the connectivity of graphs, *Amer. J. Math*, 54, 1932, 150-168.

[cZ 90]   C. Zang, Minimum cycle coverings and integer flows, *J. Graph Th.*, 14, 1990, 537-546.

[bZ 75]   B. Zelinka, Geodetic graphs of diameter two, *Czechoslovak Math. J.*, 25, 1975, 148-153.

[sZ 91]   S. Zhan, On hamiltonian line graphs and connectivity, *Disc. Math.*, 89, 1991, 89-95.

Conjecture Index

In this index we list, alphabetically using their authors' names where appropriate, all open conjectures quoted in this book. For the definitions of the terms used, the reader will need to refer to the relevant text.

Bloom, Kennedy and Quintas' Conjecture: p.73.
>   If $c(G) \geq 2$, then G has a subgraph H with $c(H) = 2$.

Bondy's Conjecture: p.293.
>   Every non-empty connected vertex-transitive graph has a 2-factor with at most two components.

Chen's Conjecture: p.294.
>   The Petersen graph is the only non-hamiltonian 3-connected graph for which $\sigma_2(G) = \frac{1}{2}n$.

Chen and Lih's Conjecture: p.301.
>   The graph $G(n,k,t)$ is hamiltonian for any admissable (n,k,t) except $(5,2,0)$ and $(5,3,1)$.

Conjecture 3.3.3: p.104.
>   Every k-snark contains a subgraph contractible to P.

Conjecture 6.2.3: p.192.
> $f(r,g) < f(r+1,g)$.

Cycle Double Cover Conjecture: p.305ff, 322.
>   Let G be a bridgeless finite graph. Then there exists a sequence of cycles in G such that each edge of G is in exactly two of the cycles.

Fiorini and Wilson's Conjecture: p.324.
> $P(9,2)$ is the only nonplanar cubic graph which is uniquely 3-edge-colourable.

Fleischner's Conjecture: p.308.
>   Every snark has a dominating cycle.

Hadwiger's Conjecture: p.68-70, 75, 154, 172, 175-176.
    If $\chi(G) \geq k$, then G is subcontractible to K_k.

Häggvist's Conjecture: p.326.
    If H is any cubic graph with $3r+1$ vertices for $r \geq 5$, then $H \mid K_{3r+1}$.

Hajós' Conjecture: p.69-70, 75.
    If $\chi(G) \geq k$, then G contains a subdivision of K_k.

Jackson's Conjecture: p.294.
    If G is a 2-connected graph on n vertices such that $\sigma_2(G) \geq \frac{1}{2}n$, then either G is hamiltonian or G is a spanning subgraph of a specified and very limited type or G is the Petersen graph.

Kneser's Conjecture: p.295, 297.
    If all the n-subsets of a $(2n + k)$-set are partitioned into $k + 1$ classes, then one of these classes contains two disjoint n-subsets.

Lovász's Conjecture (1): p.289-291.
    Every vertex-transitive graph has a hamiltonian path.

Lovász's Conjecture (2): p.292-293.
    All Cayley graphs are hamiltonian.

Meredith and Lloyd's Conjecture: p.300.
    The odd graph, $O(k), k \geq 4$, has a hamiltonian decomposition.

Parthasarathy and Srinivasan: p.309.
    Every diameter two geodetic graph of minimum degree three, contains P as an induced subgraph.

Schrag and Cammack's Conjecture: p.316.
    $P(n,k)$ is 2-extendable if and only if $n \neq 5k$.

Sheng Bau and Holton's Conjecture: p.242.
    Let G be a cubic 3-connected graph. Then there exists a cycle through any 13 vertices in G, unless G is contractible in the obvious ways to P or to the graph of Figure 7.5.5.

Strong Embedding Conjecture: p.306.
    Every 2-connected graph has a strong embedding in some surface.

Notation Index

Index

Printed in the United States
67616LVS00003B/23

9 780521 435949